Our National River Ganga

Rashmi Sanghi
Editor

Our National River Ganga

Lifeline of Millions

Springer

Editor
Rashmi Sanghi
Indian Institute of Technology
Kanpur, India

ISBN 978-3-319-00529-4 ISBN 978-3-319-00530-0 (eBook)
DOI 10.1007/978-3-319-00530-0
Springer Cham Heidelberg New York Dordrecht London

Library of Congress Control Number: 2013952900

Printed on acid-free paper

Springer is part of Springer Science+Business Media (www.springer.com)

Foreword

Ganga*ji* or the Ganges is the lifeline of India and needs no introduction. Flowing from the Himalayas, through the most heavily populated basin in the world, it is also the most sacred river. From cradle to destination, this 2,225 km long river flowing from Gangotri to Gangasagar has an integral role to play in many customs and traditions of people of India. The national river of India, Ganges, yet an endangered heritage of our country, needs special attention with respect to its utility and its deteriorating water quality. The river is not just a matter of curiosity, but also of concern even for those far away from its banks, as not only it weaves diverse communities of India but also diverse nationalities. It has become a topic of discussion, debate, deliberation and research.

Numerous databases, websites, blogs, reports, articles, books, videos, are full of facts and figures on the Ganges. It is very difficult for one author to cover the multidisciplinary aspects of the river in a single article. The idea of coming up with such a book was to be able to compile the expertise of some key persons who have worked on specific aspects of the Ganga river basin.

I hope that this book will prove to be a useful insight into the scientific world of the holy river connecting students, experts, policy makers and common man. I also humbly hope that this book will also be a medium to create mass awareness, to ultimately help reach the goal of '*Aviral*' and '*Nirmal*' *Ganga*.

Professor
Department of Computer Science and Engineering
Indian Institute of Technology Kanpur, India

T.V. Prabhakar

Preface

In his book *Discovery of India*, Jawaharlal Nehru says:

> The Ganges above all is the river of India, which has held India's heart captive and drawn uncounted millions to her banks since the dawn of history. The story of the Ganges, from her source to the sea, from old times to new, is the story of India's civilization and culture, of the rise and fall of empires, of great and proud cities, of adventures of man.

In these few lines he said it all about the great mighty river gushing from the Himalayas of Nepal and India, through the great plains of India, to the world's biggest delta, straddling West Bengal and Bangladesh, where it merges with the Brahmaputra and Meghna. The Ganges river basin is one of the most fertile and densely populated region in the world draining a basin of extraordinary variation in altitude, climate, land use, flora and fauna, social and cultural life. It occupies a special position in the ethos of the people of India, arguably being the most sacred river in the world. No doubt the Ganga river is unique rightly deserving a national status. This is why the Government of India declared the holy river *GANGA* as a "national river on 4 Nov 2008. Fourteen months later, on January 19, 2010, the critically endangered Ganges river dolphin was declared India's national aquatic animal. For the first time in the history of India, a river has been given this status.

Rivers in general play a very important part in the water cycle acting as drainage channels for water surface carrying water and nutrients. Though in India every river is sacred, Ganga river is unique. Besides providing irrigation, food, habitat, energy, transport and attracting tourism, it has long been considered a holy river and as the mother goddess. It is tradition, culture, belief, devotion and celebration for the masses. Originating in the heavens the Ganga of mythology came down to earth from the Milky Way in a lock of the god Shiva's hair, thus becoming a sacred bridge between this world and the divine thereby opening the gateway to the next life. This is embodied in the death rituals of cremation followed by most Hindus. The beliefs of the people over the years have only got firmer and are perhaps well justified as the Ganga water possess the purifying qualities saturated with viruses killing harmful bacteria.

Perception of the Ganga river Basin are always different depending on the individual and his or her mindset. No two communities would ever look at it from the same viewpoint. For a scientist it's a hub of research, and for a sage a place of worship. A connect between mythology and the physical attributes of Ganga is complex as scientists, industrialists, religious leaders all look at her from their different perspectives. A deeper and more comprehensive framework is perhaps the only way to grasp the complexities of the issues confronting this river.

The river is one of my favorite metaphors, the symbol of the great flow of life itself; an integral part of my childhood memories. Bowing to Ganga as a mark of respect came naturally to me. The very sight and touch of Ganga water fills me with immense pleasure and calm. Early years of my life whilst at Allahabad I was exposed to the serene beauty and calmness of nature's creation. The latter half of my life I was destined to be again near Ganga but at Kanpur. Here the very sight of my beloved Ganga was very disturbing. It has slowly aged and is no more able to maintain its health and clearly is in need of treatment. Age is undoubtedly catching up with me, but far more so with Ganga! I see the free flowing happy Ganga, my childhood companion for so long now, being used and abused at Kanpur. Though as a child I was mystified by the vastness of the river, today as a scientist I wonder what keeps it going despite the enormous overdose of toxins it gets along its course. I wonder what its saturation capacity is and I do wonder for how long it can keep on taking the poison. Will it be able to recuperate itself and be back to its happy flowing stage or will it keep ageing under neglect and insensitivities of its users and even worshippers?

As a scientist I may claim to understand the complexities involved with the river and may even try to shrug off the mythological beliefs attached to it. But as a Ganga user how can I deny the immense peace it gives me every time I go near it, every time I take a dip in it? There is something sacred, something mystifying, something intriguing, something special about my bond with the mighty river. I cannot reconcile the scientific aspect of Ganga with the mythological beliefs of people. In terms of understanding Ganga, it is just impossible to come to a common consensus. The paradox is How can so many millions take a 'holy dip' in the river believing that it will wash away their sins yet at the same time use it for drainage/sewage and pollute it to highly threatening levels that will eventually lead to its demise? It is both the subject of human worship as well as the object of human indifference. The problems associated with Ganga may be many but then so are the solutions, but only if realisation dawns on us. We cannot continue to use it and then abuse it with disdain.

Recently, the Ganga is facing numerous threats originating from human activity throughout its Ganga's basin. Organic pollution from domestic sewage, industrial effluents, and hazardous chemicals from non-point sources of agriculture and health sectors have seriously impacted water quality, aquatic biodiversity and human health. The Ganga ecosystem and its food-web must be viewed holistically with integrated and watershed perspectives. The river and its tributaries have not received proper attention from the researchers, policy makers and environmental managers in India.

There is a plethora of information available on this river in the form of books, blogs, articles, websites, videos and what not. But unfortunately, most of it is in a scattered form and the need to get the basic yet already available information on Ganga under one umbrella was strongly felt. This book was conceived with this objective in mind and the first introductory chapter is an effort in this direction. Part I (two chapters) gives a basic introduction to the cultures about Ganga water in the subcontinent. The existing data and available information from various sources have been compiled in a pictorial fashion in the form of knowledge models in the introductory chapter. Its cultural importance with changing times has also been discussed. It includes information on the catchment, hydrology, tributaries, water uses, and environmental features such as river water quality, aquatic and terrestrial flora/fauna, natural resources, ecological characteristics, sensitive environmental components and many other features. Part II (three chapters) looks at the rich biodiversity of Ganga basin. It gives a detailed description of the major floral and faunal biodiversity with special emphasis on the two aquatic indicator species Gharial and Dolphin. In particular, understanding the role and ecological importance of Sunderbans, the largest mangrove wetland in the world, with respect to impact and management of land and water resources is discussed in details. Part III (four chapters) examines 'The Ganga Water as it flows'. The water quality as well as the challenges associated with the specific ecological qualities of the Ganga system. Part IV (four chapters) looks at the complexities of issues confronting the river 'Ganga in changing times', be it snowmelt runoff, river bank erosion hazard and hydropower assessments with 'Population, Poverty and Pollution' being an integral part of this big change. Part V (two chapters) touches some aspects of the livelihoods and tourism along the river. The intention here is to explore the complex relationships between river resources and livelihoods with interrelations and connections across various Ganga aspects. Part VI includes 'Ganga as perceived by Some Ganga Experts'. This is the voice of some Ganga lovers who have worked on issues concerning the river and feel passionately bonded to the river.

Ganga system has inadvertently also become the source for intense debates about river benefits, ownership, entitlements and even legal jurisdiction. Through the development of new concepts, significant advances have been made towards understanding the dynamics of the river system. Today the major challenge is to integrate the secondary dataset with current findings in order to develop a holistic understanding of the river dynamics. Though India's culture and commonsensical folk wisdom could suggest solutions that are more in tune with the Indian psyche, as well as costing a great deal less money, it remains to be seen whether "new engineering" can offer different possibilities.

This compilation is just an effort to bring together the work of some experts on one common platform. Through the lens of basin interdependence, it has been a challenge to map the relative informational and conceptual Ganga river resources and bring together the experts in their respective areas of expertise. By evaluating the common present realities of accepting that river flow is a geo-morphological and dynamic process, the experts have also explored different strategies for harnessing the Ganga system by providing information relevant to their area of expertise. Though it was not possible to cover all the aspects of the vast river system, an attempt has been made to cover the major aspects ranging from cultural, biodiversity, hydrology to livelihood and tourism.

As late as the 1950s, the Thames flowing through London was an open sewer; called "The Great Stink" yet, by mid-1970s, the river was restored to its pristine glory, after the clean-up campaign that began in 1964. This book is put together and conceived not only to bring to clarity the issues and challenges facing our national river but to also inspire us to join hands as 'One' to restore, protect, and maintain Mother Ganga and Her tributaries, our national heritage and identity. If Thames can do it why not Ganga which is worshipped by millions. The future of our Ganga Ma is in our hands and we cannot just let it disappear in the times to come. The authors views are reflected in their respective chapters as they have tried their best to come up with updated contents which will help the readers to increase their knowledge database on many aspects of the river.

Kanpur, India Rashmi Sanghi (Editor)

Acknowledgements

The time has come to express my gratitude to all those who were the motivating factors behind the making of this book. It is an outcome of the blessings of many people, and it may not be even possible to name and thank all of them. During my long 'Ganga' journey I met many interesting personalities and Ganga lovers, who truly inspired me. They have been contributing to the cause of protecting 'Ganga' in many ways; by taking on the government, cleaning the dirty ghats, organizing rallies and outreach programs, and even going on a fast. As for my contribution via this project, I took an easier road perhaps. But at the end of this journey, I feel charged to take up more difficult challenges in the future. I can already visualize 'Aviral and Nirmal Ganga' and the role played by me in the near future.

I owe my presence in this project to Dr. Vinod Tare who is the man entrusted with the Ganga mission. His positive energies were like a recharge for me, every time I met him. Prof. T. V. Prabhakar is the mentor behind the inception and conception of this book. Without his contribution to the design and development of this book, it would not now exist, and just mere thanks and acknowledgement are not enough. I am intrigued by his spontaneity and multitasking which is what makes him a visionary. He was always there to guide me, encouraging and pushing me forward with his great ideas. At this point, I just hope I have lived up to his expectations, and if not I wish I get another chance to do so.

The contributors of this book are more than those visible in the book. Some could contribute in the form of chapters, but some who could not made contributions in the form of suggestions and requisite feedback. I wish to thank all my contributors for their time, patience and understanding. But overall, I received a very enthusiastic positive response from all those I approached. There are some whom I regret for not being able to approach at the right time. It would be a matter of great honor to be able to work with them later at some point of my life.

My special thanks to Mr. Suresh Babu and Mr. Nitin Kaushal (WWF) for their unflinching support and timely inputs. I also thank Julian for his extra critical comments and zeal with which he reviewed my chapter selflessly. Mr. Rakesh Jaiswal (Eco friends) was equally supportive, as also Ms. Archana Tadinada, whom I really wanted to rope in this project and who I think would have done full justice to the theme of the book. She was with me at the beginning of the project, and oblivious to her, I managed to extract her positive energies which were much needed in the long journey I had undertaken. The Parmath Niketan (Rishikesh) family, full of love not just for Ganga and her lovers, deserves a special mention. I observed the sevaks and sevikas, both with amazement and wonder and imagine myself as one of them someday serving Ganga selflessly. They all were a big help in preserving and refining my vision of the book.

Most importantly, I wish to thank to my Gangapedia team Neelam and Shashank for carrying out all the odd jobs at all the odd times. Meeta, Shalini, Tulika and my other office colleagues for their assistance and encouragement in the making of the 'Gangapedia'. They were all constantly providing strong background support to me.

Last, but not the least to my family. My husband for his endless patience and for holding the fort at home. Surabhi my 14 year old daughter, her inputs to beautify the book; she even offered to write a poem for the same. My son Udit (10 years) was like a sudden storm, making me jump out of the seat, I was so often glued to, and come back to reality. My parent's blessings I always take for granted and I hope to get their prayers answered by making them proud one day. My sister Sweety, who was instrumental in inculcating this love for the mighty river. My friend Prof. Sudhir Jain for his very useful inputs and critical comments which helped me improvise immensely. His relentless persuasion and belief in my abilities is treasured.

I do hope this book will be a good connect between the Ganges and its people.

Rashmi Sanghi (Editor)

Contents

The mighty Himalayas: As viewed from Kedarnath (Courtesy Rashmi Sanghi)

Part I
Introduction to Ganga

Gaumukh: Terminus of the Gangotri glacier with Bhagirathi peaks (Courtesy Rashmi Sanghi)

Chapter 1
Introduction to Our National River Ganga via cmaps

Rashmi Sanghi and Nitin Kaushal

Abstract The Government of India declared the holy river ***GANGA*** as a "national river" on 4 Nov 2008. For the first time in the history of India, a river has been given the status of national river. The Ganga river basin is the lifeline for millions sprawling across Himalayas of Nepal and India, through the great plains of India, to the world's biggest delta, straddling West Bengal and Bangladesh. The unique importance of this multifaceted Ganga basin stems from its geographical, historical, socio-cultural, environmental and economic value. It has been facing serious threat due to discharge of increasing quantities of sewage effluents, industrial effluents and other pollutants on account of rapid urbanization, industrialization and agricultural growth. The challenge is compounded by the competing demands on river water for irrigation, domestic purposes, industrial use and power. Due to the complexities of the issues confronting the Ganga river, it has inadvertently become the source for intense debates about river benefits, ownership and even entitlements with a indistinct line between its use and abuse.

Though there is a vast array of information available on the Ganga river in the form of write-ups, books, articles, blogs, it was felt that there was a need to present the information in a simpler and comprehensive way. Based on the available information from various sources, this chapter presents the salient features of the profile of the Ganga Basin in the form of knowledge models (cmaps). This chapter encompasses a holistic view of the river, – be it physiographic, ecological, cultural, spiritual or hydrological. It also touches briefly on the various challenges leading to the current state of affairs for the river also. The chapter concludes with the discussion on the establishment of National Ganga River Basin Authority (NGRBA) and the preparation of Ganga River Basin Environment Management

R. Sanghi (✉)
Media Labs, Indian Institute of Technology Kanpur, Kanpur, India
e-mail: rsanghi@iitk.ac.in; rsanghi@gmail.com

N. Kaushal
River Basins and Water Policy, World Wide Fund for Nature India, 172-B, Lodi Estate, New Delhi-110003, India
e-mail: nkaushal@wwfindia.net

R. Sanghi (ed.), *Our National River Ganga: Lifeline of Millions*,
DOI 10.1007/978-3-319-00530-0_1,

Closer look of Gaumukh: The origin of the mighty Ganga river (Courtesy Rashmi Sanghi)

Plan (GRB EMP) by the consortia of IITs (Indian Institute of Technology) and its partners. With a view to an easier understanding, several cmaps have been used throughout the chapter.

Keywords Ganga river • Tribuataries • Biodiversity • Endangered • Festivals • Water resources

1.1 Introduction to the Ganga River Basin

Since times immemorial, river Ganga has been the cradle of civilization of India. The river has unique status in the cultural and spiritual lives of people of, not only India, but also to the millions of people across the globe. The river is revered by a majority of Indians, to the extent that they consider it as 'mother' and have conferred the status of Goddess on to it. There have been numerous examples of vitality and greatness of river Ganga in ancient *hindu* mythology. The Ganga river system is the lifeline for more than 40 % of population of the country. The dependency on the Ganga river system is ever increasing for various reasons including – population growth and that too in a highly concentrated manner in the Ganga basin, rising standards of living and exponential growth of industrialization and urbanization. Ganga drains a basin of extraordinary variation in altitude, climate, land use and cropping patterns [1].

1.1.1 Source and Location

The source of Ganga water is the melting of snow in the Himalayas and monsoon rains. Mythologically it is believed that Gaumukh (cow's mouth) at an altitude of 4,000 m is the source of Ganga, though factually Ganga has its origin in different small streams at an altitude of more than 6,000 m or above.

The Ganga basin lies between longitudes 73°30 and 89°0 East and latitudes 22°30 and 31°30 North, covering an area of 1,086,000 km^2, extending over India, China, Nepal and Bangladesh. It has a catchment area of 8,62,769 km^2 in India, which means about 80 % of the basin area falls in Indian territory. The basin area in India constitutes 26 % of the country's land mass [2] and supports about 43 % of the population (448.3 million as per 2001 census). Out of the many river basins (12 river basins, and 14 minor and desert river basins) in India (Fig. 1.1), the Ganga basin – considered as the part of composite Ganga-Brahmaputra-Meghna basin, is the largest in India and fifth largest in the world [3]. The Gangetic plain is the most thickly populated and the oldest settled plain in the world and agriculture sustains that population. The average population density in the basin is over 550 individuals per square kilometer which in the delta zone rises to over 900 individuals per square kilometer [4].

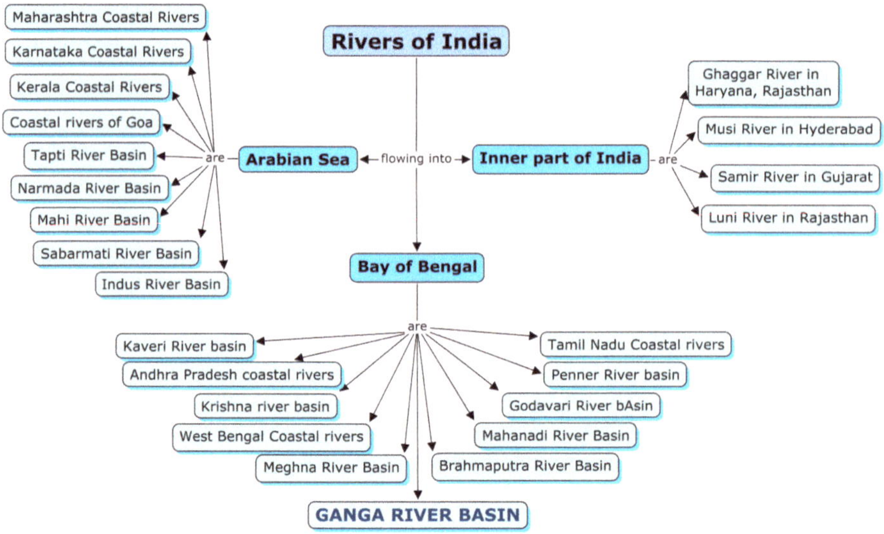

Fig. 1.1 River Basins in India (Source: Compilation from Wikipedia)

The Northern mountains, Indo-Gangetic Plains and Central Highlands together form the Ganga river basin (Fig. 1.2) which is bounded on the north by the Himalaya mountain ranges; in the west the Ganga basin borders the Indus basin and then the Aravalli ridge; on the south it is bounded by the Vindhyas and Chhotanagpur plateaus and on the east by the Brahmaputra ridge. The Ganga basin consists of hilly terrains of the Himalaya with dense forest, sparsely forested Shiwalik hills and the fertile Ganga plains [5]. The Ganga is the major river of the Indian subcontinent and most important perennial river. During its 2,525 km course, gouging a distance of about 220 km in the Himalayas, it enters the plain at Haridwar and flows southeast through the Indian states of Uttarakhand, Uttar Pradesh, Bihar, Jharkhand and West Bengal ultimately joining the sea at the Bay of Bengal [6] (Fig. 1.3). Of this total length, 1,450 km falls in Uttarakhand and Uttar Pradesh, 445 km in Bihar and Jharkhand and 520 km in West Bengal. The mean annual water discharge is the fifth highest in the world (18,700 m^3/s). The distribution of the geographical and catchment area of the Ganga River by state along with its population within India, is given in Table 1.1.

1.1.2 Headwaters

Although many small streams comprise the headwaters of the Ganga, the six longest headstreams: the Alaknanda, Dhauliganga, Nandakini, Pindar, Mandakini, and Bhagirathi rivers and their five confluences known as the Panch Prayag – all along the Alaknanda – are considered sacred. The five confluences in downstream

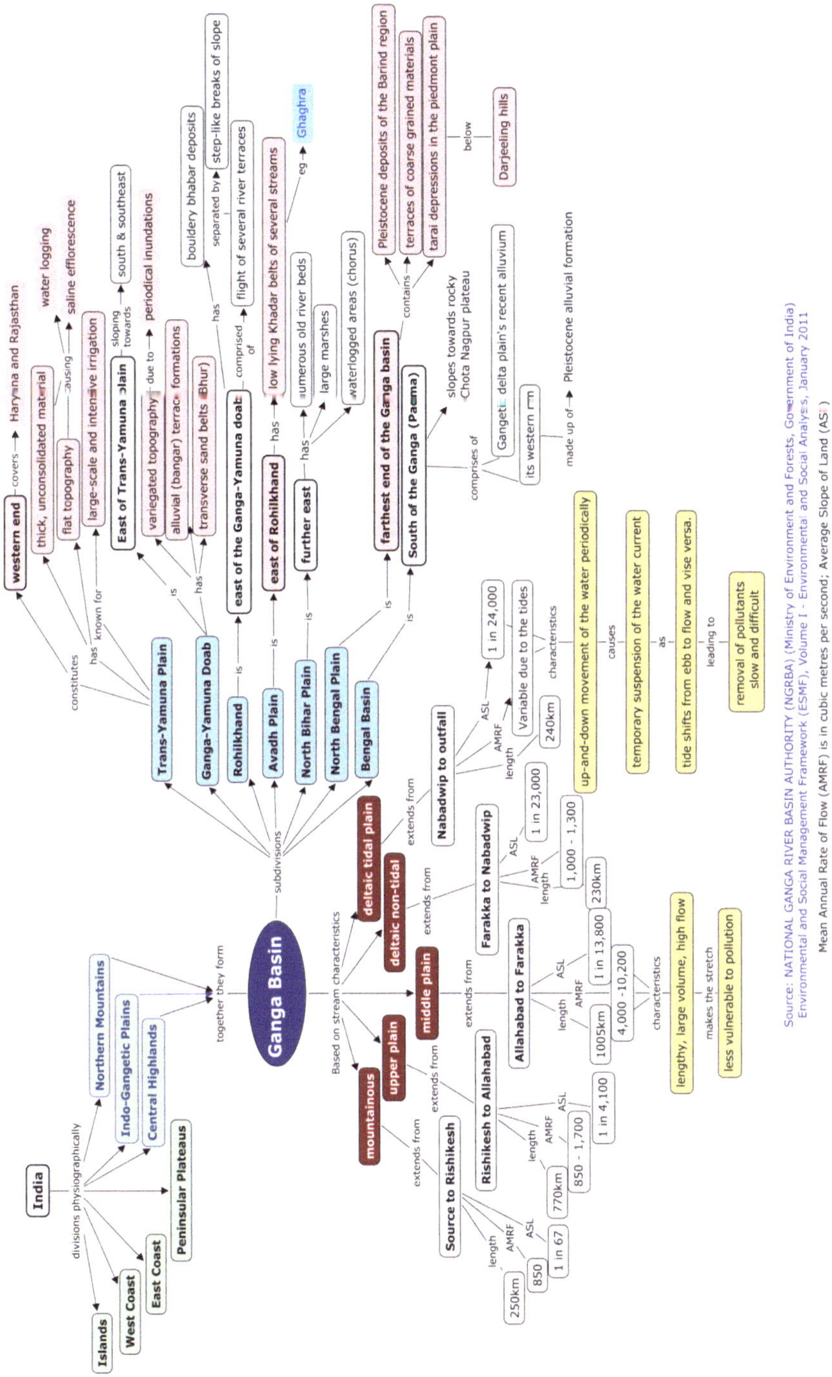

Fig. 1.2 Physiographical divisions of Ganga river

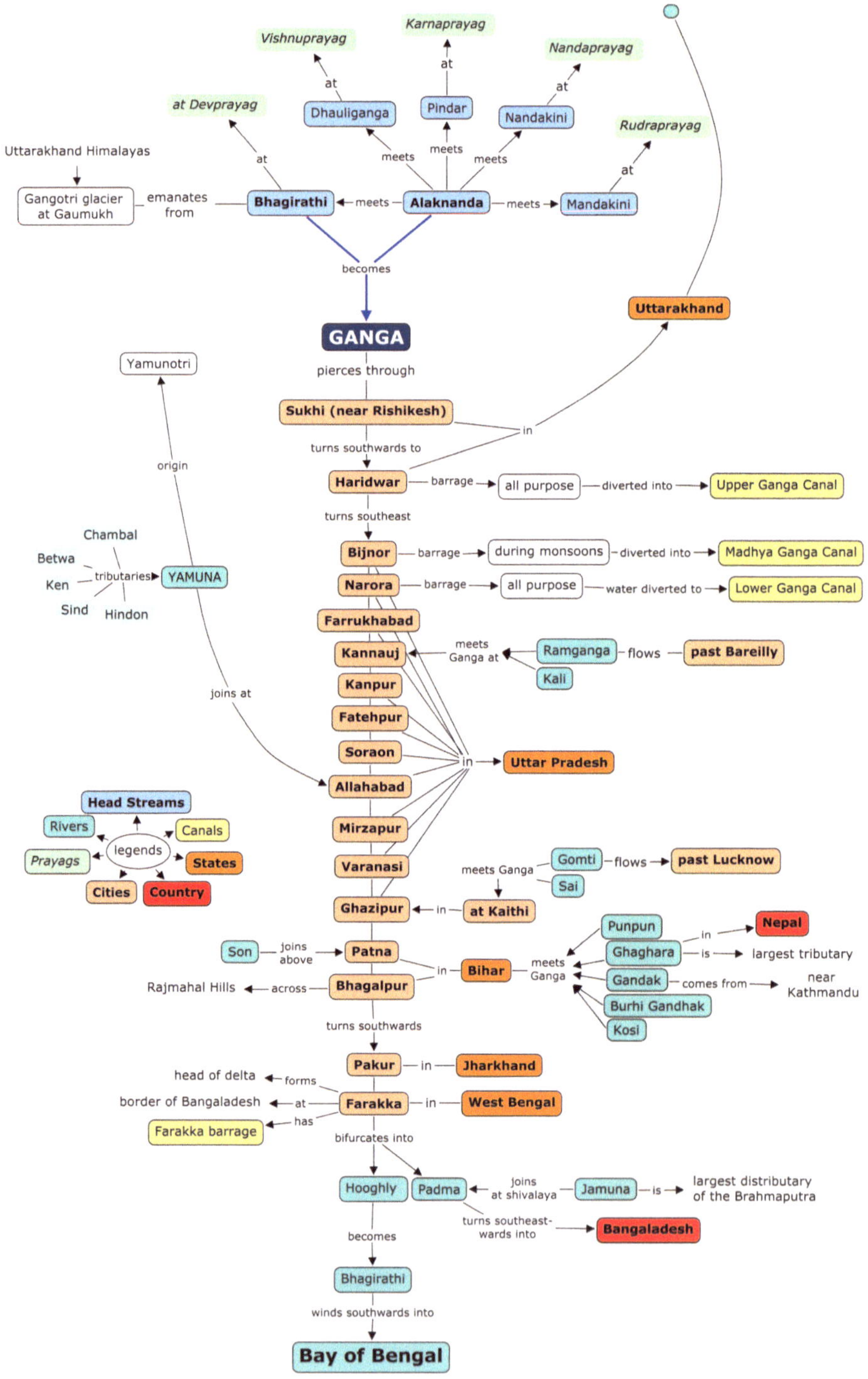

Fig. 1.3 Course taken by the river Ganga

Table 1.1 State-wise distribution of the drainage area of Ganga River Basin in India

State	Total geographical area (km^2)	% of total geographical area	Catchment area (km^2)	Population (in lakh)
Uttar Pradesh & Uttarakhand	2,94,364	34.2	2,94,411	1,391.12
Madhya Pradesh	1,98,962	23.1	2,00,252	379.42
Bihar & Jharkhand	1,43,961	16.7	1,36,925	803.26
Rajasthan	1,12,490	13.1	1,15,845	214.87
West Bengal	71,485	8.3	63,136	547.93
Haryana	34,341	4	33,422	129.66
Himachal Pradesh	4,317	0.5	6,201	4.61
Delhi	1,484	0.2	1,483	94.21
Ganga Basin (Total)	**8,61,404**	**100**	**8,51,675**	

Source for Geographical Area: Environmental and social management framework [7]; **Source for Catchment Area and population:** Forest survey of India, 1995

order are: Vishnuprayag, where the Dhauliganga joins the Alaknanda; Nandprayag, where the Nandakini joins; Karnaprayag, where the Pindar joins, Rudraprayag, where the Mandakini joins; and finally, Devprayag, where the Bhagirathi (which is considered to be the source stream of Ganga) joins the Alaknanda to form the Ganga River (Fig. 1.3).

1.1.3 Tributaries

The Ganga has a large number of tributaries (Fig. 1.4) with considerable water wealth, both in the Himalayan region before it enters the plains at Haridwar, and further downstream before its confluence with the Bay of Bengal. The river system covers cold upland streams and warm water stretches, including deltaic habitats. The freshwater flow in the river system is mostly from its tributaries and therefore, the water availability greatly varies from 59 billion m^3 at Allahabad, before the confluence with river Yamuna, to 459 billion m^3 at Farakka in the lower stretch [8]. The important tributaries to the left of Ganga are Mahakali, Karnali, Koshi, Gandak, Ghaghra, Mahananda, Gomti, Ramganga and to the right are Yamuna, Tons, Son and Punpun. Out of its 17 main tributaries Yamuna, Sone, Ghagra and Kosi contribute over half of the annual water yield of the Ganga. The Yamuna, although a tributary of the Ganga, is virtually a river by itself. Yamuna River originates at Yamnotri glacier, to the east of the Bhagirathi River (60 km away from Kedarnath) and runs 1,376 km parallel to Ganga and shares a common flow with Ganga only from Triveni Sangam at Allahabad. Its major tributaries are the Chambal, the Sindh, the Betwa and the Ken.

At Pakur, near Farakka, Ganga starts diverting to form its distributary 'Bhagirathi-Hoogly' which goes to form the Hoogly River. The main branch of Ganga enters Bangladesh and is known there as Padma River. It is fed by Jamuna and Meghna (distributaries of Brahmaputra) before merging into the Bay of Bengal. In central

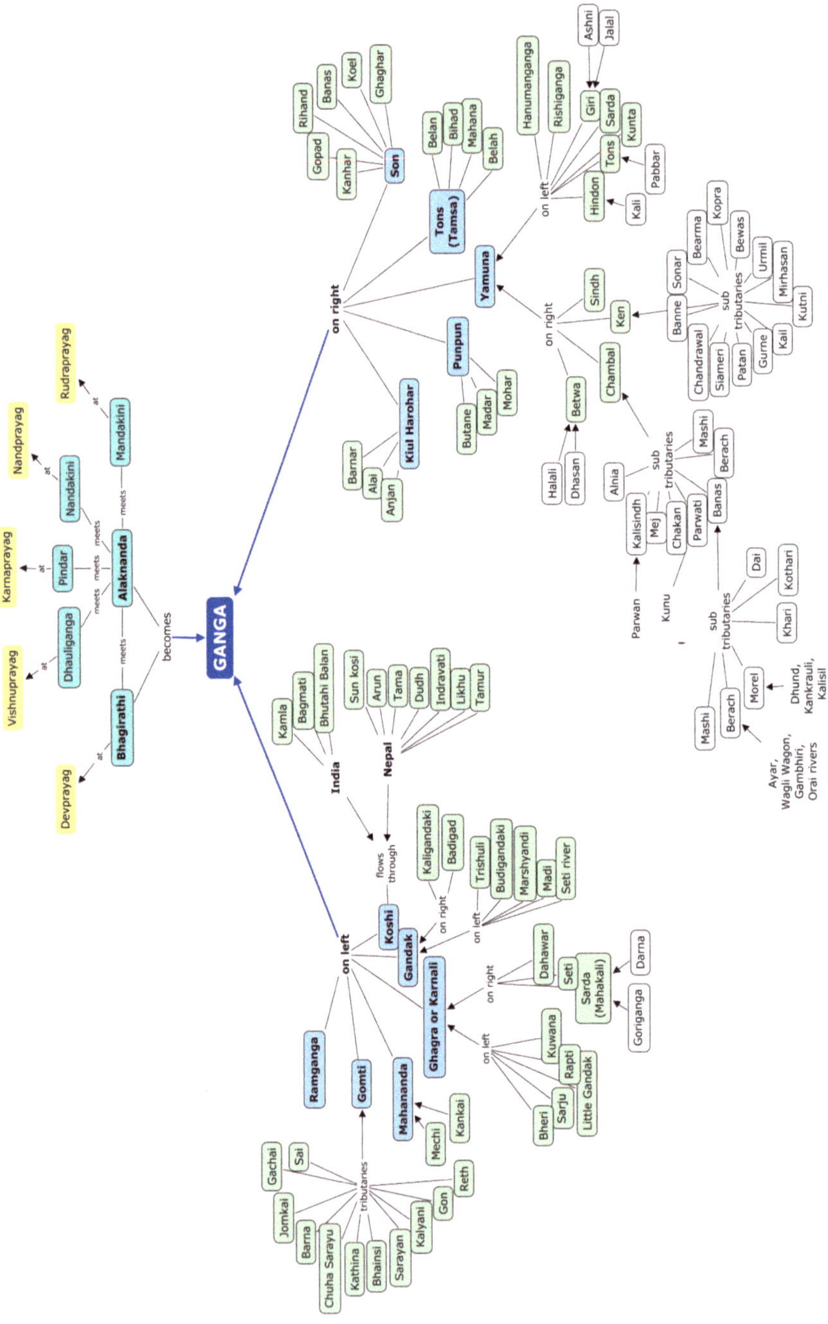

Fig. 1.4 Tributaries of the river Ganga

Bangladesh it is joined by the Brahmaputra river and Meghna rivers. Their combined waters (called the Padma River) empty into the Bay of Bengal and form a delta 354 km wide, which is shared by India and Bangladesh.

1.2 Key Features of River Ganga

The river Ganga is the 20th longest river in Asia and the 41st longest in the world occupying a special position in the ethos of the people of India. The Ganga is unique among several other rivers of the world because of its delta – the Sunderbans – which is the world's largest delta. Ganga has extraordinary variation in altitude, climate, flora and fauna, land use and cropping pattern. The Ganga was once famous for its ability to retain oxygen as it was the only river in the world which had 12 ppm of oxygen [9]. It has great importance in respect of culture, economy as well as ecology. This river is vastly used for agriculture, irrigation, power generation, human and cattle consumption, fish production, tourism, pilgrimage and recreation.

1.2.1 Ganga Water (Gangajal)

The Ganga water has an extraordinarily high rate of oxygen retention, allowing it to remain fresh during long storage periods. The amount of organic waste that goes into Ganga should have already exhausted the amount of oxygen. But this has not happened. Its water, widely known as *Gangajal*, has certain chemical qualities enriched with many extraordinary healing properties (Fig. 1.5). Indian environmentalists have confirmed that owing to its self cleaning properties *Gangajal* does not deteriorate or loose its special values even if it is kept in a closed vessel for years. There is some material present in Ganga that replenishes the oxygen content and kills the pathogens. Interestingly, boiled Ganga water does not have the capability to kill pathogens. Similarly, Yamuna, a river which joins Ganga, does not contain the special material present in Ganga (website 1). See Chap. 6 for a detailed description of the Gangajal, by an expert.

Subtle Quality of Ganga Water: Accept Until Proven Otherwise Rather Than Not Accepting Until Established Scientifically

अस्या जल्स्य गुणा : शीतत्वम्, स्वादुत्वम,स्वछत्वम, अत्यन्तरुच्यत्वम्, पथ्तत्वम्,
पावनत्वम्, पापहारित्वम्,तृष्णामोहध्वंसत्वम्, दीपनत्वम्, प्रज्ञाधारित्वंच, इति राजनिर्घणटः

Coolness, sweetness, transparency, high tonic property, wholesomeness, potability, ability to remove evils, ability to resuscitate from swoon caused by dehydration, digestive property and ability to retain wisdom.

– By Professor Vinod Tare from Indian Institute of Technology, Kanpur

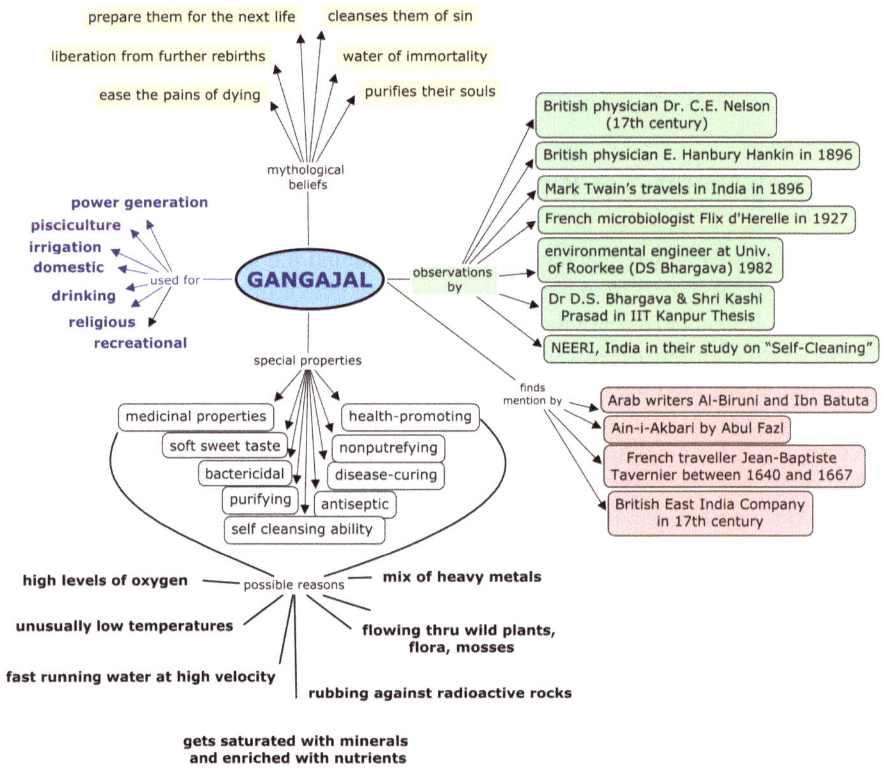

Fig. 1.5 Compilation on Gangajal

1.2.2 Soil Type

The perennial river Ganga originates in the grand Himalayas and the geology of Ganga River shows a wide variety in composition. The Ganga basin is characterized by a wide variety of soils, some of which are highly susceptible to erosion (Fig. 1.6). The Ganga river basin is one of the most fertile basins in the world and the important soil types found in the basin are sand, loam, clay and their combinations such as sandy loam and silty clay. The abundance of clay minerals like illite, chlorite, smectite and kaolinite in the Ganga basin has recorded varied degrees of physical and chemical weathering with dominance of physical weathering being prevalent in early post-glacial deposits.

To preserve such soil resources, adequate conservation measures and appropriate land management interventions are required, keeping the turbidity levels of the surface water within tolerable limit. If managed properly, the alluvial soils are highly fertile soils, capable of producing the highest possible yields of major crops central to the agricultural economies of India and Bangladesh and to feed the

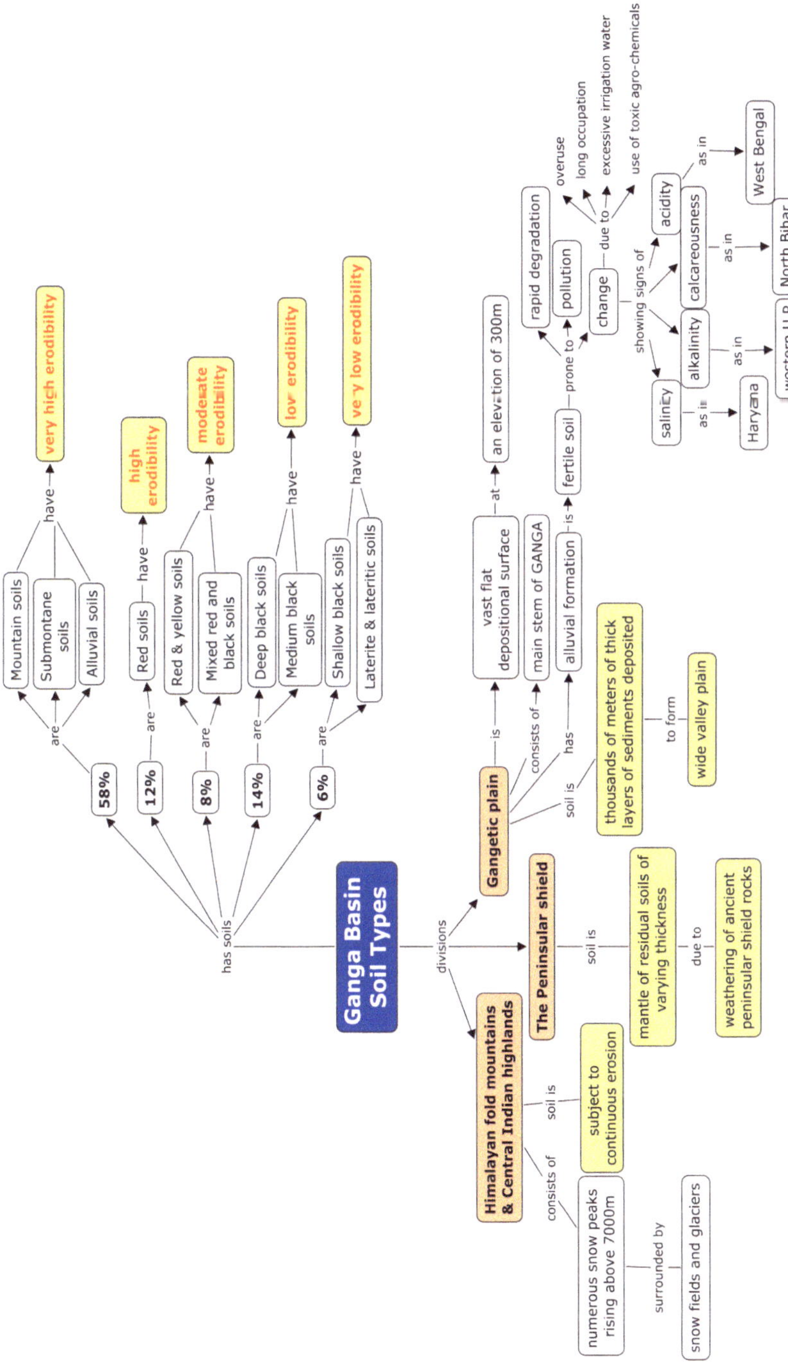

Source: STATUS PAPER ON RIVER GANGA, State of Environment and Water Quality National River Conservation Directorate; Ministry of Environment and Forests August, 2009; Alternate Hydro Energy Centre, Indian Institute of Technology Roorkee

Fig. 1.6 Soil types of river Ganga

millions. The Ganga and its tributaries provide a constant source of irrigation over an extensive area that produces a wide variety of crops. The major ones cultivated include rice, lentils, sugarcane, potatoes, oil seeds and wheat. Near the banks of the river, the existence of swamps and lakes provide a rich fertile area for crops like legumes, chilies, sesame, mustard, sugarcane, and jute.

1.2.3 Vegetation

The vegetation type of the Ganga basin is largely comprised of tropical moist and dry deciduous types, but also includes a few additional varieties. The vegetation type of Ganga basin is illustrated in Fig. 1.7 [7].

1.2.4 Biodiversity

Rivers basins are ecosystems, the way in which nature gathers and delivers water for human use providing – habitat for diverse fauna and flora, transport, recreation and tourism and also the power generation. The Ganga river basin is known for rich biodiversity, which sustains diverse group of flora and fauna supporting abundant biological wealth (see Chap. 3). The faunal resources of the river Ganga can be categorized into three distinct zones: the Upper Ganga in hilly terrain flowing within Uttarakhand; the middle Ganga, which flows through Uttar Pradesh, Bihar, Jharkhand and West Bengal and finally the lower Ganga in deltaic tract [10]. 1960s and 1970s were periods of unprecedented destruction of aquatic life and its habitat [11]. Though the upper portion is the cleanest and contains a wide range of biodiversity and fragile ecosystems, the middle section is the most polluted, due to human actions, while in the lower segment, the sediment flowing in from Nepalese tributaries cause severe problems.

The Ganga river basin is characterized by its rich fisheries and faunal diversity. Apart from being the original abode of the most prized carp species of the subcontinent, *viz., Catla catla, Labeo rohita, Cirrhinus mrigala* and *Labeo calbasu*, the river sustains fisheries of large catfishes, mahseer, hilsa and other fish species. Ganga is also the major source of riverine spawn, which meets 30 % of the carp spawn requirements of the fish culture sector in India. Over the years there has been a measurable shift in the hydrology and water quality of the river. The fish catch from the river has declined significantly and species composition changed more in favour of non major carp and miscellaneous species. Exotic fishes have invaded some stretches of the river. Hence the status of the ecology and fisheries of the river Ganga has been a serious concern [12]. There are over 140 fish species, the richest freshwater fish fauna in India [13, 14], 90 amphibian species, and five areas supporting birds found nowhere else in the world. The basin is home to five species of freshwater cetaceans including the endangered Ganga River Dolphin which faces an annual mortality rate of 10 % [14] and the rare freshwater shark, *Glyphis gangeticus* [15].

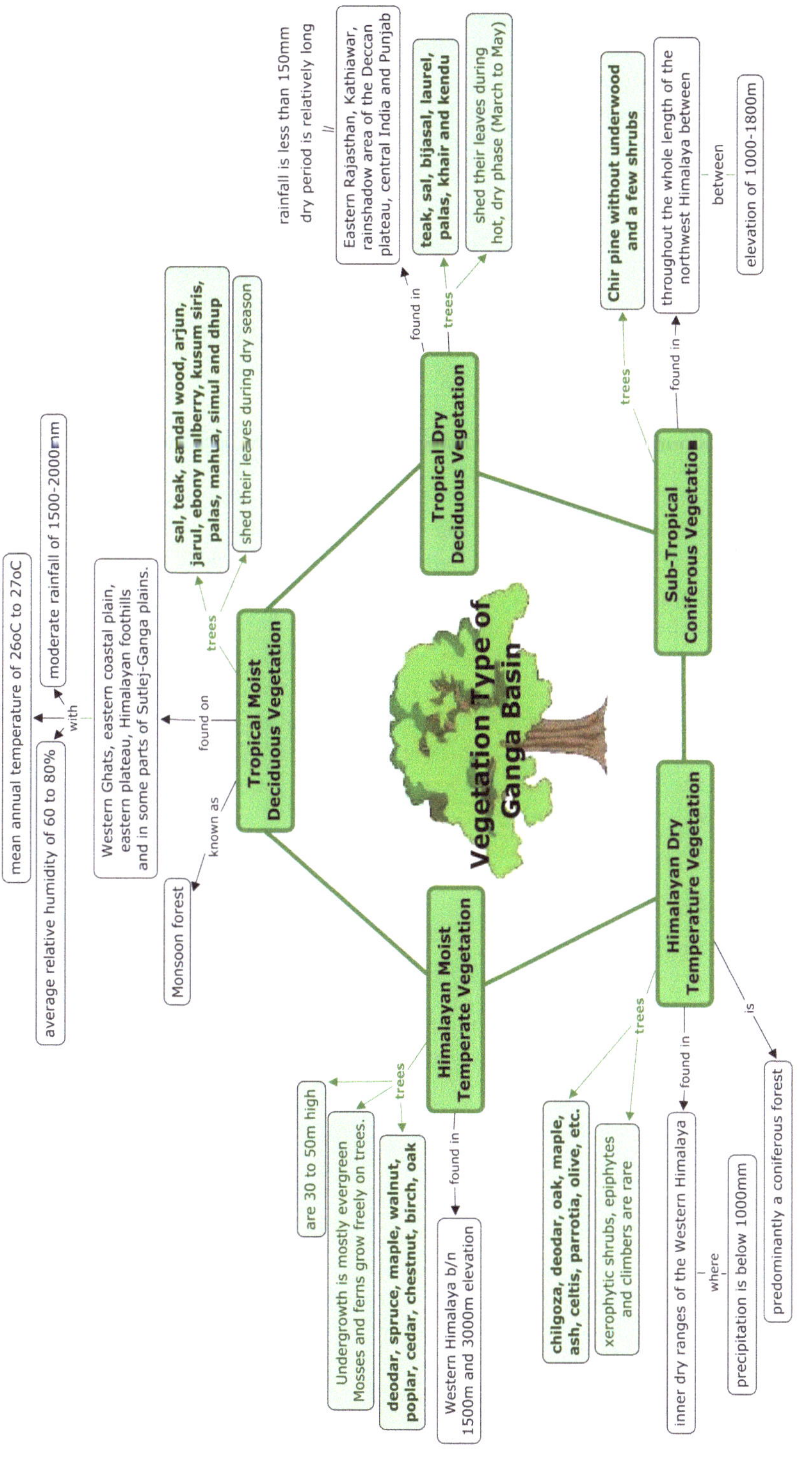

Fig. 1.7 Some vegetation types of river Ganga

Being the only mammalian predator of the Ganga, the dolphin occupies the apex of the food chain and plays a vital role in maintaining and balancing the ecosystem (see Chap. 4). As is rightly said "as tiger is to forest, dolphin is to Ganga". In spite of a species representing an ecosystem is in need of conservation, its status as a flagship is a matter of serious concern. Practices like sand mining destroy the dolphin habitat by smothering the bottom fauna and lowering the river productivity. Construction of barrages restricts the movement of dolphins in water which may inhibit genetic, social as well as ecological interactions among individuals, thereby limiting the gene flow, increase their vulnerability to natural catastrophes and ultimately lead to their extinction. Poaching (its oil has very high medicinal value), siltation, extensive fishing and excessive water extraction are the major factors leading to its disturbed existence. Gangetic dolphins have been included in the Schedule-I of Wildlife (Protection) Act, 1972.

Anthropogenic activities are impacting the rich and diverse biodiversity of the river. Ever growing population causes the change in water and land use patterns and irregular water flow from the reservoirs in the upper reaches. Irrigation canals and barrages on the Ganga decrease the water flow and the river is too shallow for dolphins to navigate. Over fishing and sand mining activities lead to habitat destruction of turtles and disturbs the nesting and basking ground for the crocodiles. Some endangered species of Ganga are shown in Fig. 1.8.

Aquatic species have been heavily exploited in the last few decades, which have pushed them near to extinction and are highly endangered due to many reasons. Some major ones being habitat destruction and defragmentation [16], water abstraction for irrigation, industries and private use which leads to reduced flows and also compounds the problems of pollution [17–20], exotic species introduction [21], pollution [22] due to discharges from the industries, cremation activities and agricultural activities, high sensitivity to the quantitative and qualitative alteration of aquatic habits [23, 24].

1.2.5 Irrigation Systems

This river system has been facing major disruptions since the 1850s. The first and oldest canal system on this river was commissioned in the year 1854.

> The Ganga Canal (now called Upper Ganga Canal) is the first irrigation system on the Ganga, which was commissioned during 1854. The construction of this canal system was done under the supervision of the British engineer, Sir Proby Thomas Cautley. The main diversion structure of this system is at Bhimgoda, Haridwar. The objective of this canal system was two pronged; one is to provide irrigation and to initiate navigation through the canal. It is for this reason, it was recommended that a discharge of 6,750 cusec (191 cumec) would be adequate to irrigate 5,96,474 ha area and to carry one main line of navigable canal from Kankhal (near Haridwar) to Kanpur [25].

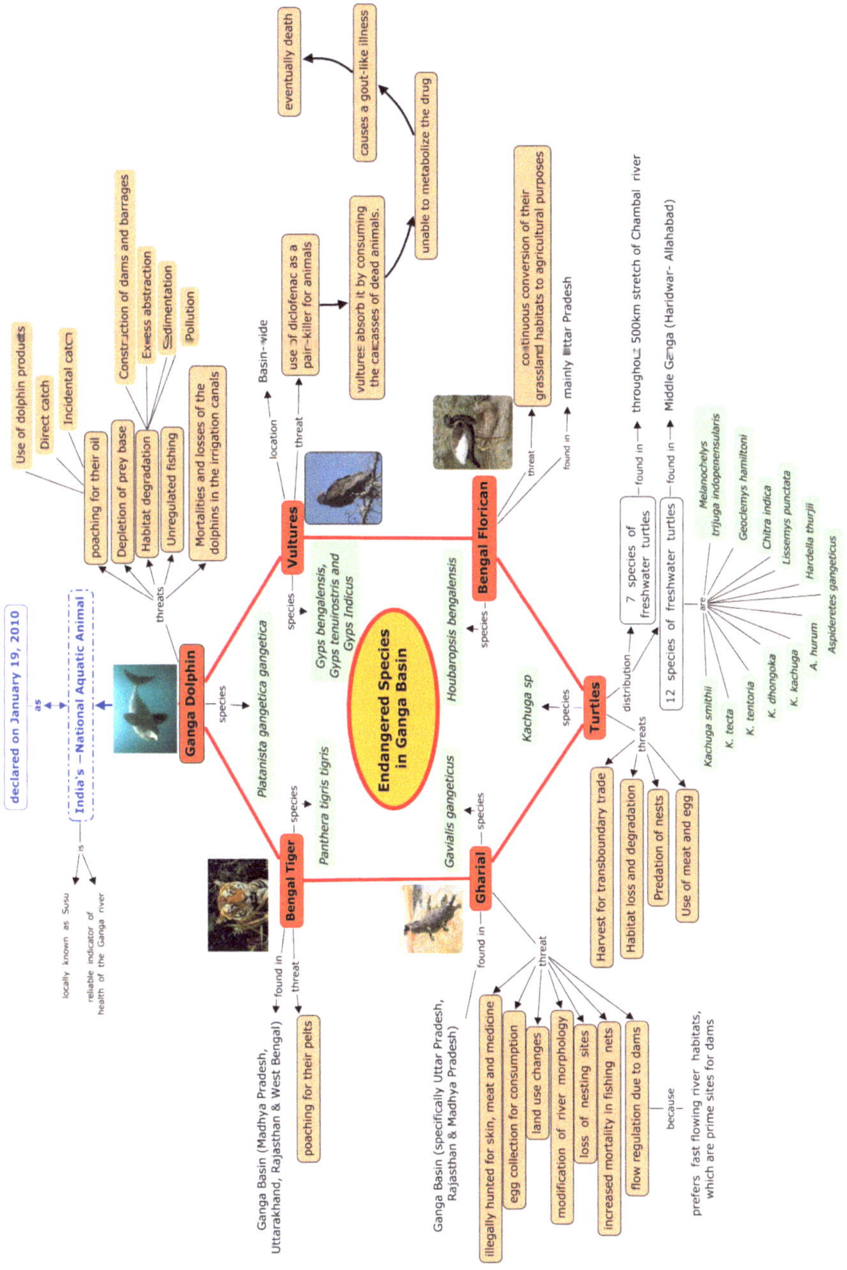

Fig. 1.8 Some endangered species in Ganga basin

Later on in year 1878, the Lower Ganga Canal was commissioned (Source: UP Irrigation Department). More recently, the Madhya Ganga Canal and the Eastern Ganga Canal were also built and commissioned. Whilst the Ganga river system is one of the major contributors of the prosperity of the region, it also caters for drinking and other domestic needs of the people living on its banks. Lately, the plentiful water resources of the river have also led to proliferation of industries at its banks. Consequently, these developments have posed various threats to the very sustainability of the river.

1.2.6 Environmental Habitats in Ganga Basin

Ganga basin has a diversity of biological wealth distributed in its forests, wetlands, freshwater river channels and in its marine areas. These constitute the ecologically fragile zones and need to be protected from pollution and abatement of flow rates in various sections of the river. At present about 31 % of the National parks in India are located in Ganga Basin along with 15 % of wildlife sanctuaries protecting some of the most endangered species such as Bengal Tigers, Gangetic Dolphins and Vultures. Mangroves which are amongst the more resilient and very unique ecosystems on the planet are also located in the Ganga basin in Sundarbans (West Bengal) [7].

- Out of 15 Biosphere reserves in India, 2 are within Ganga Basin – Nanda Devi Biosphere and the Sundarbans National Park (Fig. 1.9)
- Out of 94 National parks in India 29 are within Ganga Basin (Fig. 1.10)
- Out of 502 animal sanctuaries in India, 75 are located in the Ganga basin
- Out of 28 tiger reserves, 10 tiger reserves are found in Ganga basin (Fig. 1.11)

1.2.7 Sundarbans – A Unique Entity

The Sundarbans are the largest littoral mangrove belt in the world, stretching 80 km into the Bangladeshi and Indian hinterland from the coast (see Chap. 5). The Sundarbans has been declared a UNESCO World Heritage Site (website 2) (Fig. 1.12). The forests are not just mangrove swamps though, they include some of the last remaining stands of the mighty jungles which once covered the Gangetic plain. They cover an area of 10,500 km^2, of which about one-third is covered in water/marshy area. Since 1966 the Sundarbans have been a wildlife sanctuary, and it is estimated that there are now 400 Royal Bengal tigers and about 30,000 spotted deer in the area (website 3). The Sundarbans National Park, a Tiger Reserve, and a Biosphere Reserve are located in the Sundarbans delta in the Indian state of West Bengal. Sundarbans South, East and West are three protected forests in Bangladesh. This region is densely covered by mangrove forests, and is one of the largest reserves for the Bengal tiger.

The Sundarbans mangrove forest, one of the largest such forests in the world (140,000 ha), lies on the delta of the Ganga, Brahmaputra and Meghna rivers on the Bay of Bengal. It is adjacent to the border of India's Sundarbans World Heritage site

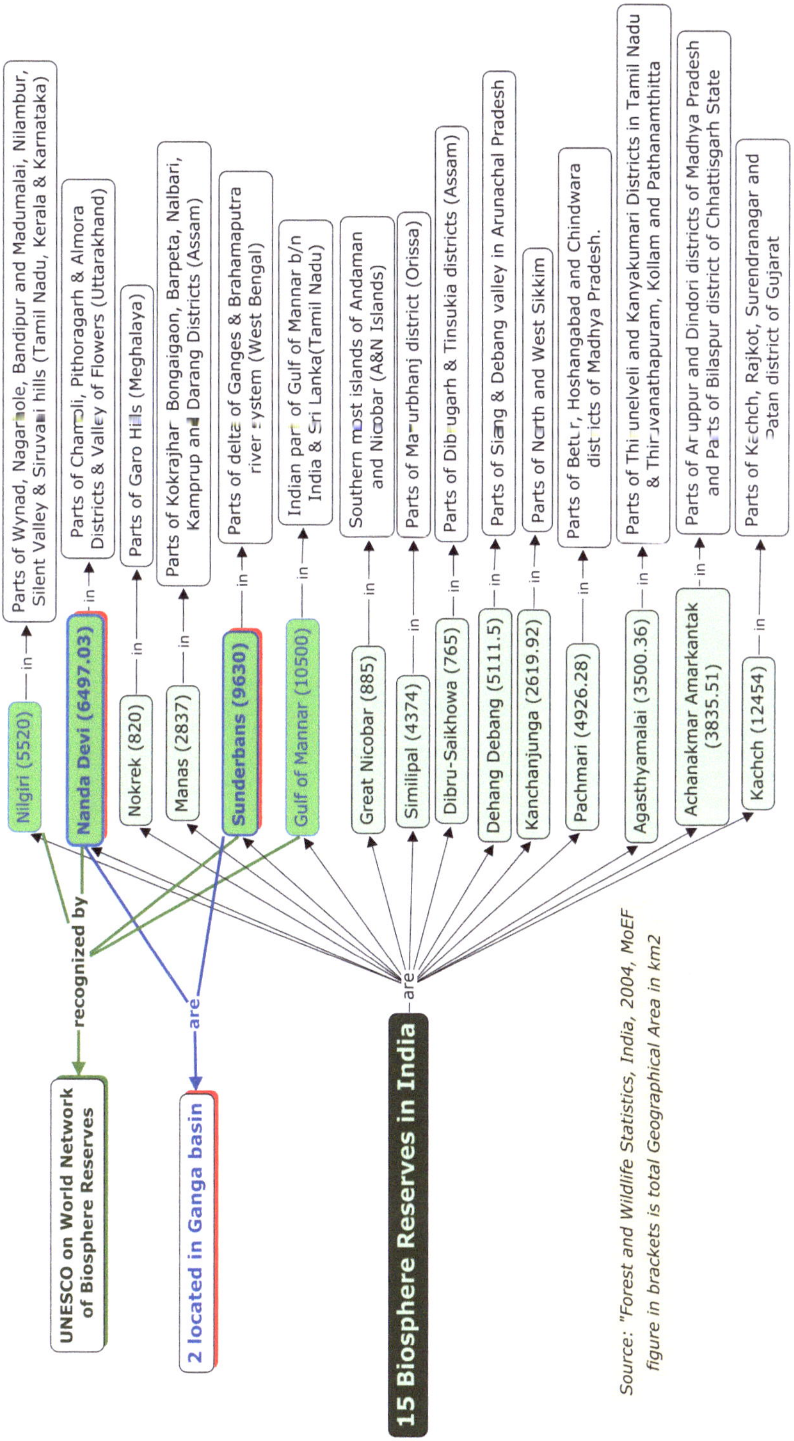

Fig. 1.9 Biosphere reserves in India

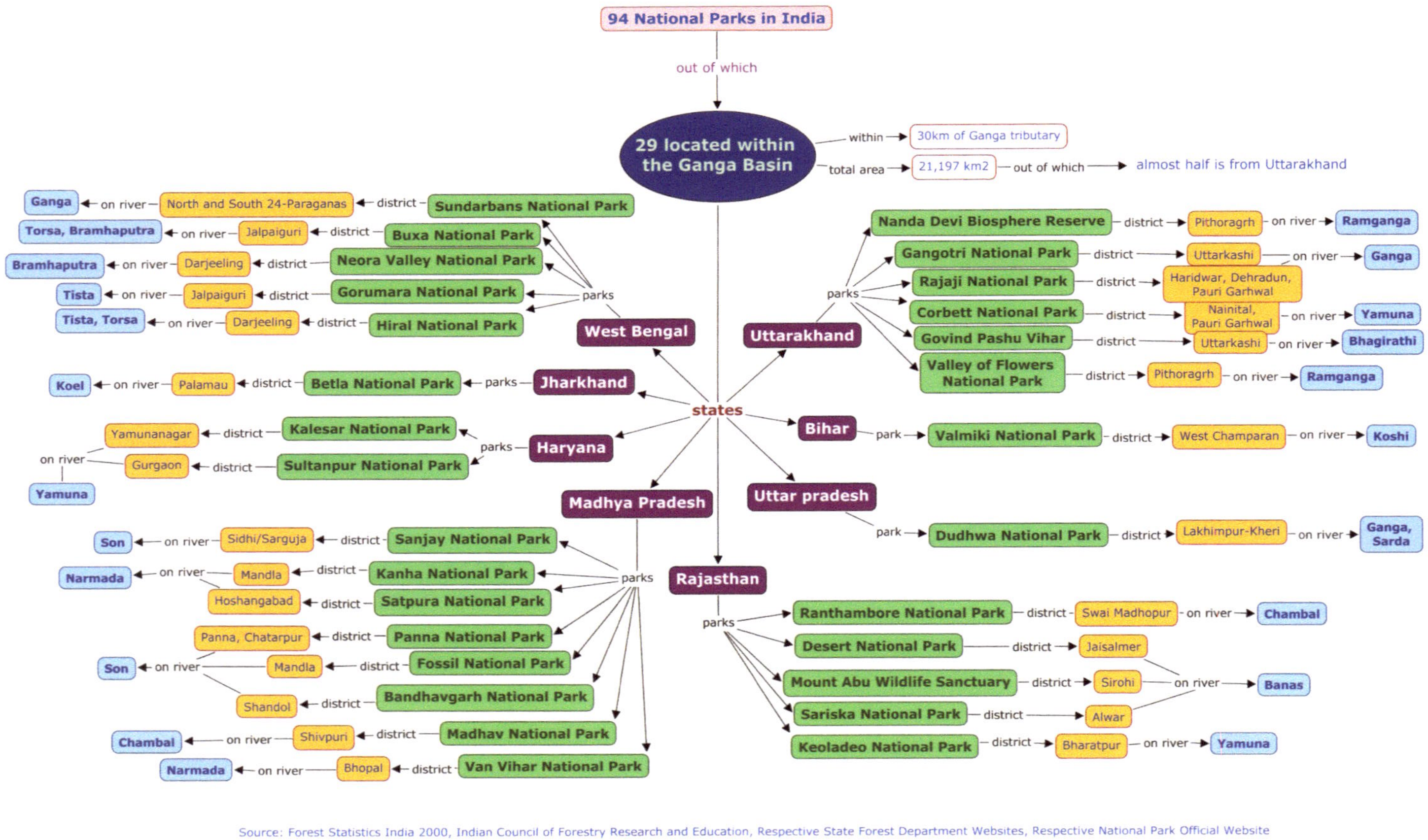

Fig. 1.10 National Parks located within Ganga Basin

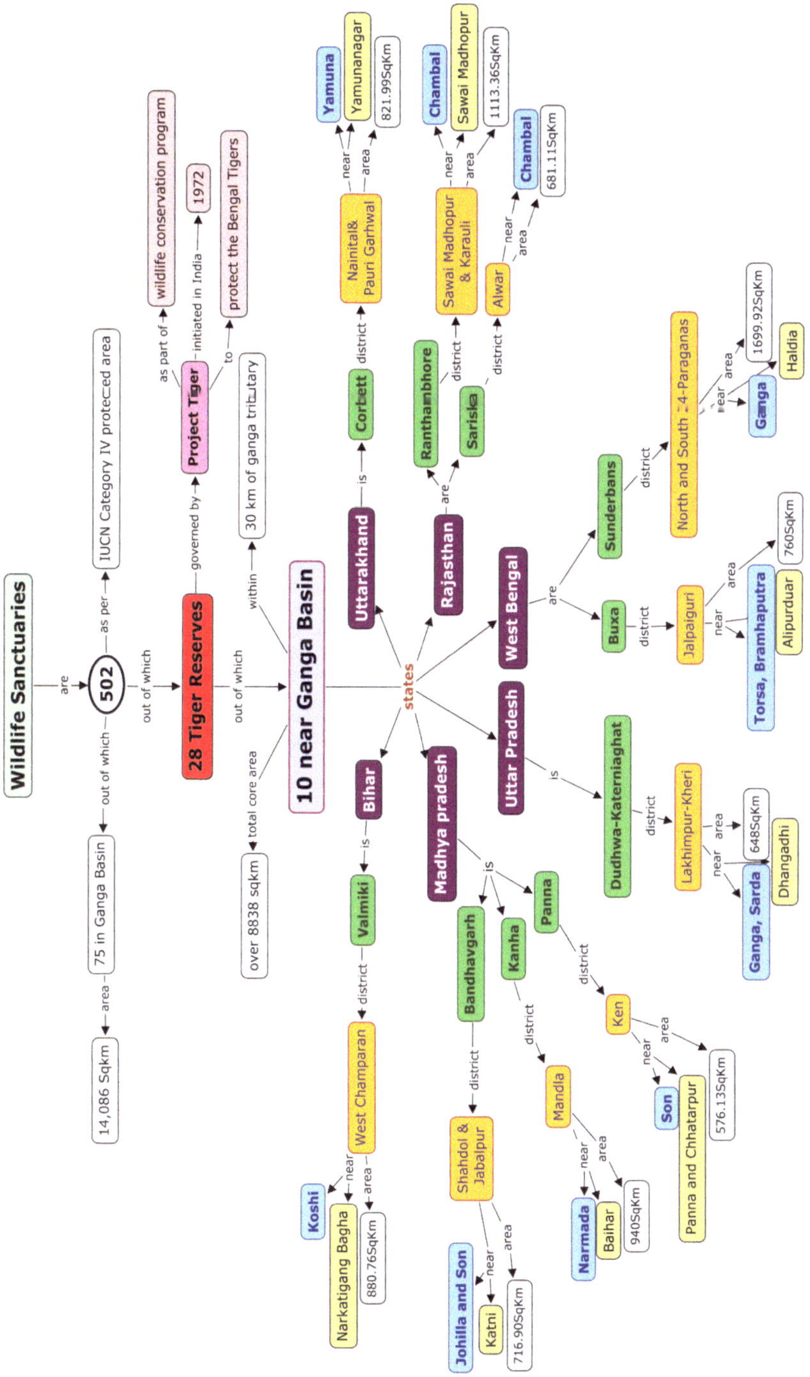

Source: Project Tiger Task Forest Reports (2004, 2010), Project Tiger Website

Fig. 1.11 Tiger reserves in Ganga Basin

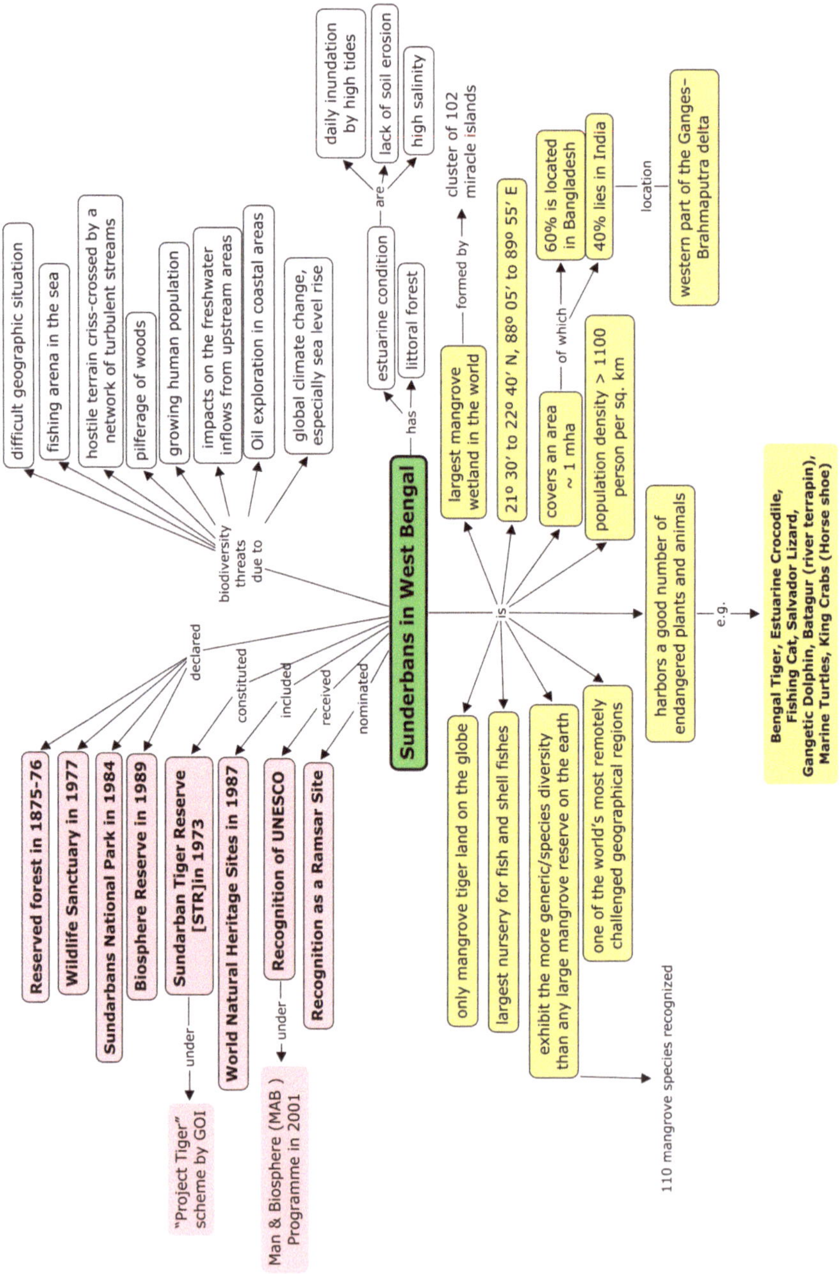

Fig. 1.12 Sunderban mangroves in Ganga Basin in India

inscribed in 1987. The site is intersected by a complex network of tidal waterways, mudflats and small islands of salt-tolerant mangrove forests, and presents an excellent example of ongoing ecological processes. The area is known for its wide range of fauna, including 260 bird species, the Bengal tiger and other threatened species such as the estuarine crocodile and the Indian python (website 3). Together the Brahmaputra and Ganga watersheds span 10 biomes and contain the widest diversity of all large river systems as classified by Nilsson et al. [26].

According to the study 'Snow and Glaciers of the Himalayas' conducted by the Union Ministry of Environment and Forests in association with the Ahmedabad-based Space Application Centre of ISRO over a period of 4 years from 2004–2005 to 2007–2008, there are 32,392 glaciers in the Indus, Ganga and Brahmaputra basins that drain into India [27]. The country alone has 16,627 glaciers, covering an area of 40,563 km^2. The Ganga basin has 6,237 glaciers occupying 18,392.90 km^2 of glaciated area. There are seven glaciated sub-basins in the Ganga basin. This study is of great significance as 2,500 glaciers have been monitored to estimate glacial advance. It is the first time that such a detailed inventory with the help of satellite data has been published that contains figures indicating the advance or retreat of glaciers.

1.3 Cultural Significance of River Ganga

River Ganga is known for its physical and spiritual sustenance with powers of healing and regeneration. It holds an immense sacred position in the *hindu* religion and finds its mention in many ancient Indian religious books such as the *Vedas*, the *Puranas*, the *Ramayana* and the *Mahabharata*. River Ganga is mentioned in the *Rig Veda*, the earliest of the *Hindu* scriptures. It is found mentioned in the *Nadistuti*, which lists the rivers from east to west. River Ganga is known by many names and is associated with many legends and myths and is one of the few living Goddesses in the Hindu pantheon (see Appendix). Ganga is worshipped by Hindus, who believe that bathing in the river causes the remission of sins and facilitates liberation from the cycle of life and death. Pilgrims travel long distances to immerse the ashes of their kin in the waters of the Ganga, bringing their spirits closer to *Nirvana* (salvation).

In Hindu mythology, the river is a symbol *par excellence* of purity and the purifying power to all that she touches. Her entire course is a pilgrimage route for the millions of *Hindu* devotees who visit the preeminent *tīrtha*s that mark her path. Emotional attachment to the river and to the centers of pilgrimage (Rishikesh, Haridwar, Allahabad, Varanasi and Patna) on its banks runs deep and long in the

Indian history. In reverence '*Ganga snan*' – a holy dip in the river – is considered to rid one of the sins committed. It is believed to be the gateway to the heavenly abode for the pious souls, its banks have been places of worship to Gods for soul searching and cleansing. Its emotional, spiritual, socio-cultural, and historical place in Indian civilization has given the Ganga its uniqueness. Pilgrims visit these places to bathe in the Ganga, to drink her water, to worship the river, to perform various religious and cultural functions and to chant her holy name. More than 60 million people came to the city of Allahabad for pilgrimage in January 2007, making it the largest gathering in the world. On important *Hindu* festivals/occasions, millions of people converge on the river in select cities to pray and bathe in the waters, and for them a clean Ganga holds great value. It is also a lifeline to millions of Indians who live along its course and depend on it for their daily needs.

Kumbh Mela is a mass Hindu pilgrimage attracting the world's largest congregation of religious pilgrims. The pilgrims gather at the Ganga and river Godavari, where bathing for purification from sin is the gateway to *moksha* (Salvation). The *Kumbh Mela*, the greatest of the Indian cultural and religious fairs commemorates the legend of the struggle between the Gods and Demons over the *Amrita Kumbha* (Pot of Nectar of Immortality). It is believed that during the battle, Lord *Vishnu* flew away with the *Kumbha* of elixir, drops from which spilled over Allahabad (*Prayag*), Haridwar, Ujjain and Nashik which are now the pilgrimage places (see Fig. 1.13). The Mela takes place on a rotational basis in four of the most holy Hindu places in India – on the banks of the Godavari river in Nashik (Maharastra), the Shipra river in Ujjain (Madhya Pradesh), the Ganges river in Haridwar (Uttarakhand), and at Prayag in Uttarpradesh – convergence of the Ganges, Yamuna, and Saraswati rivers.

The Maha Kumbh Mela 2013 was described as the "greatest socio-cultural show on earth" with over 100 million people visiting the city of Allahabad, Uttar Pradesh on the banks of the River Ganga from January 14 to March 10, 2013. Millions of cultural tourists and pilgrims visit the confluence of Ganga, Yamuna and Saraswati (called Sangam) and take a holy dip for satisfaction of their spiritual aspirations in hope of attaining salvation. Kumbh is like a magnet attracting people not just from different parts of India but from across the globe who travel a long way to witness the mega gathering of humanity on the banks of the Ganges in Allahabad. Kumbh mela is a meeting of mystical minds and brings holy men together to discuss their faith and disseminate information about their religion. It is a conference of holy saints and religious leaders to share their knowledge under one common platform.

The *Chota Char Dham* is an important *Hindu* pilgrimage circuit in the Indian Himalayas located in the Garhwal region of the state of Uttarakhand (see Fig. 1.14). The circuit consists of four sites – Yamunotri Gangotri, Kedarnath, and Badrinath. Each of these sites is unique in its own way, and are viewed together as *Chota Char Dham* in pilgrimage practice. Originally, ***Char Dham*** was the umbrella term for India's most popular pilgrimage circuit, the four important temples-Puri, Rameshwaram, Dwarka and Badrinath-clubbed together by the eighth century saint *Shankaracharya* into the all-India pilgrimage circuit to the four cardinal points of the country.

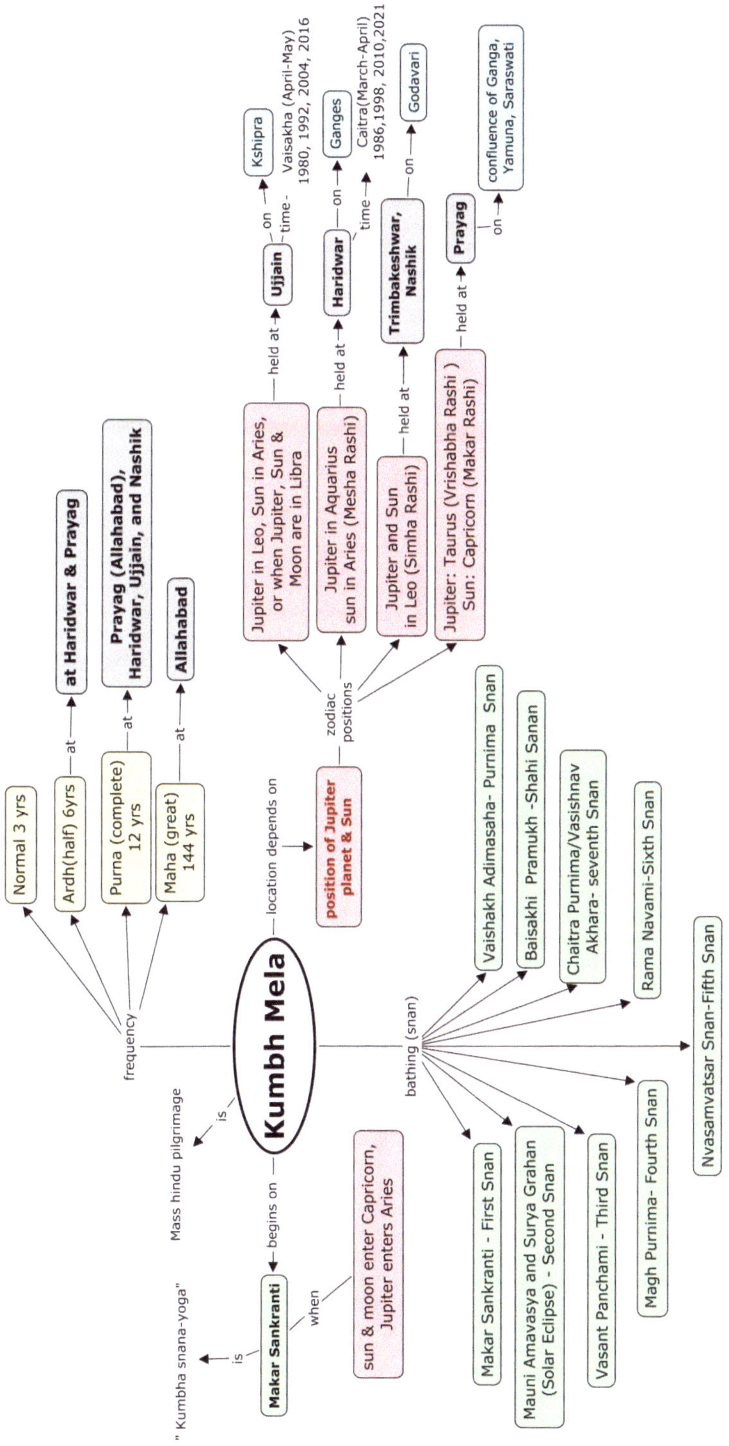

Fig. 1.13 More about Kumbh Mela

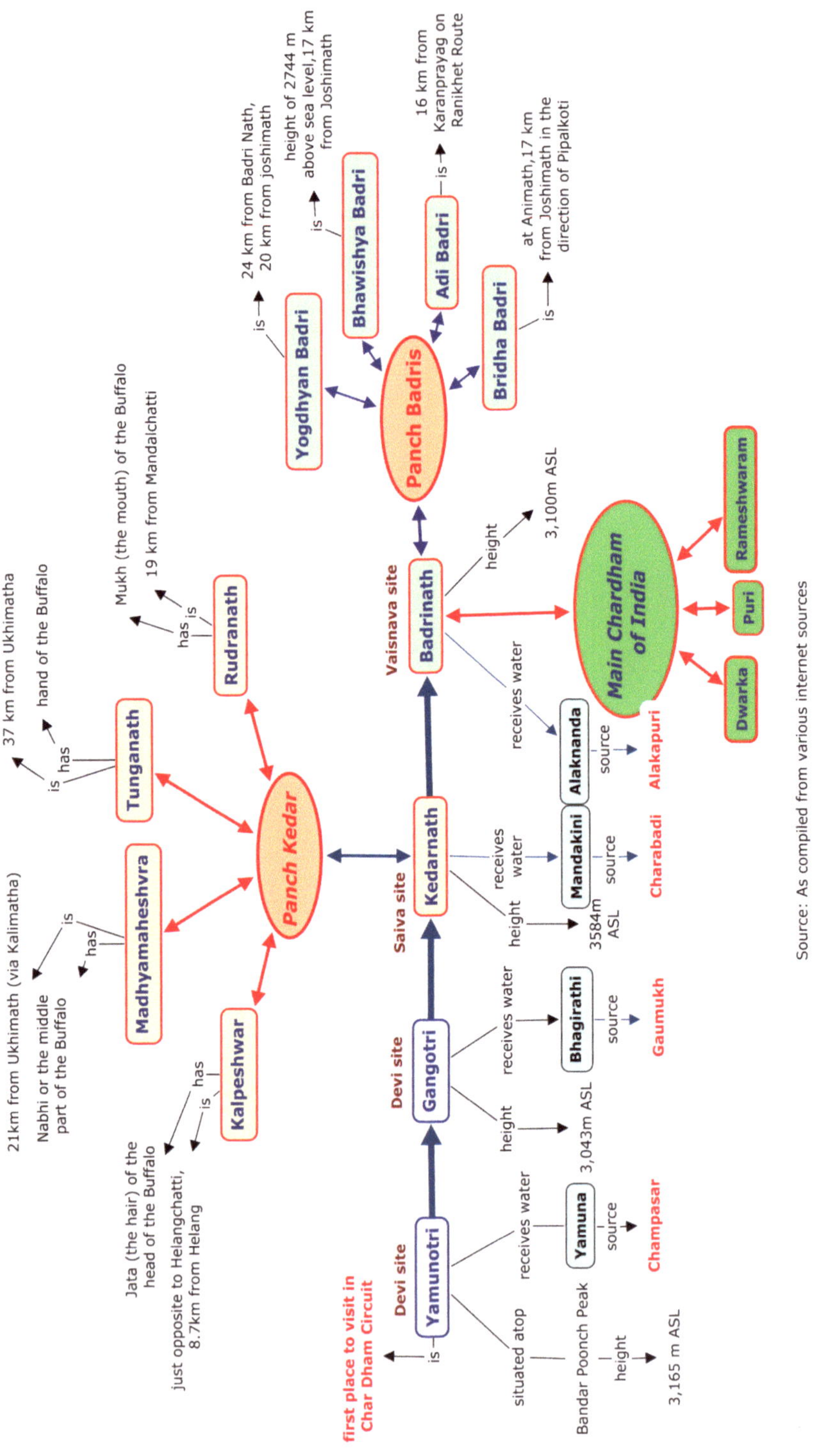

Fig. 1.14 The chota chardham circuit

As per Professor Vinod Tare from Indian Institute of Technology (Bulletin GBP/IIT/QNL/5-6, Nov 4, 2011), seven of the Ten Criteria of the UNESCO's Revised Operational Guidelines for the Implementation of the World Heritage Convention (2005) considered to be met by the river Ganga are:

1. Exhibit an important interchange of human values, over a span of time or within a cultural area of the world, on developments in architecture or technology, monumental arts, town- planning or landscape design;
2. Bear a unique or at least exceptional testimony to a cultural tradition or to a civilization which is living or which has disappeared;
3. Be an outstanding example of a traditional human settlement, land – use or sea – use which is representative of a culture (or cultures), or human interaction with the environment especially when it has become vulnerable under the impact of irreversible change;
4. Contain superlative natural phenomena or area of exceptional natural beauty and aesthetic importance;
5. Be outstanding examples representing major stages of earth's history, including the record of life, significant on-going geological processes in the development of landforms, or significant geomorphic or physiographic features;
6. Be outstanding examples representing significant on – going ecological and biological processes in the evolution and development of terrestrial, fresh water, coastal and marine ecosystems and communities of plants and animals;
7. Contain the most important and significant natural habitats for in – situ conservation of biological diversity, including those containing threatened species of outstanding universal value from the point of view of science or conservation.

Note: Fulfillment of the above criteria is considered sufficient for declaration of Ganga as a World Heritage Site. Addition of any such recognition may provide impetus for conservation of river Ganga.

1.4 Gloomy State of Ganga

Ganga is the lifeline for Indians who rely on the river for agriculture and irrigation of crops. Agriculture is the major livelihood activity of majority of rural population in the Ganga Basin and is thus the major consumer of water in the basin area. However, due to rapidly increasing population, rising standards of living and exponential growth of industrialization and urbanization leading to alarming pollution level, developmental activities, and unsustainable collection and poaching, the living resources in the Ganga are threatened. As a result, there is a strong demand and competition for natural resources, especially water for domestic use and irrigation, and several key tributaries of the basin are regulated by barrages.

A new global study by World Wide Fund for Nature (WWF) has included River Ganga in the world's top 10 most endangered rivers at risk (Table 1.2) [28] threatening the livelihoods of millions of people. According to this report, the over-abstraction of water is the key reason for such state of river Ganga; however the problem further compounded due to increasing domestic and industrial pollution load into the river.

The environment of Ganga basin is also deteriorating with time due to severe natural episodes of periodic floods and storms as well as anthropogenic factors such as population growth, deforestation, agricultural activities, urbanization, fertilizer and fossil fuel consumption and construction activities such as dams and bridges. All these have inconceivable adverse impacts on the health and natural regeneration capacity of the river basin. The presence of micro pollutants in water and sediments of this river turns the system into being unsustainable to the biota [29]. Ganga, is gradually losing its oxygen absorption and retention capacity and today, its oxygen has reduced from 12 ppm to 4–8 ppm. Ganga is under great threat – from pollution, but above all from a rapidly modernizing India whose appetite for Ganga's water far outstrips the river's capacities [30].

Point and non-point sources of pollutants such as industrial effluents, sewage, hospital wastes, radioactive materials, toxic chemicals released from various sources, solid wastes of non-degradable building materials, pesticides and insecticides, number of dead body disposal, pollutants released from cremation grounds, temple wastes, quantity of plastics etc are all causes for its deteriorating state (Fig. 1.15). Thousands of people bathe in the river every day. About 95 % of this pollution is caused by sewage discharge into the river from the point sources of pollution. The level of this pollution can be quantified by BOD (Biological Oxygen Demand) in mg/l of the river water, which is a measure of organic matter

Table 1.2 Summary of threats as per WWF for the ten most endangered rivers of the world

River Basin	**Threat**
Salween-Nu	Infrastructure–Dams
Danube	Infrastructure–Navigation
La Plata	Infrastructure–Dams and Navigation
Rio Grande-Rio Bravo	Water Over-extraction
Ganga	**Water over-extraction and pollution**
Indus	Climate change
Nile-lake Victoria	Climate change
Murray-Darling	Invasive species
Mekong-Lancang	Over fishing
Yangtze	Pollution

Source: World's top 10 rivers at risk (2007). WWF – International, Gland Switzerland [28]

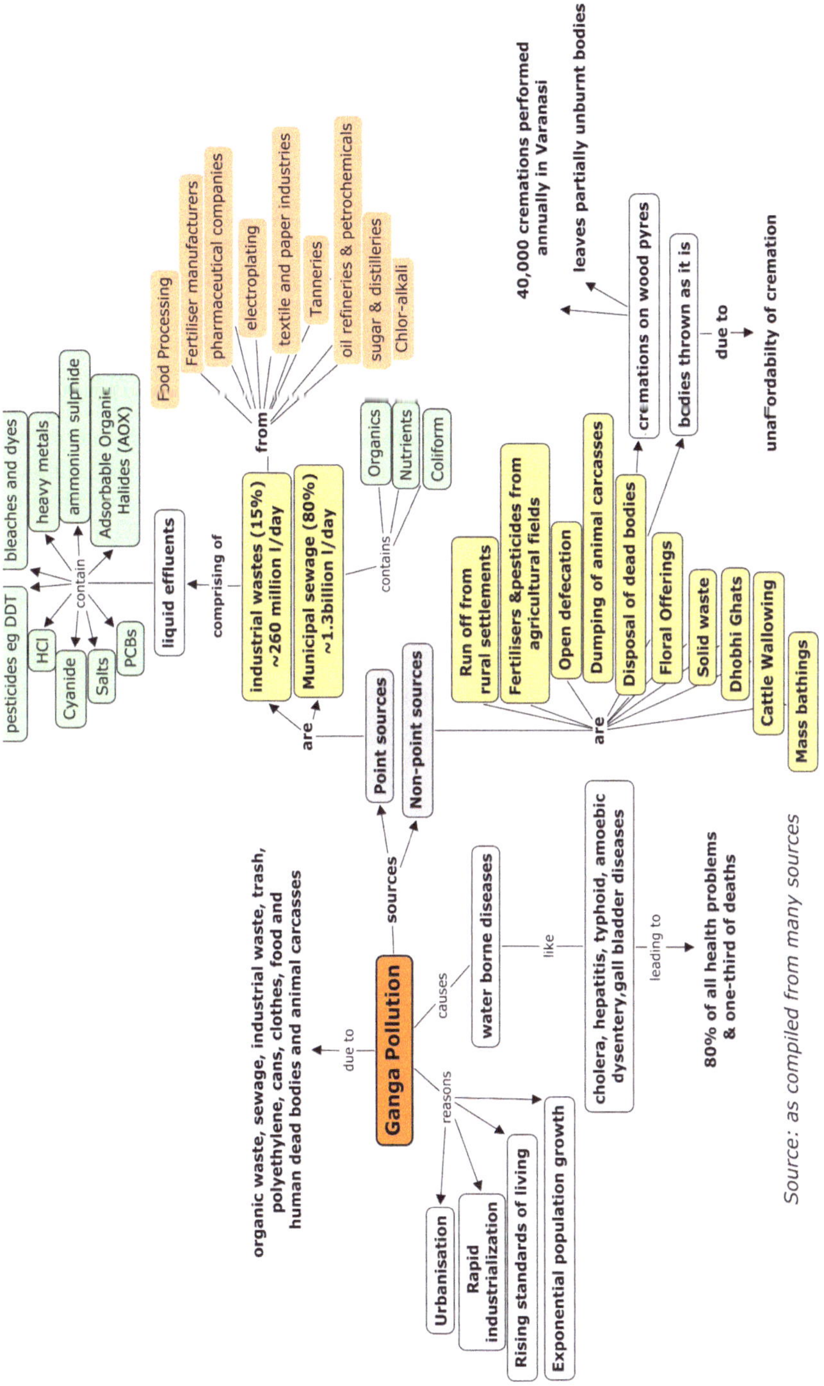

Fig. 1.15 Ganga pollution sources

present in water, and the number of FCC/100 ml of water sample, indicating the presence of faecal coliform bacteria in water, the root cause of practically all water-borne diseases.

The deterioration of river Ganga's health is mainly due to various anthropogenic Activities. However the impacts of climate change are adding another set of challenges. The 27 km long Gangotri, the feeding glacier of Ganga, is receding at nearly 21 m/year. In 1935 the glacier was melting by 7 m/year, in 1990 the meltdown had increased to 18 m/year and in 2006 the key Himalayan glaciers have shrunk by almost 21 % causing water shortage in the coming years due to melting of the glaciers. Data for 61 years (1936–96) show that the total recession of Gangotri glacier is 1,147 m, with the average rate of 19 m/year [31]. However, over the last 25 years of the twentieth century it has retreated more than 850 m (34 m/year) [32].

The increasing sand bed in the Ganga basin, which defines the ecosystem, is an indicator of the gloomy future of the Ganga. According to Uday Kant Chowdhary, a professor of civil engineering, and coordinator of Ganga Research Laboratory, Institute of Technology, Banaras Hindu University (BHU), sand bed in the Ganga is increasing 5–6 m in width and 8–10 cm in height annually. It means the width of the river is reducing in proportion to the increase in sand bed. Factors like increasing pollution, over-extraction of water and reduced flow of the Ganga are the key reasons for this situation and are thus leading to cause slow death to the river. The water holding capacity of the Ganga river and its tributaries are reducing due to the high rate of siltation, resulting in devastating floods at regular intervals during the recent years. It is imperative to initiate immediate preventive measures to lower the pollution rate, otherwise the very livelihood crisis will deepen further.

1.5 Water Resources Projects in Ganga Basin – A Special Mention

Historically hydropower is considered to be the most environment friendly mode of power generation in comparison to other traditional modes including the coal based one; however with advancing technologies, new means of power generation are emerging, such as – nuclear, wind and solar energy. The very notion that, the hydropower development is the most environmentally benign mode of power generation is highly contestable for various reasons. The available worldwide literature on consequences of dam development reveals that the impact of dams on ecosystems are profound, complex, varied, multiple and mostly negative. By storing or diverting water, dams alter the natural distribution and timing of stream flows. This in turn, changes sediment and nutrient regimes and alters water temperature and chemistry resulting to impacts on ecosystems and biodiversity elements that these streams support and on their attendant socio-economic aspects. These ecosystem impacts may result in consequent changes in freshwater biodiversity, which is already threatened by other factors. It is well understood that land clearing, anthropogenic disturbances in the landscape, combined with increasing water withdrawals and alterations of river systems can result in adverse effects to the sustainability of natural resources. Within a basin, the greater number of dams leads to greater fragmentation of river ecosystems. While it is acknowledged that energy is essential for economic progress and well-being of the people, the loss of biodiversity cannot be compensated by economic growth. It is essential to ensure that water demands for energy and irrigation do not become a cause for the reduction of forested areas, receding wildlife habitats and loss of biodiversity resources.

In the context of Ganga basin, given the wealth of water resources within the state which offers numerous suitable sites, hydropower development has been considered on a massive scale by the state, to not only improve the erratic and 'yet-to-reach-all-villages' power scenario but also to attract the industries with better infrastructure for the overall development of the state.

Within the Ganga basin, there are three types of Hydro Electric Projects: Diversion (Run-off) the river; Pump storage and Reservoir (see Chap. 12). The water resources projects, be it hydroelectric ones or the irrigation projects, are bound to have environmental implications, nevertheless they are necessary to meet the power and food demand of ever-growing population of the country. However, careful and environmentally benign planning for such projects could lead to minimized impact onto the ecology in specific, and the environment in general.

The status of water resources projects in the Ganga basin varies in different sources and reports. It's very difficult to comprehend the exact number of projects at any given moment; the numbers in this regard appears to be dynamic all the time. This could be due to the fact that there are several projects which are under different stages of construction, i.e. planning/development/construction/completion/commissioning.

Based on the NRCD's 2009 [1] data, Fig. 1.16 informs about the number of hydropower projects and major/medium irrigation projects including their respective command areas as a percentage of total area in the states falling within the Ganga Basin. The WII 2012 report [33], takes into account the cumulative impact of 70 Hydro Electric Projects of, which 17 are existing, 14 are under-construction and 39 are proposed in two river basins *viz.* Alaknanda and Bhagirathi which fall in the north-western part of Uttarakhand (Figs. 1.17 and 1.18). The report is an outcome of the study conducted in response to the directives of the Ministry of Environment and Forests (MoEF). The focus of the study included assessment of the baseline status of rare, endangered and threatened (RET) species of flora and fauna, identification of the critically important habitats for RET species in the two basins, delineation of river stretches critical for conservation of RET aquatic species, and assessment of the minimum flows for ecological sustainability of the two rivers. Of the 1,121 km long stretch of rivers that flow in the entire Alaknanda-Bhagirathi basins, a minimum of 526.8 km long river stretch is expected to be affected, if all proposed HEP are implemented. This is 47 % of total rivers stretch in the entire Bhagirathi-Alakananda basin. Therefore, significant area of the fish habitat would either be modified or lost due to proposed Hydro Electric Projects in the basin. This means that about 87 % of fish species would be affected, if all proposed Hydro Electric Projects get implemented in the basin.

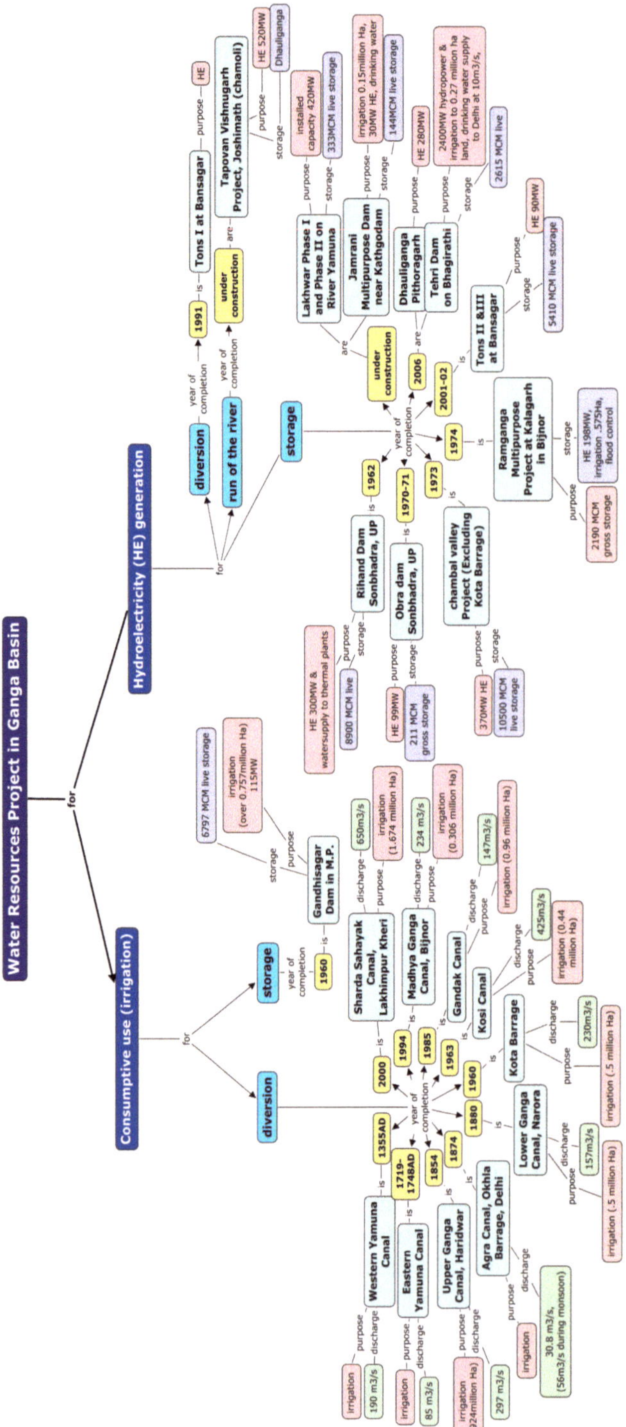

Source: NATIONAL GANGA RIVER BASIN AUTHORITY (NGRBA)(Ministry of Environment and Forests, Government of India) Environmental and Social Management Framework (ESMF) Volume I - Environmental and Social Analysis, January 2011 Prepared by The Energy and Resources Institute Consultant

Fig. 1.16 Water Resources Projects in the Ganga Basin

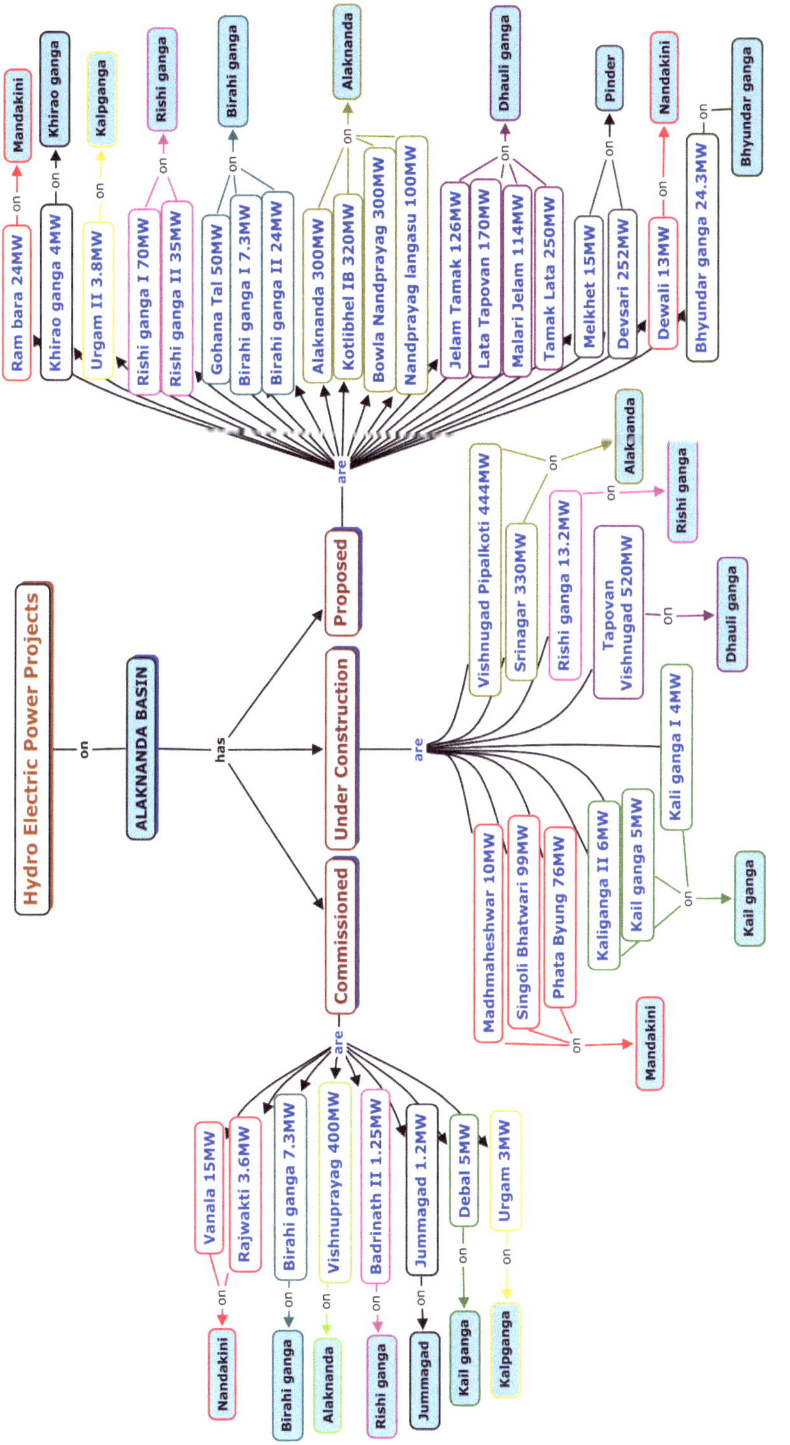

Source: Report by Wildlife Institute of India 2012 on Cumulative Environmental Impact Assessment

Fig. 1.17 Alakananada Hydro Electric Power Projects

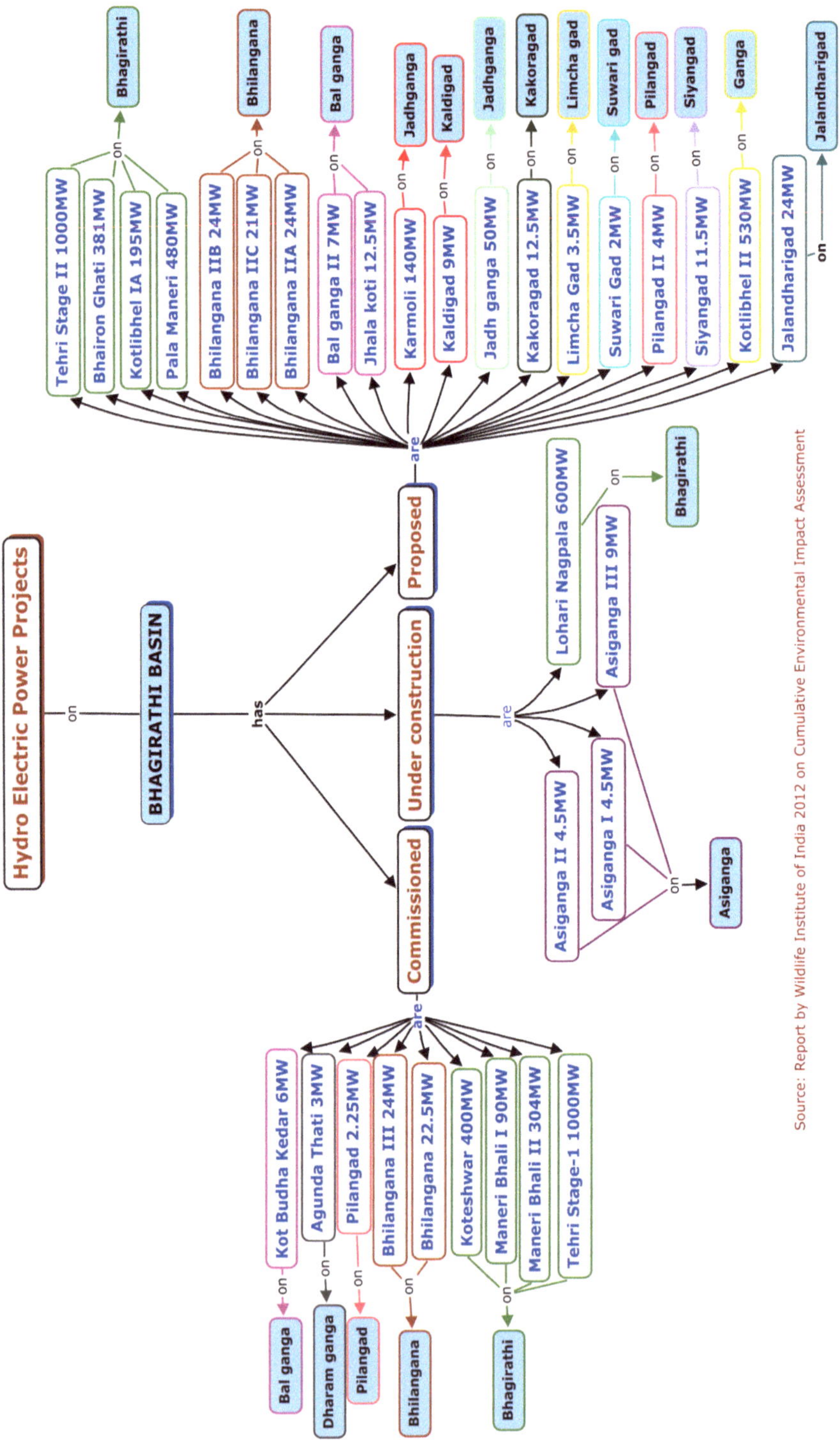

Fig. 1.18 Bhagirathi Hydro Electric Power Projects

Conflicts over dams have heightened in the last two decades largely due to the social, ecological and environmental impacts of dams that were either ignored in the planning process or were not adequately evaluated [34]. The very subject of water resources projects is controversial in the context of Ganga river basin, however in a way this has helped in making the governmental entities and communities more vigilant about the impacts of such projects. The recent movement to save river Ganga headed by Swami Gyanswroop Sanand (the eminent scientist Prof. GD Agrawal) has led to scrapping of three hydropower projects on the Bhagirathi, which is considered to be the source stream of river Ganga. These projects include 600 MW Lohari Nagpala, the 480 MW Pala Maneri and the 350 MW Bhairo Ghati. It was mounting environmental and social concerns which prompted the union government to halt work on the Lohari Nagpala plant and other state government projects at a meeting of the National Ganga River Basin Authority in November 2010. In fact, the move to scrap the 440 MW Vishnusagar projects in Chamoli district too sparked outcry among those in favour of hydel projects in the state [35]. The report by Wildlife Institute of India [33], has recommended that 34 dams on the Alaknanda and Bhagirathi rivers should not be allowed to come up as they will cause irrevocable harm to biodiversity in Uttarakhand. The big projects in the "red list" include the 530 MW Kotlibhel II, the 250 MW Tamak Lata on Dhauli Ganga, the 320 MW Kotlibhel IB on Alaknanda, the 381 MW Bharon Ghati and the 195 MW Kotlibhel IA on Bhagirathi.

The government also decided to declare the river course of the Ganga (Bhagirathi) in initial 135 km stretch from Gaumukh to Uttarkashi as India's first "eco-sensitive" zone on a major river. This will ensure that this stretch of the river cannot be disturbed for any projects [36].

1.6 Improving the Health of River Ganga

Over the past three decades or so, the pollution of the Ganga has reached serious levels, leading to local and national campaigns and actions to restore the health of the sacred river. The Ganga action plan (GAP) was launched by Shri Rajeev Gandhi, (former Prime Minister of India) on 14 January 1986 as its main objective pollution abatement, and to improve the water quality by Interception, Diversion and Treatment of domestic sewage and present toxic and industrial chemical wastes from identified grossly polluting units. The Government of India then extended this model with suitable modifications to the national level through a National River Action Plan (NRAP). However, due to fast urbanization these efforts were not adequate to achieve the clean Ganga target (see Fig. 1.19).

More than two decades after Rajiv Gandhi conceptualized the GAP, Prime Minister Manmohan Singh declared the holy river ***GANGA*** as a "national river" on 4 Nov 2008. Thereafter, on October 5, 2009, the critically endangered Ganga river dolphin was declared as "national aquatic animal". For the first time in the history of India, a river has been given this status. This landmark event is linked to the research done by Swami Gyanswroop Sanand (Prof. GD Agrawal), a former dean of IIT Kanpur whose

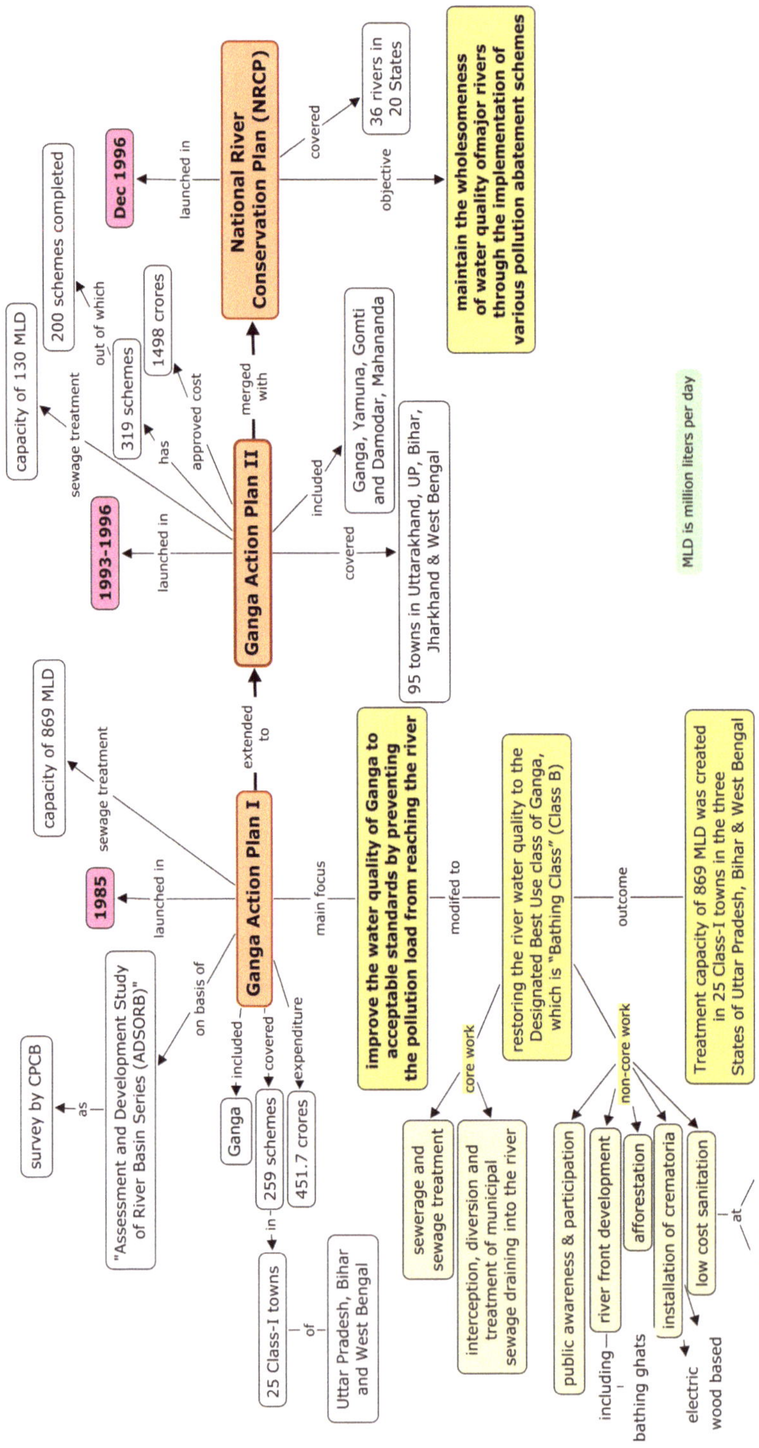

Fig. 1.19 Ganga action plan

contention was that unless the hydropower projects between Gangotri glacier that feeds the Ganga and Uttarkashi were stopped, the Ganga would be reduced to a series of dry riverbeds, small lakes and tunnels carrying water for electricity. In June 2008, he threatened to fast unto death unless all development work on such projects was halted and eventually his persistence to save the Ganga paid off partially. This has been explained in the previous section.

On 4 Nov 2008, the Government of India declared river ***GANGA*** as the "national river".

On 19 January 2010, the critically endangered Ganga river dolphin was declared India's national aquatic animal.

1.6.1 NGRBA and GRB EMP

The status of a "National River" for Ganga, facilitated the constitution of a National Ganga River Basin Authority (NGRBA) an empowered planning, financing, monitoring and coordinating body on 20 Feb 2009 under the Environment (Protection) Act, 1986 for cleaning of the river Ganga. The Authority is chaired by the Prime Minister and has as its members the Union Ministers concerned and also the Chief Ministers of the States through which Ganga flows, viz., Uttarakhand, Uttar Pradesh, Bihar, Jharkhand and West Bengal etc. The objective of NGRBA is to ensure pollution prevention and conservation of the river Ganga by adopting a river basin approach. The National Ganga River Basin Authority (NGRBA) is expected to provide the framework for a long-term institutional setup in the management of the environmental sustainability of the Ganga (Fig. 1.20).

In the Mission 'Clean Ganga' which aims to ensure that no untreated sewage enters the River Ganga by the year 2020, a Memorandum of Agreement (MoA) was signed by the Ministry of Environment & Forests (MoEF) on 6 July 2010, with a consortium of seven IITs (Bombay, Delhi, Guwahati, Kanpur, Kharagpur, Madras and Roorkee) for the development of Ganga River Basin Environment Management Plan (GRB EMP), implementation of which is expected to result in the restoration of the 'wholesomeness' of all the rivers in the Ganga basin. It is for the first time that a consortium of 7 IITs has come together for a project of such national importance and magnitude. For the assignment of this nature, the consortia has also partnered with other technical and scientific institutions, like – IT BHU, CIFRI and some NGOs, like – WWF-India.

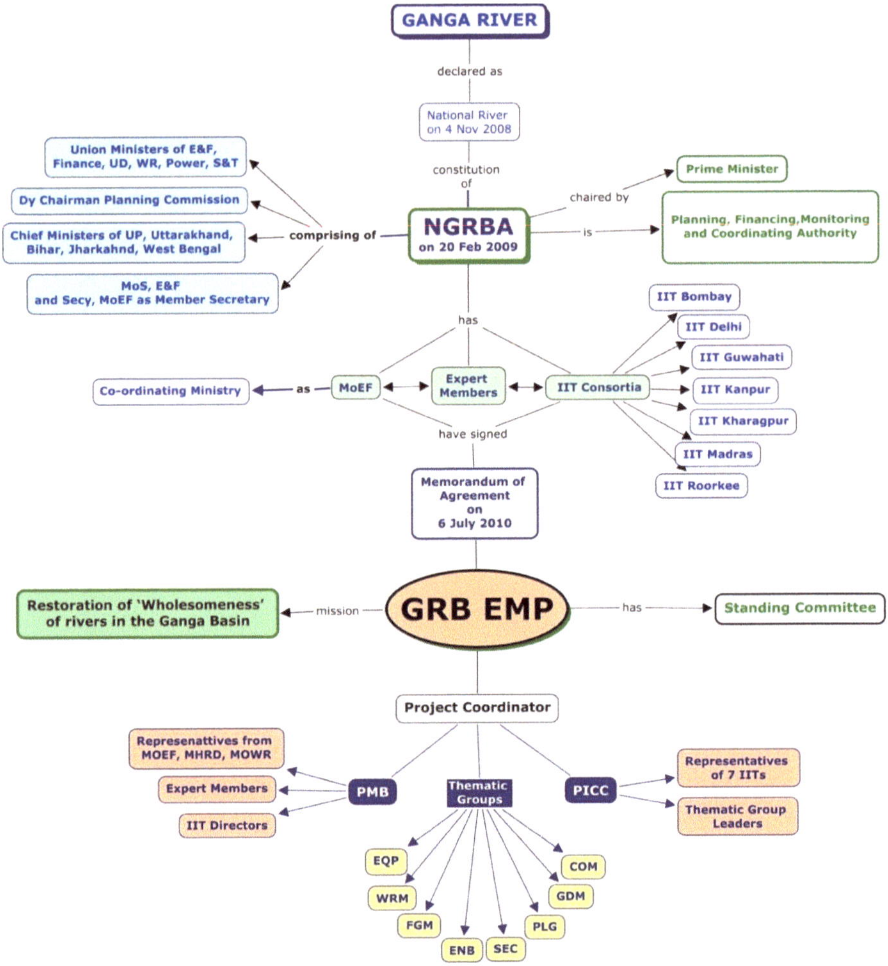

Fig. 1.20 The structure of Ganga River Basin Environment Management Plan (GRBEMP)

GRB EMP will aim to have adequate provision for water and energy in the Ganga basin to accommodate the pressures of increased population, urbanization, industrialization and agriculture while ensuring protection of the sanctity of the fundamental aspects of the river system. These include:

- river must continuously flow (***Aviral Dhaara***)
- river must not be seen as a carrier of waste loads (***Nirmal Dhaara***)
- river must have longitudinal and lateral connectivity
- river must have adequate space for its various functions
- river must function as an ecological entity

The Ganga river basin is a multifaceted system and requires multidisciplinary and interdisciplinary management approach. A river system is not just a physical entity but a living dynamic and a continuing resource system. For integrated management of the Ganga basin several aspects ranging from religious and socio-economic to highly technical ones, are required to be considered. Keeping in view the need for a holistic approach, the GRB EMP with the aim to improve the health of Ganga river was segregated under eight different themes, which led to formation of eight thematic groups coordinated by a Project Implementation and Coordination Committee (PICC):

Theme I: Environmental Quality and Pollution (EQP)
Theme II: Water Resources Management (WRM)
Theme III: Fluvial Geomorphology (FGM)
Theme IV: Ecology and Biodiversity (EBD)
Theme V: Socio-Economic-Cultural (SEC)
Theme VI: Policy, Law and Governance (PLG)
Theme VII: Geo-spatial Database Management (GDM)
Theme VIII: Communication (COM)

However, soon it was realized that since the Environmental Flows for river Ganga is an equally important aspect which is required to be considered while making the GRB EMP, so an additional theme was added and the E-Flows thematic group was created, therefore the total number of themes became nine. Each theme is important in its own way and is working independently as well as in coordination with other themes.

The Prime Minister, whilst chairing the third National Ganga River Basin Authority (NGRBA), said urgent steps should be taken to save the river, deemed sacred, and said that "time is not on our side. Every day about 2,900 million litres of sewage is discharged into the main stream of the river Ganga from municipal towns located along its banks. The existing infrastructure has a capacity to treat only 1,100 million litres per day, leaving a huge deficit" [37].

1.6.2 Gangapedia as the Communication Platform

The integration and co-ordination of activities with the involvement of public at large, in monitoring the effectiveness of implementation of GRB EMP would have a positive impact on the water quality and health of river Ganga. It is important to mobilize the general public/NGOs as watchdogs, strengthen the statutory status of community-based organizations and to empower them for participation in GRB EMP. Effective communication support is essential when dealing with complex multi thematic processes that may have small to large overlap spanning through several institutions and organizations involving many contributors. There is a need to have a repository of data and information which is easily accessible to anyone. Though the quantum of data on the river system is voluminous there are numerous lacunae in the information that still needs to be plugged. The immense amount of data generated by many institutions, organizations need to be screened meticulously, arranged methodically and evaluated for presentation in a comprehensive way. It is hoped that the numerous organizations involved in different aspects of Ganga river studies will be assisted by the GRB EMP efforts.

As 'Ganga basin information and communication network', the website 'Gangapedia' (www.gangapedia.in) was launched. The objective of Gangapedia website is to foster coordination amongst people with the interest in issues related to Ganga river basin. It also brings together experts on this subject, so that they can share their knowledge and experience.

Gangapedia aims to consult, discuss and take feedbacks from the national and state level stakeholders, policy makers as well as the general public at large on the status, issues and challenges to implement the GRB EMP and to narrow down the knowledge gaps among institutions and key stakeholders. The structure, function and legalities of this website is based on this philosophy and the information provided through the website is not only about the river but also for the users to optimally utilize its resources.

1.6.3 Need for Communication

While disseminating information, it is important to see whether the information is provided to keep the users curiosity at bay or is it just to highlight the current status of the river which has a past as well as a future. In the past, the pristine state of the river had abundance of resources and less number of users depicting a river friendly livelihood and traditional way of using resources; whereas, the present state of the river has certain stretches of river, which are at the verge of extinction and where science and technological help is needed to improve the state of river.

The future of the Ganga river would be the combination of its past and present. In order to do so, one has to utilize the information on the past and present state of the Ganga basin. The past pristine state cannot be revived in totality, however, with the application of participatory approach and present scientific technology, one can ensure that it's ecological and cultural integrity improves. This by no mean advocates zero usage of the resources but actually calls for optimal usage. The idea is not merely for promotion of sustainable development but also for saving the civilization, by making them aware about various facets about the river and set of challenges being faced by it.

1.7 Conclusions

The growing recognition of the Ganga as an environmental resource is reflected in the endless warnings by scientists, environmentalists, fasts by seers, efforts of people, billions spent by the government. The serious nature of the threats it is facing has led many NGOs, youth clubs, welfare associations, women's groups and groups of pilgrims, religious saints and local administration to participate in the Ganga cleaning activities either directly or indirectly. Several agencies like the National River Conservation Directorate, Ministry of Environment and Forests, Central and State Pollution Control Boards; several research establishments, National laboratories and scientific organisations, Government and non-governmental organizations have done or are doing considerable work under various projects.

Despite all that, today the Ganga river which has nurtured and protected the Indian civilization since ages, is in need of protection itself. Over the centuries, the river with all its past and present symbolisms has become a great symbol of national unity and has the potential to serve as a catalyst for change at the national level. If the British could rejuvenate a polluted Thames and the Europeans should save Danube can we not save our precious Ganga from disappearing?

Water Quality: The Mantra: जल गुणवत्ता: मंत्र:

गंगां पुण्यजलां प्राप्य त्रयोदश विवर्जयेत्। शौचमानं सेकं निर्माल्यं मलघर्षणम्।
गात्रसंवाहनं क्रीड़ा प्रतिग्रहमथोरतिम्। अन्यतीर्थरतिंचैवः अन्यतीर्थ प्रशंसनम्।
वस्त्रत्यागमथाघातं सन्तारंच विशेषतः।।

ब्रहमादन्दपुराण (८00ई0)

Thirteen actions are prohibited on approaching the scared waters of the Ganga, namely:

1. Defecation
2. Ablutions
3. Discharge of wastewater
4. Throwing of used floral offerings
5. Rubbing of filth
6. Body shampooing
7. Frolicking
8. Acceptance of donations
9. Obscenity
10. Offering of inappropriate praises or even hymns in a incorrect way
11. Discarding of garments
12. Beating and
13. Swimming across, in particular

Acknowledgements The author (RS) is thankful to Prof TV Prabhakar who conceptualized the portrayal of the Ganga river in the form of cmaps and encouraged her to take it up. All the concept maps have been designed and developed by RS. The author (NK) is thankful to its parent organization, WWF-India for his involvement in 'Living Ganga' and 'Rivers for Life' Programme which helped him take up this project.

References

1. National River Conservation Directorate (NRCD) (2009) The status paper on Ganga Action Plan. Ministry of Environments and Forests, Government of India. Available at http://ahec.org.in/Status
2. Central Water Commission (2009) About the basin. http://cwc.gov.in/regional/patna/welcome.html
3. Welcomme RL (1985) River fisheries. Food and Agriculture Organization, Technical Paper 262, Rome
4. Behera SK (1995) Studies on population dynamics, habitat utilization and conservation aspects of Gangetic dolphin (Platanista gangetica) in a stretch of Ganga River from Rishikesh to Kanpur. PhD thesis, School of Studies in Zoology, Jiwaji University, Gawalior
5. Misra AK (2011) Impact of urbanization on the hydrology of Ganga Basin (India). Water Resour Manage 25:705–719
6. Krishnamurti CR, Bilgrami KS, Das TM, Mathur RP (eds) (1991) Ganga: a scientific study. Northern Book Center, New Delhi, 246
7. NGRBA (Ministry of Environment and Forests, Government of India) Environmental and Social Management Framework (ESMF), vol. I – Environmental and social analysis, January 2011
8. Joshi DM, Kumar A, Agrawal N (2009) Assessment of the irrigation water quality of river Ganga in Haridwar district. Rasayan J Chem 2(2):285–292
9. The Times of India, Varanasi, Nov 4 (2009), First anniversary of Ganga being declared as National River of India. http://articles.timesofindia.indiatimes.com/2009-11-04/varanasi/28103392_1_national-river-ngrba-ganga-action-plan
10. Ghosh AK (1991) The Ganga. A profile and biological resources. In: Jairajpuri MS (ed) Faunal resources of Ganga, Part I. Zoological Survey of India, Calcutta, pp 1–12
11. Rao RJ (2001) Biological resources of the Ganga River, India. Hydrobiologia 458:159–168
12. Vass KK, Mondal SK, Samanta S, Suresh VR, Katiha PK (2010) The environment and fishery status of the River Ganges. Aquat Ecosyst Health Manage 13:385–394
13. Jones T, Phillips B, Williams CE, Pittock J (eds) (2003) Managing Rivers Wisely: lessons from WWF's work for integrated river basin management. WWF International, Gland
14. World Resources Institute Annual Report (2003) Ideas into Action
15. Martin R, Aidan (2003) The mysterious, endangered river sharks (Glyphis spp.) ReefQuest Centre for Shark Research. http://www.elasmo-research.org/conservation/river_sharks.htm
16. Fu Cuizhang, Wu Jihua, Chen Jiakuan, Wu Qianhong, Lei Guangchun (2003) Freshwater fish biodiversity in the Yangtze River Basin of China: patterns, threats and conservation. Abstract submitted to the Second International
17. Dawson TP, Berry PM, Kampa E (2003) Climate change impacts on freshwater wetland habitat. J Nat Conserv 11:25–30
18. Gibbs JP (2000) Wetland loss and biodiversity conservation. Conserv Biol 14(1):314–317
19. Ricciardi A, Rasmussen JB (1999) Extinction rates of North American freshwater fauna. Conserv Biol 13:1220–1222
20. Szollosi-Nagy (2004) In: Proceedings of the United Nations seminar, 25–26 November 2004, Delft, Netherlands
21. Copp GH, Bianco PG, Bogutskaya NG, Eros T, Falka I, Ferreira MT, Fox MG et al (2005) J Appl Ichthyol 21(4):242–262

22. Lima-Junior SE, Cardone IB, Goitein R (2006) Fish assemblage structure and aquatic pollution in a Brazilian stream: some limitations of diversity indices and models for environmental impact studies. Ecol Freshw Fish 15(3):284–290
23. Kang B (2009) Aquaculture of detritivorous fish in China: status and perspective. CAB Rev Perspect Agric Vet Sci Nutr Nat Resour 4(10):1–7
24. Laffaille P, Pétillon J, Parlier E, Valéry L, Ysnel F, Radureau A, Feunteun E, Lefeuvre JC (2005) Does the invasive plant Elymus athericus modify fish diet in tidal salt marshes? Estuarine Coastal Shelf Sci 65(4):739–746
25. Ganga Canal Works report by Sir Proby T Cautley (1864) Chapter-1
26. Nilsson C, Lepori F, Malmqvist B, Törnlund E, Hjerdt N, Helfield JM, Palm D et al (2005) Forecasting environmental responses to restoration of rivers used as log floatways: an interdisciplinary challenge. Ecosystems 8(7):779–800
27. 'Snow and Glaciers of the Himalayas' (2011) Study by Space Applications Centre (SAC), Indian Space Research Organisation, Ahmedabad on joint project of the Ministry of Environment and Forests and Department of Space
28. Wong CM, Williams CE, Pittock J, Collier U, Schelle P (2007) World's top 10 rivers at risk. Report by WWF International, Gland, March 2007
29. Sarkar SK, Bhattacharya A, Bhattacharya B (2003) The river Ganga of northern India: an appraisal of its geomorphic and ecological changes. Water Sci Technol 48(7):121–128
30. Journey to the Source of Ganga. http://www.ibaradio.org/India/ganga
31. Naithani AK, Nainwal HC, Sati KK, Prasad C (2001) Geomorphological evidences of retreat of the Gangotri glacier and its characteristics. Curr Sci 80(1):87–94, http://www.ias.ac.in/currsci/jan102001/87.pdf
32. Sharma MC, Owen LA (1996) Quaternary glacial history of the NW Garhwal, Central Himalayas. J Quat Sci Rev 15:335–365
33. Wildlife Institute of India (2012) 'Assessment of Cumulative Impacts of Hydroelectric Projects on Aquatic and Terrestrial Biodiversity in Alaknanda and Bhagirathi Basins, Uttarakhand'
34. World Commission on Dams, Nov (2000) WCD report 'Dams and Development'. Earthscan Publications
35. IANS Nd (2012) Business/Economy, Environment/Wildlife, Tuesday, 29 May 2012 IANS
36. NIDM Newsletter – Disaster Update. Issue No.: 1958. Dated 3rd November 2010
37. Hindustan Times Article (2012) New Delhi, 17 Apr 2012

Bhagirathi joins Alakananda river (Sangam) at Devprayag and forms the mighty GANGA (Courtesy Nitin Kaushal)

Chapter 2
Ganga – Our Endangered Heritage

Subhajyoti Das

Abstract Ganga, the Holy river of India, provides life sustaining succor for environment and ecology. It has a historical bondage with our civilization. But the river water which was famous for its self-purifying capacity is now highly polluted due to discharge of untreated municipal and industrial waste waters and effluents directly into the Ganga and Yamuna. The problem is compounded by its depleted flow due to diversions caused by construction of dams and barrages. Deforestation in its catchment, sand mining and stone quarrying in its river bed and bank are doing enormous harm to its environment and ecology. Ganga is now rated as one of the ten most endangered rivers of the world. Climate change and global warming syndrome is also likely to add to the crisis. The restoration of the river to its original pristine quality needs concerted actions from all sections of the society and administration. The article gives an overview of the state of the Ganga and possible measures to restore its quality and flow.

Keywords Ganga • Yamuna • Environmental flows • Glacier melting • Pollution • Ganga Action Plan • Arvari river

2.1 Introduction

Ganga, the Holy river originates in the snow-capped Himalayas flowing out of gorges and narrow valleys in Uttarakhand, descends on to the plains at Hardwar, winding its course through 11 states and murmuring into the Bay of Bengal at Ganga Sagar in West Bengal. As mythology goes, Sage Gautam and Bhagirath after

S. Das (✉)
Formerly with Central Ground Water Board, Flat # M 901, HM Tambourine,
28 Jaraganahalli, J. P. Nagar 6th Phase, Kanakpura Road, Bangalore 560078, India
e-mail: subhajyoti_das@hotmail.com

R. Sanghi (ed.), *Our National River Ganga: Lifeline of Millions*,
DOI 10.1007/978-3-319-00530-0_2, © Springer International Publishing Switzerland 2014

rigorous penance to Lord Siva brought down the River Ganga from the heavens to the earth for salvation of 60,000 sons of King Sagar doomed by Sage Kapila's fury of anger. Hence a holy dip in the river (Ganga Snan) is believed to free the suffering souls from the bondage of sins on earth leading to the heavenly abode. Along its banks are many places of worship and pilgrim cities for soul searching and cleansing. It is on its banks that 'Amrit Kalas' was hidden to save it from the demons. Through ages whenever misfortune or calamity has befallen the humanity, masses have joined in congregation on its bank for confabulations, to invoke heavenly bliss to win over inimical forces of destiny, or for guidance to the paths of enlightenment. Thus started the tradition of festivals: Purna Kumbh and Ardha Kumbh. Ganga has thus age-long emotional, spiritual, cultural, religious and historical bondage with our civilization, immortalized in the folk lore, ancient scripts, art, culture and literature starting from the Vedas, Puranas, Ramayana, Mahabharata to Nehru's *Discovery of India*.

Ganga is revered as incarnation of the Goddess for providing the life- giving and life- sustaining succor for the environment and ecology. To keep it free from pollution and its flow unimpeded, people traditionally used to abide by certain rules and customs laid down by the community and spiritual leaders. These age old traditions or customs which ensured its purity and unobstructed flow have now been discarded with disdain. In the last 30–40 years our life style, culture, and social values have drastically altered. Economic liberalization, privatization have all posed new challenges rendering the natural resources saleable commodities for profit-making, leading to its overexploitation, encroachment, diversion through dams and barrages, and pollution. Human greed has gripped the flow to near-stagnation and to the peril of his future and progeny. The ecology and environment are the two essential elements of a river regime. Hence the following chapters aim at understanding these dynamics of the river regime.

2.2 Physical Setting

The mighty Ganga gushes out of Gaumukh in Gangotri glacier of the Himalayas. Ganga follows a long course of about 2,525 km, and along with its tributaries winds through the states of Uttarakhand, Uttar Pradesh, Bihar, West Bengal, and draining parts of Himachal Pradesh, Haryana, Rajasthan and Madhya Pradesh before debouching on to the Bay of Bengal (Fig. 2.1). It is one of the largest river basins of the world covering 26.3 % of the country's total geographical area. It has a number of tributaries, namely Yamuna, Gomati, Ghaggar, Kosi, Son, Gandak and subtributaries, the Yamuna being the largest of all. Its perennial flow is fed mainly by monsoon precipitation and partially by the Himalayan glacier melt. Its total catchment area measures 861,404 km^2, out of which Yamuna's share is 366,223 km^2. The Yamuna rising at Yamunotri joins the Ganga at Allahabad. Roughly the land use pattern of the basin is: total forest cover 116,973 km^2, cultivable area 509,944 km^2, net sown area 427,991 km^2, and gross sown area 556,488 km^2.

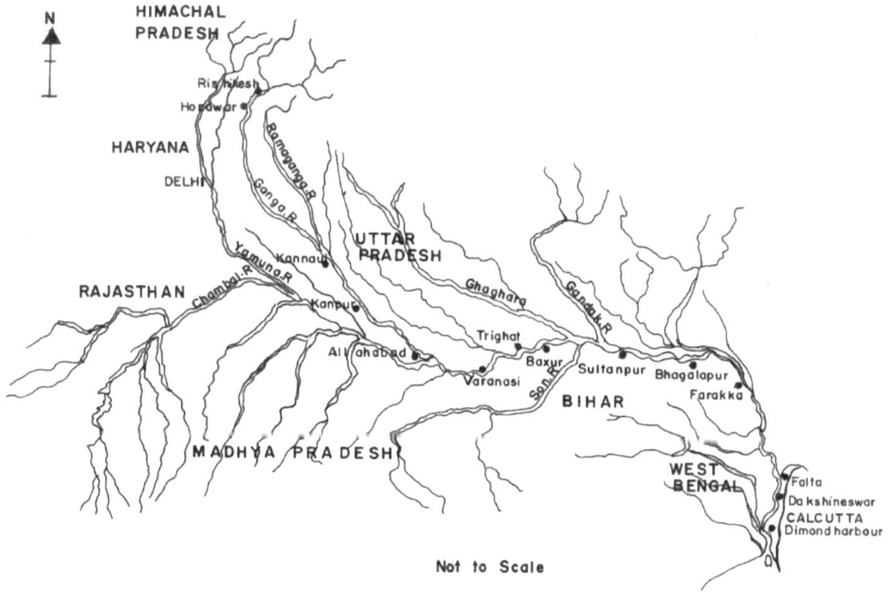

Fig. 2.1 The Ganga River Basin Line drawing (After Central Board for the Prevention and Control of Water Pollution [1])

The rainfall in the Ganga basin varies widely from nearly 160 cm in the Bengal Delta in the east and southeast to less than 40 cm in the west in Haryana, but nearly 230 cm in the Himalayan area. Three months of monsoon rainfall (July–September) provides the water surplus and about 85 % of its runoff which sometimes overreaches the banks to wreak flood havocs.

Geologically the Ganga river basin occupies the Himalayan foredeep. This crustal warp formed between the mountains in the north and Peninsular plateau in the south in the geological past, is filled with a thick pile of alluvial sediments, 1,000–1,500 m or more in places in a sinking basin, and forming one of the largest groundwater basins in the world. These are the products of floodplain deposits of torrential streams. The geological set up and rock types are diverse. The river divides the terrain into two parts, – the northern part, overwhelmingly an alluvial terrain of Quaternary age, and the southern part mainly composed of consolidated crystalline and sedimentary formations of Archaean to Palaeozoic ages. This is reflected in the mineralogical composition of the sediments in the valley. This geological set up controls the soil characteristics and groundwater hydrology. The soils in the northern parts are mainly alluvial derived from active erosion of the Himalayas being rich in nutrients and supporting rich agricultural produce. But in the southern parts red and black, black, and lateritic soils predominate being residual soils, the products of in situ weathering of rocks. Red and yellow soils have high fertility and serve as paddy soils par excellence. The Ganga basin is known as the rice bowl of India.

2.3 Flow Regime

Reliable river flow data and flow statistics are hardly available in the public domain except for piecemeal information from various sources. The long-term account of river flows in India by K. L. Rao [2] is treated as the most reliable information subject to only minor alterations. Quoting from the Report of Central Board for Prevention and Control of Water Pollution (CBPC) [1], the Ganga and some of its major tributaries have annual yields as: *Yamuna 76,000, Son 22,420, Ghaghara 94,400, Ramganga 15,258, Damodar 12,200, Kosi 61,560, Gandak 52,200, and Ganga 468,700 million* m^3 *(revised as 525,023* Mm^3 *in CWC [3])*. The river flow varies significantly from month to month and year to year due to seasonal rainfall, which too is capricious in nature, as also due to large scale withdrawal of water for canal irrigation. Reportedly more than 100 billion cubic meters of water are annually withdrawn from different streams of the Ganga.

Except the Yamuna, Son, Ghaggar, Gandak and Kosi, the other tributaries have but limited flows. Being fed by glacial melt Ganga is one of the largest perennial rivers of the world. The live storage capacity of the barrages and dams in the Ganga basin, completed and under construction totals 60.6 BCM only, leaving a huge balance in the river flow [3], but this is hardly the real picture.

Recent river flow analysis by Central Pollution Control Board [4] records the discharge of the Ganga outside the peak periods of monsoon as meager as 90–386 m^3/s at Kanpur, 279–997 m^3/s at Allahabad, and 278–1,160 m^3/s at Varanasi. Upstream of Varanasi Ganga flow has virtually vanished at many places. At Varanasi the flow has lost depth from an average 60 m to a mere 10 m at places. Its largest tributary, the Yamuna, also records reduction in the flow due to diversion of large volume of river water into the irrigation canals.

The sediment load carried by the river is the highest compared to other major rivers of the world, ranging from 1,085 to 2,400 million tons [5].

2.3.1 Dams and Barrages

Dams and barrages have been constructed on the River Ganga and its tributaries mainly for flood control, irrigation and hydropower generation. They divert the flow through canals or tunnels. This impacts the flow and the quality of river water downstream. There are several major canal projects on the Ganga-Yamuna river system which cause voluminous diversions of the flow. The live storages of some of the major and medium projects are: Upper Ganga (2,613 Mm^3); Sarda (634 Mm^3);

Matalia (780 Mm^3); Betwa (84 Mm^3); Ken (60 Mm^3); Ramganga (1937 Mm^3); Son (8,971 Mm^3); Damodar (4,226 Mm^3); Mayurakshi (555 Mm^3); Kangsabati (972 Mm^3); Chambal (8,500 Mm^3). The largest storage reservoir is Rihand reservoir holding 10.6 BCM of water, and Tehri Dam holds 2.6 BCM. In addition Tehri Dam commissioned in 2006, is a massive storage structure on the Bhagirathi for hydroelectric generation. Reportedly 600 more dams are on the anvil awaiting government approval. The Western and Eastern Yamuna Canals are the two major canal projects on the Yamuna.

A massive diversion of its water by the Upper Ganga Canal Network off the head works at Hardwar reduces the flow to only 15 billion m^3 per annum at Balawali [1]. The Lower Ganga Canal is another large irrigation network, which diverts near Aligarh whatever flow is regenerated between Hardwar and Aligarh, thereby causing a critical fall of dry weather flow downstream and downgrading its water quality. There is very little dry weather flow in the Ganga at Kannauj and Kanpur where there is a heavy discharge of pollutants in the river.

2.3.2 *Environmental Flows*

"The question is to find an agreeable trade off in which the diversion of water will be socially acceptable and ecologically sustainable and all the damages to livelihoods and ecosystems dealt with" [6]. Environmental flow implies strategically releasing water downstream of dams and reservoirs to protect the services it provides and maintain their regime [7]. The concept is still in infancy. It is important to note that the river flows help rivers to purify themselves, sustain aquatic life and vegetation, effect groundwater recharge, support livelihoods, facilitate navigation, preserve estuaries, prevent incursion of salinity, and enable rivers to play their role in people's cultural and spiritual lives. Richter et al. [8] listed five aspects of hydrological regimes which will be altered by dam or barrage construction such as magnitude of flow, timing of flow, frequency and duration of flow events, and rate of change between types of flow events. Low flow means lower populations of fish and wetland wild life, reduced pollution assimilation, and protection of downstream water quality. The timing and frequency affect seed dispersal, signaling mating periods, flushing out sediments, cycling nutrients. Duration and rate of change of flow pattern determine connectivity between stream and flood plain, and in turn how long inundated riparian zone is available as breeding and nursery grounds by fish and flora.

Quantifying the amount of water a river must have is difficult as a river's ecological needs have not been defined yet [9]. In India various authorities are still working on it. Ministry of Environment and Forests suggests 20 % of average flow of four lean flow months to be released downstream of the dam. Central Water Commission reportedly suggests a minimum flow of not less than 2.5 % of dependable annual flow for Himalayan Rivers. There is urgency of implementing the policies of environmental flow, keeping in view impacts of climate change on river flows and aquatic biodiversity [9].

Table 2.1 Recession of Gangotri Glacier

Period	Total retreat (m)	Rate of retreat (m/year)	Total length change (m) along central flow line	Total area vacated at snout[a]	Average area vacated at snout[b]
1965–1968	−17.7 ± 12.8	−5.9 ± 4.2	+77 ± 12.8	22.9 ± 6.0	7.6 ± 2.0
1968–1980	−323.2 ± 21.5	−26.9 ± 1.8	−465 ± 21.5	163.9 ± 10.1	13.7 ± 0.8
1980–2001	−441.0 ± 26.2	−21.0 ± 1.2	−537 ± 26.2	215.4 ± 12.3	10.2 ± 0.5
2001–2006	−37.0 ± 20	−7.4 ± 4	−6 ± 20	16.39 ± 9.4	3.2 ± 1.8
Total	−818.9 ± 14	−19.9 ± 0.3	−1,085 ± 14	418.5 ± 37.8	10.2 ± 0.9

Source: Bhambri et al. [10]
[a] 10^3 m^2
[b] 10^3 m^2/year

2.3.3 *Climate Change and Glacier Melting*

Climate change is a global environmental challenge with severe impacts on hydrological system, ecosystem, food production, freshwater supply, health, and sea level rise. Climate models show a clear trend of rising temperature in the Ganga basin: 1–2 °C by 2050. The hydrological cycle is likely to alter with severity of droughts, and intensity of floods are likely to increase. Crop yield may decrease with rise in temperature. Imminent fallout is melting of the Himalayan glaciers (Asian Water Tower) affecting flows of the mighty Ganga among other Asian rivers. Ganga basin has 968 glaciers spreading over an area of 2483.91 km^2, and having a volume of 213.72 km^3. Based on the study of high resolution satellite data from 1965 to 2006, Bhambri et al. [10] concluded retreat of Gangotri glacier up to 819 ± 14 m, and loss of 0.41 ± 0.03 km^2 $year^{-1}$ at its front (Table 2.1, Fig. 2.2). With a 2 °C increase by 2050, 35 % of the present glaciers will disappear, and runoff will increase, peaking between 2030 and 2050, but ultimately decrease as the glaciers disappear by 2109 (Shripad [11]: In: Thakkar [12]). The rising sea level will cause saltwater flow into the Ganga River, threatening its ecosystem and turning vast farm lands barren.

However, based on the study of the climate models Eddy Moore [13] infers that as the total precipitation rises due to increasing temperature, snow fall in the Himalayas will reduce by 30 %, spring melt will occur earlier. People living downstream will be impacted more by changes in monsoon rain due to rising temperature than by glacial melting. The climate change will impact food and drinking water of about 550 million people in the basin.

2.4 River Water Quality and Pollution

Since time immemorial the river water of the Ganga is famous for its purity and its unique capacity to purify itself against all odds (*Charak samhita*), which makes it distinct from other rivers in the world. It is the major source of drinking water, as no germ, bacteria or fungi can thrive in it. The self-purifying capacity of the Ganga is

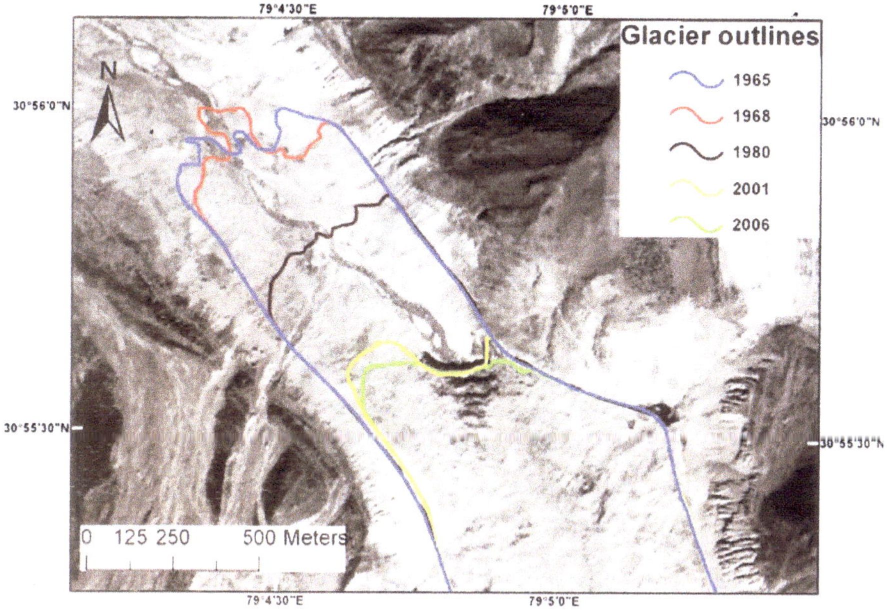

Fig. 2.2 Glacier outlines derived from high resolution satellite data overlaid on carto Sat 1 (2006) imagery [10]

derived from its unusually high ability to retain dissolved oxygen (DO) inherited from its environment. This capacity to regenerate its quality by killing germs owes to the geology, soils, wind, sun shine, climate and rainfall, trees and vegetation in the drainage basin. Bhargava and Agarwal [14], Bhargava [15], Sharma et al. [16], Sharma and Ghosh [17] reported large amount of well adapted microorganisms and radioactive radon in slightly higher concentrations in Ganga water (86.5 pci/l). This may be an important factor for fast killing of bacteria and regenerating capacity.

Sarin and Krishnaswamy [18] studied the major ion composition of the Ganga. The values of the major ionic composition of the Ganga (September 1982) at Allahabad, Kanpur and Hardwar were reported as below:

Hardwar	Na 59, K 38, Mg 153, Ca 353, HCO_3 894, Cl 23, SO_4 121, *SiO_2 109* (μmol/l), TDS 94 (mg/l)
Kanpur	Na 354, K 82, Mg 242, Ca 565, HCO_3 1829, Cl 78, SO_4 103, *SiO_2 157* (μmol/l), TDS 173 (mg/l)
Allahabad	Na 533, K 83, Mg 257, Ca 471, HCO_3 1751, Cl 146, SO_4 112, *SiO_2* 112 (μmol/l), TDS 170 (mg/l)

The river water quality of the Ganga in different sections has been analyzed by several workers, and summarized by Ghose and Sharma [19], Priyodarshi (American Chronicle, July 2009) and Das [20]. The Central Pollution Control Board (CPCB), too, is monitoring and evaluating the water quality at various monitoring stations. The quality evaluation has been done mainly based on Biochemical Oxygen

Demand (BOD), Chemical Oxygen Demand (COD), Faecal Coliform count (FC), Dissolved Oxygen (DO), electrical conductivity, chloride, sulphate, sodium, calcium, magnesium, hardness and alkalinity. Sewage and industrial wastes containing oxidisable organic matter exert a demand on dissolved oxygen, thereby reducing DO content of the river. BOD indicates biochemically degradable organic content at the expense of dissolved oxygen. COD indicates total organic content of water. In a healthy river adequate reserve of oxygen supports biochemical oxidation, in the lack of which anaerobic breakdown of the organic matter causes evolution of malodorous products, termed as septic condition of the river.

2.4.1 Scholastic Studies

Pahwa and Mehrotra [21] studied the stretch of the river Ganga from Kanpur to Rajmahal, reporting maximum turbidity (1,100–2,170 ppm) in monsoon and minimum (<500 ppm) in January to June, pH 7.45–8.30, D.O count 5.0–10.5 ppm (minimum in monsoon), Cl 4.0–35.4 (minimum in monsoon, maximum in Jan–May), PO_4 0.05–10.21 ppm, NO_3 0.08–0.22 ppm, SiO_3 4.0–20.3 ppm. Saxena et al. [22] reported that BOD in Ganga water at Kanpur varied from 5.3 ppm in winter to 16.00 ppm in summer. Jayaraman [23] reported pH value of Ganga water increases owing to increase in HCO^{-3} ion concentration along the entire downstream stretch of the river. The river is highly supersaturated with lime and magnesium carbonate. Bhargava and Agarwal [14] and Bhargava [15] surveyed the total length of river Ganga and found that Ganga is rich in polymers excreted by bacteria. These polymers remove turbidity by coagulation and settle the suspended matter. Nath [24] reported 4.0 ppm of DO concentration in River Hooghly along its 35 km stretch from Falta Water Works to Garden Reach in Calcutta. The river being estuarine in nature does not disperse pollutants effectively, causing increase in pollutant load. Ghose and Sharma [19] reported that Ganga receives total Biochemical Oxygen Demand load of 30,483 kg/day through 315 mld of domestic wastes at Buxar, Patna, Munger and Bhagalpur.

Ghose and Sharma [19] studied the chemistry of Ganga river water at Patna. The chemical pollution is not alarming because of high dilution. The pH values varied from 8.7 (winter) to 7.1 (summer). The higher values were attributed to increased production of lotic and lentic aquatic ecosystem. High temperature in summer enhances microbial activity causing excessive production of CO_2 and reduction in pH. The average DO was 7.4. The average hardness and alkalinity of Ganga at Patna were 114.2 and 135 mg/l respectively. Values of alkalinity higher than hardness indicate that Ganga water was having carbonate and bicarbonate hardness. The average concentrations of Ca and Mg were 29.4 and 21.3 mg/l respectively. Iron concentration is up to 0.22 mg/l at Bansghat, but 0.1 mg/l at Kurjighat, the higher value being due to concentration of ashes of human bodies at the crematorium. Sewage discharge, too, increases iron content up to 0.3 mg/l. Further, Ganga water contains an average BOD value of 2.3 mg/l. Average COD value is 21 mg/l. The average ammonical nitrogen of Ganga water at Patna is 0.12 mg/l exceeding tolerance limit and reflecting its polluted nature. The average albuminoid nitrogen content is 0.2 mg/l. The range of chloride content is 2–62 mg/l. The sulphate value of Ganga water at Patna is 15.7 mg/l. The phosphate value is on average 0.21 mg/l, silicate value is 8–30 mg/l being maximum in monsoon due to increased volume of transport of siliceous sediments. The fluoride concentration is 0.6–1.5 mg/l, while the Iodine concentration in the Ganga after the confluence of the Gandak, a tributary of Ganga, is low (av. 4.2 mg/l) causing the disease of goiter. Average conductivity value is 373.6 micromhos/cm. Bacteriological pollution is very high due to human excreta. However, in their view the Ganga at Patna is comparatively less polluted than at Buxar and Allahabad upstream.

Studies at Rishikesh, Rajmahal and Dakshineswar indicate that Ganga water is of alkaline earth bicarbonate type (Handa [25]).

2.4.2 *Report of CBPC [1]*

In a detailed report of Central Board for the Prevention and Control of Water Pollution (CBPC) [1] on the quality and pollution of the Ganga water, first of its kind, based on river water monitoring, CBPC has presented the pollution status of the river at various monitoring stations during summer (March–June) 1980–1982 as summarized in Table 2.2 below.

According to CBPC 1 the total dissolved solids (TDS) in the Ganga water are generally low from Kannauj to Falta and downstream (167–370 mg/l), but thereafter it rises under the influence of sea water to 4,590 mg/l at Diamond Harbour in summer. The alkalinity is 100 mg/l (August) to 120 (May) mg/l at Kannauj, 200 mg/l (May) at Allahabad, 100 mg/l (August) at Rajmahal, 70–80 mg/l downstream of Farakka. Cl concentration is 10 mg/l at Hardwar in May, 34.3 at Mirzapur, 9.5 mg/l at Patna, but rises to 1,670 mg/l at Diamond Harbour in the tidal zone. SO_4 values in summer vary from 2 mg/l at Hardwar to 505 mg/l at Diamond Harbour. Na concentrations are in the range of 14–28 mg/l in the stretch of Trighat to Allahabad, but it rises to 2,632 mg/l at Diamond Harbour in summer. Incidentally, "Ganga accounts for 2.5 % of global flux of sodium to oceans" [19]. Ca concentration varies from 26.2 to 30.9 mg/l in summer, and 35–40 mg/l in winter. BOD values are higher than tolerance limit (3 mg/l) from Hardwar to Trighat during the period October to June. The BOD values at Dakshineswar and Diamond Harbour are 15.23 mg/l (winter) and 15.58 mg/l (summer) respectively. From October to June in the stretch from Hardwar to Trighat BOD concentration is higher than the maximum tolerance limit (3 mg/l) for drinking water. BOD concentration is significantly high in Kanpur stretch. During the dry period (October to June) Ganga from Kannauj downstream station to Diamond Harbour has in general a total coliform count much higher than the maximum tolerance limit (5,000 MPN/100 l) permissible for drinking. DO has lowest recorded average of 4.35 mg/l in summer at Kanpur, and highest average of 11.75 mg/l at Varanasi in winter. However, DO level in the Ganga is generally higher

Table 2.2 Major Ionic Composition (mg/l) of the Ganga at various monitoring stations

Station	EC	pH	Na	Ca	Cl	SO_4	Total Kjedahl N	NO_3	BOD	DO	COD	Total coliform (FC)
1	661	8	4.2	43.2	8.3	2.0	2.45	0.22	2.0	5.7	8.6	2,400
2	399	8	NA	91.4	18.6	16.8	43.0	NA	15.3	6.5	28.4	
3	439	8	23.1	75.2	24.9	15.6	4.70	2.04	9.3	7.4	22.0	31,250
4	332	9	22.6	78.8	25.4	15.7	4.59	1.26	9.7	6.4	20.0	12,075
5	NA	NA	NA	27.4	35.0	19.5	NA	NA	NA	NA	NA	NA
6	325	8	16.4	69.4	14.3	21.0	0.62	0.16	0.03	8.1	16.2	125,500
7	430	7	16.0	78.6	13.8	27.4	0.70	0.12	2.0	6.0	19.8	240,000
8	2,060	7	2,632.3	145.7	1,370.7	505.3	4.37	0.41	8.0	7.1	185.4	8,250

EC in mhos/cm at 20 °C, FC – MPN/100 ml
1 Haridwar, 2 Kanpur D/S, 3 Allahabad D/S, 4 Varanasi (Dasaswamedh Ghat), 5 Buxar, 6 Farakka, 7 Daksineswar, 8 Diamond harbour
NA not available

than the minimum limits required for support of fish fauna, for bathing and for use as drinking water. In June total nitrogen level reaches the highest value of 43 mg/l at Kanpur due to large quantity of chemical effluents from the factories. Such concentration of total kjeldahl nitrogen may provide enough concentration of ammonia to kill the fishes. At Diamond Harbour in June the total nitrogen value is 4.37 mg/l. Further, in post monsoon months the value of total nitrogen is more than 6.85 mg/l from Allahabad to Trighat. Hardness is maximum in summer and minimum in monsoon. pH of Ganga water is relatively high in January – February and becomes as low as 7.1 in monsoon and summer. Hardness, alkalinity, dissolved salts and nutrients in Ganga water increase with intensity of pollution. Summarizing, the fairly large volume and high flow of the Ganga between Allahabad and Farakka render this stretch relatively less vulnerable to pollution, while the sections upstream of Allahabad and downstream of Farakka are prone to pollution.

2.4.3 Recent Reports of CPCB [4]

In the recent reports of CPCB [4], the latest pollution status of the Ganga river water has been comprehensively summarized based on crucial indices of FCC, BOD, DO. The following is a vivid narration of the current state of the river along its entire course from the catchment to its confluence with the Bay of Bengal. Monitoring of the river water quality undertaken by CPCB [4] clearly shows that the pure unpolluted river water at Gangotri with BOD 0.0–2.0 mg/l, DO 7.2–10.0 mg/l, 2–340 MPN/100 ml, sharply deteriorates at the pilgrim city of Hardwar with DO 6.0–9.2 mg/l, BOD 2.2–5.4 mg/l, due to disposal of sewage and half burnt dead bodies. At Kanpur, an industrial hub, the water is generally alkaline, but FCC and BOD far exceed permissible limits. Further BOD and FCC values up to 16.3 mg/l and 1,48,333/100 ml respectively have been reported during the period in the river at the holy city of Varanasi, much beyond the permissible limits for drinking and bathing. Pathogenic organisms in river water pose a veritable threat to the residents of the city. Downstream the Ganga turns black and septic, and ripples are foamy at many places, emitting methane bubbles, with floating unclaimed bodies of the dead, corpses of animals, half cremated remains of the babies thrown to the river to attain eternal peace, floral offerings, garlands and garbage. At Buxar (Unnao) further downstream values of BOD, COD and H_2S rise beyond tolerance limits. At Kolkata FCC and BOD are precariously high, while DO levels decrease. Though dissolved oxygen is sufficient, it shows a clear down trend all along the river course. Figure 2.3 reveals the polluted state of Ganga at Kanpur.

CPCB [4] has observed that the ambient water quality (Class ‘D’) between Kannauj and Trighat (border of Uttar Pradesh and Bihar), as also between Kolkata and Diamond Harbour is suitable for only wild life and fisheries, and unfit for human use.

Table 2.3 depicts the latest pollution status of the Ganga and Yamuna (Source: CPCB [26]).

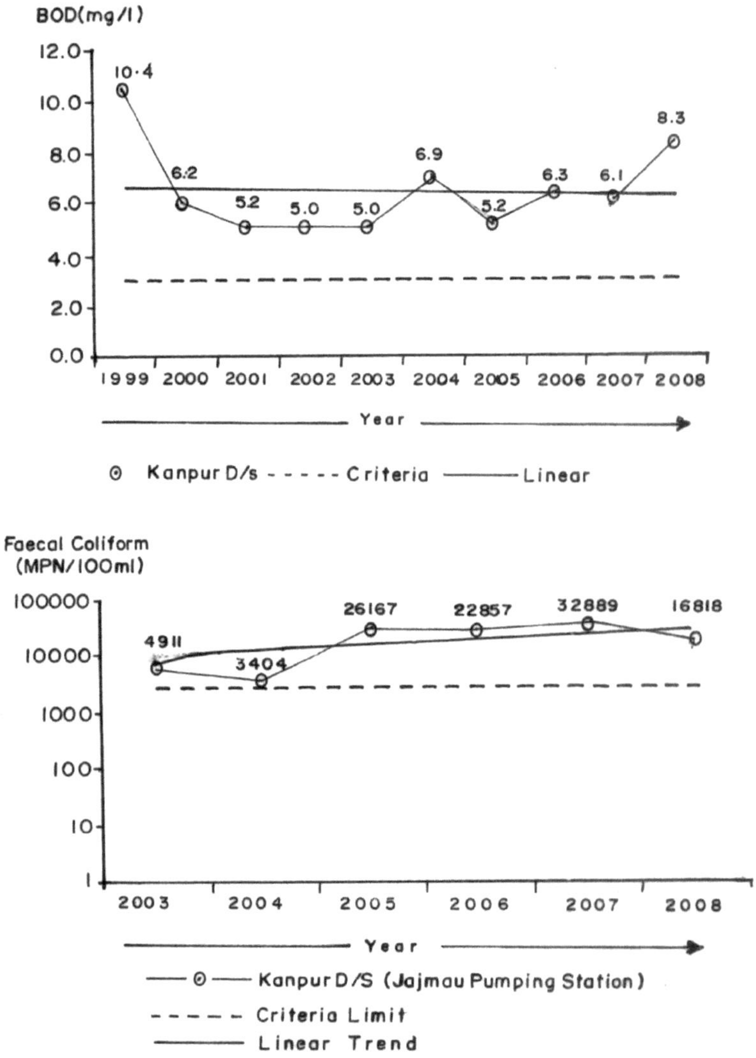

Fig. 2.3 Water Quality (BOD & Faecal Coliform) trend of the River Ganga at Kanpur (After CPCB [4])

2.4.4 *The Yamuna*

Rising from the Yamunotri glacier in the Himalayas, the River Yamuna flows through seven states before joining the Ganga at Allahabad. The Yamuna is the largest tributary of the Ganga bifurcating in to Western and Eastern Yamuna Canals at Hathnikund barrage, about 40 km upstream of Yamunanagar. The detailed major ionic composition of the river as studied by Sarin and Krishnaswamy [18] at the

Table 2.3 Water Quality Data of River Ganga and Yamuna (2010)

Locations	D.O. (mg/l)			pH			Conductiviity (µmhos/cm)			B.O.D. (mg/l)			Fecal coliform (MPN/100 ml)			Total Coliform (MPN/100 ml)		
	Min	Max	Mean	Min	Max	Mean	Min	Max	Mean	Min	Max	Mean	Min	Max	Mean	Min	Max	Mean
Ganga at Kanpur	1.6	9.9	5.6	7.4	8.8	8.0	67	860	479	3.8	21.0	8.3	1,500	46,000	16,818	15,000	240,000	94,250
Ganga at Allahabad Sangam	6.5	9.7	8.0	8.1	8.4	8.3	270	758	469	3.3	5.4	4.0	1,700	11,000	5,167	3,500	17,000	8,875
Ganga at Varanasi	6.5	7.9	7.3	8.4	8.9	8.6	264	397	353	9.2	12.4	10.5	18,000	110,000	74,500	94,000	220,000	144,833
Ganga at Dakshi-neshwar (Kolkata)	4.1	8.2	6.1	7.4	8.2	7.8	201	378	316	1.4	6.0	4.2	35,000	350,000	283,333	55,000	1,400,000	605,417
Yamuna at Nizamuddin (Delhi)	0.0	4.4	0.5	7.3	8.0	7.6	410	1,590	1,153	2.0	55.0	25.7	170,000	1,990,000	1,064,545	2,300,000	17,900,000	8,918,182

Source: CPCB [26]

monitoring stations of Agra and Allahabad comprised Na 1401 & 556, K 130 & 57, Ca 794 & 547, Mg 395 & 190, HCO_3 1921 & 1753, Cl 1172 & 222, SO_4 412 & 67, SiO_2 170 & 164 μmol/l and TDS 287 & 173 mg/l respectively. The effects of pollution on the river water quality at Agra are quite perceptible.

Center Science and Environment [27] at Delhi has worked on the current pollution status of the river. It is reported to be relatively clean till it enters Delhi at Wazirabad barrage turning into a sewer. It is only after 958 km of traverse that regains its freshness after confluence of Chambal river. The pollution hotspots are recorded as: Sonepat (BOD 2 mg/l, DO 6.8 mg/l, FC 29,678), Palla (BOD 2 mg/l, DO mg/l 8.2, FC 7892), Nizamuddin Bridge (BOD 24.3 mg/l, DO 0.3 mg/l, FC 175,54,444), Agra canal (BOD 14.3 mg/l; DO 0.6 mg/l; FC 1,824,583), Majhawali (BOD 22.5 mg/l; DO 4.7 mg/l; FC 1,942,500), Mathura (BOD 8 mg/l, DO 6.9 mg/l, FC 74892), Agra (BOD 15 mg/l, DO 6.3 mg/l, FC 256,833), Bateswar (BOD 11 mg/l; DO 13.5 mg/l, FC 12,423, 333), Etawah (BOD 10.5 mg/l, DO 13.4 mg/l, FC 1,37,823). DO levels are dipping with increasing organic pollution. At Tajewala it has dipped to 8 mg/l in 2005. At Delhi (Nizamuddin) DO is 0.3 mg/l.

Table 2.3 gives the pollution status of the river as in 2010.

The total nitrogen contribution downstream of Okhla barrage increases from 36 to 53 mg/l, thereafter falls to 25 mg/l at Majhauli. Between Hasanpur and Vrindaban total nitrogen increases from 25 to 36 mg/l. Further with the same total nitrogen level from Okhla downstream to Agra, any rise in free pH level may lead to rise in free ammonia levels and toxicity killing fishes. Mathur [28] found ammonical and albuminoid nitrogen ratio considerably high in Yamuna water.

According to a report of CPCB, monitoring of the Yamuna at Nizamuddin (Delhi) showed that the lowest level of faecal coliform was 4.4 lakh per 100 ml as measured in May 2009, owing to the discharge of untreated sewage and industrial wastes through drains (*The Times of India*, 2nd Feb 2010, p 10). Figure 2.4 shows the polluted Yamuna near Noida, Uttar Pradesh.

2.5 Polluting Factors

Pollution of the Ganga water is from three major sources: Discharge of untreated municipal sewage and industrial effluents; agricultural runoffs or return flow of leachates from fertilizers and pesticides. Lack of adequate 'dissolved oxygen' leads to septic condition and rise in biochemical oxygen demand (BOD) in the river. According to Markandya and Murty [30], 1.3 billion liters of sewage, 260 million liters of industrial waste, runoff from six million tons of fertilizers and 9,000 tons of pesticides used in agriculture, and large quantities of solid wastes, are daily released into the river. The major water polluting industries include chemicals, textiles, pharmaceuticals, cement, electrical and electronic equipment, glass and ceramics, pulp and paper board, leather tanning, food processing, and petroleum refining. But the sewage treatment capacity is 44 % in the Class I cities, and 8 % in Class II cities as on 2011. Reportedly the discharge of untreated sewage/sullage into the Ganga is

Fig. 2.4 River Yamuna: Polluted and Foaming near Noida (Source: Srikantia [29])

responsible for 75 % of its pollution with nearly 3,000 mld (million liters per day) of sewage generated in the towns along the Ganga. At Kanpur 45 tanneries, 10 textile mills and several other industrial units discharge 37.15 mgd of waste waters. According to Agarwal et al. [31] 3,000 half burnt human bodies, 6,000 carcasses, 140–2,000 tons of flesh, 200–3,000 tons of ash (produced from burning 11,000 tons of firewood) in addition to 925 mld of domestic sewage and industrial effluents are added every year to the Ganga at Varanasi. In the industrialized belt (92 km) on the bank of the Hooghly estuary, the river is tampered by heavy discharge of liquid wastes from industrial sources. The Hooghly receives maximum pollution load of suspended solid from the pulp and paper mills [24]. Reportedly in CMDA area (Kolkata) 122,849 kg/day of industrial BOD load and 2,74,090 kg/day of domestic BOD are discharged into the river Hooghly [19]. This quantity is too large for self-purification by the Ganga. Nearly 50 % of waste waters are discharged untreated. In addition, lakhs of devotees take daily holy dips in the Ganga, the number swelling many times more during the festive seasons polluting the river further.

A study of the Yamuna river water quality [32] showed that the river carries 300,000 cubic meters of sewage and 20,000 cubic meters of toxic industrial wastes per day produced from the textiles, chemicals, soaps, rubber plastics, engineering goods. According to the Center of Science and Environment [27] the waste water discharge into the river varied from 3,556 to 4,657 mld. Reportedly Delhi contributes 68 % of BOD load to the river followed by Agra 4.27, Faridabad 4.37 and Ghaziabad 5.51 %. Domestic sewage is the source of high nitrogen pollution.

Markandya and Murty [30] estimated high levels of toxic metal in waste waters discharged into the Ganga at Kanpur and Varanasi such as Cd 0.05 and 0.16; Cr 6.45

and 8.12; Cu 0.88 and 0.16; Fe 8.80 and 3.32; Mn 0.55 and 0.47; Ni 0.22 and 0.14; Pb 0.19 and 0.15; Zn 1.82 and 1.58 mg/l respectively. Mehrotra [33], Ghose and Sharma [19] reported presence of toxic metals like Hg, Pb, Cr, Ni, Cu, Zn, Co in the river sediments at Varanasi and Patna.

2.5.1 *Sand Mining and Catchment Degradation*

Sand mining and quarrying of boulders from the bank and river bed are also causing immense harm to the river bed, aquatic habitat and water quality. Sand mining within and near the river bed affects the river's physical characteristics, biodiversity, and groundwater recharge. It has already created precarious situation in Uttarakhand and Haryana. Equally disastrous is the deforestation of the catchment areas which has caused soil erosion, increase in silt generation and deposition in the river, reduction in groundwater recharge and increase in runoff. Resting and nesting grounds of many migratory birds are lost.

2.6 Impact of Flow Impediment and Pollution

2.6.1 *Flow Impediment*

If the river flows are obstructed by artificial constructions, its water quality, downstream ecology, sedimentation pattern and river morphology are susceptible to change. Ganga water has thus been deoxygenated and lost its purity. A naturally flowing freshwater river provides the habitat of hundreds of diverse life forms, meets the needs of people on its banks, recharges groundwater, and meets requirement of livelihoods of fisher folks, boatmen, farmers. The silt deposited by the river improves soil fertility. Most of these benefits are annulled by disruption of the flow. Downstream for long stretches there will be effectively no river or only a dirty polluted nallah will exist. This is the state of the River Ganga. The entire ecology has been degraded with the reduction or gradual disappearance of fishes, and near extinction of dolphins or other aquatic fauna in places along its course. If the contemplated 600 dams, mostly in Uttarakhand are constructed, an entire 935 km stretch of Ganga will be dry, with disastrous effect on environment and ecology.

Reportedly large hydropower projects in Uttarakhand including Tehri Dam are critically depleting the discharge of the river and its tributaries. The impacts of Hydro Power Projects on the environment and ecology of the Ganga River has been a matter of utmost public concern for life, health and wellbeing. "Since the dam (Tehri) was completed in 2006 the natural spring that once fed Pipola has dried up. ....several in Pipola (in Uttarakhand) had to give up farming for lack of water...left the village to work in a hotel bakery outside of New Delhi, hoping to earn enough to feed their five children. .... It's a crisis brought on by India's relentless push to modernize, as water ...is increasingly put in service of big hydroelectric dams, big cities, and big agriculture...." (*Time Magazine*, July 19, 2010, p 18). Short of enough water to fill in the dams the river water downstream is pumped back upstream for reuse to meet the demand for power. This belies 'Run of the River myth' of the hydropower projects.

Recent studies show that during the 12-year period 1991–1992 to 2003–2004, notwithstanding massive expenditure for increasing canal irrigated areas, there has been actually a reduction of 3.18 Mha of canal irrigated areas. Similarly most of the hydropower projects perform at 50 % of the promised generation and in the last 16 years generation as per installed capacity has reduced by 20 % [12]. Silting up of reservoirs and reduced flow (or overdevelopment?) of the river are cited as major reasons for deficiency in power generation.

2.6.2 Ecosystem Health

According to a World Bank Sponsored Study (State of Environment Report-U.P.) (In: Mallikarjun [34]), pollution levels in the Ganga are contributing 9–12 % of total disease burden in Uttar Pradesh. The coliform bacteria levels are in excess of 2 lakh MPN as against the national water quality standard of 5,000 [34]. The report estimated total health damage on account of water pollution in Uttar Pradesh to be around 6.4 million DALYS (Disability Adjusted Life Year). The study of Indian Toxicological Institute, Lucknow during 1986–1992, detected Hg in river water, sediments, benthic fauna, fish, soil and vegetables at Rishikesh, Allahabad and Kolkata (Dakshineswar), which finds its way to food and water [35]. Ajmal et al. [36] reported heavy metal in harmful concentrations in submerged plants and some fishes in the Upper Ganga basin around Aligarh in Uttar Pradesh.

2.6.2.1 River Dolphins

The existing population is in the form of small fragmented sub-populations. The species is having serious threats for its survival due to construction of dams and barrages (habitat fragmentation), pollution, mining of sand and stones (habitat degradation) apart from poaching etc. Due to usage of 2,500 tons of pesticides and 1.2 million tons of fertilizers annually in the catchment area, the river has been most polluted. Sand mining, siltation and excessive water extraction have degraded the river ecosystem. Sand extraction destroys dolphin habitat. Construction of barrages at Bijnor, Narora, Farakka and Kanpur have restricted the movement of dolphins, which may inhibit their genetic, social and ecological interactions among individuals, limit the gene-flow, increase their vulnerability to natural catastrophes' and ultimately lead to their extinction. Sadly, only 1,800 dolphins now survive in the Ganga and its tributaries [37]. Recent studies show high levels of organochlorines and heavy metals in the tissues of dolphins and fishes in the Ganga water, thereby affecting vital organs and processes causing extinction in the long run. The fish and amphibian populations, too, are facing extinction.

2.7 Groundwater

The subsurface regime of groundwater is the unseen part of the water cycle which too discharges to the river and hence is partially responsible for the lean month flow in the river. Hence health of the groundwater regime is another crucial aspect of the Ganga basin. The coarse clastic sediments (gravels and sands) form highly potential aquifers yielding up to even 150 cubic meters per hour or more. The annually replenishable part of recharge is utilized through wells to meet various requirements, mainly for irrigation. However this is also the only major source for drinking water. As per latest reports [38], the total utilizable ground water resources in the Ganga basin states have been estimated as 195 BCM and total draft as 105.7 BCM. Groundwater is overexploited in the states of Rajasthan (135 %), Haryana (127 %), Punjab (170 %) and NCT Delhi (138 %) and parts of Uttar Pradesh (Muzaffarpur, Moradabad, Saharanpur districts: 110–133 %), sometimes reaching a stage of more than 250 % of its natural recharge. Parts of Uttar Pradesh, and Uttarakhand, too, in its upper catchment are on the verge of overexploitation (74–92 %). The water levels in summer months dip pretty low (in cases down to even 20 m or more below ground level) precluding or reducing the scope of dry month base flows in the Ganga. The river flow in lean months is practically lacking from the upper catchment down to the holy city of Varanasi affecting its capacity for self-purification and aggravating pollution of the river during non-monsoon months.

Groundwater in the Ganga basin is generally fresh, suitable for all uses, being alkaline in nature and of earth bicarbonate type, with low mineralization, and low to medium hardness. Electrical conductivity is generally less than 750 μmhos/cm at 25 °C. But groundwater is also vulnerable to pollution from various surface sources through porous alluvium. High salinity is a characteristic of groundwater in arid terrains of Rajasthan, Haryana and in the coastal tracts, where EC as high as 3,000–10,000 μmhos/cm is also reported. Groundwater with high arsenic (0.05–3.20 mg/l) in the Lower Ganga delta of West Bengal as also in parts of Bihar and Uttar Pradesh, consumed in drinking water has caused the largest catastrophy of water pollution in the world. Fluoride contaminated ground water (2–20 mg/l) also occurs widespread in Rajasthan, Haryana, and many other parts of the basin being the reason for the dreaded disease of fluorosis if ingested through drinking water. High nitrate (up to 1,000 mg/l) is reported in groundwater in scattered pockets throughout the basin. If used in baby feed, nitrate rich groundwater may cause blue baby syndrome. Iron, fluoride and arsenic are all from geogenic sources, while nitrate is derived from leachates of fertilizers used in agriculture, as also from domestic and municipal wastes. Cyanides, heavy metals (Cd, Cr, Pb) in toxic concentrations in groundwater are also carcinogenic, and sourced from industrial effluents in all major urban centers. It assumes importance as groundwater is the principal source of drinking water, and regenerates Ganga in summer. As groundwater is contributing 60 % of total irrigation in the country, alien chemicals may enter into our food cycle. With climate change and global warming groundwater use will increase but recharge will decrease, leading to more overexploitation and pollution.

2.8 River Restoration

A river's ecology and environment depends on its volume of flow in different seasons, as also on source, nature and quantity of pollutant load, and other related factors. Hence any plan for conservation of river is virtually a tradeoff between economic potential and its environment (*Time*, July 19, 2010). Rajeev Gandhi's Ganga Action Plan (GAP) launched in 1984, was the first gigantic exercise undertaken to cleanse the Ganga. It was initially designed and implemented in Varanasi and some other cities on the bank of Ganga. It was aimed at stoppage of sewage disposal in to the river by renovation of sewage pumping and treatment plants, and existing sewerage system, as also installation of new sewage treatment plants (STP), extending the existing sewerage systems to the unsewered systems in parts of the cities, prevention of throwing dead bodies into the river, and regulated use of pesticides and insecticides for agriculture. Trunk sewers parallel to the river were already laid for interception of sewage, leading it to the sewage treatment plants. The effluents from the sewage treatment plants could be treated as rich source of energy (say biogas) and manorial matter. But sadly, Ganga Action Plan Phases I and II met with only partial success due to various factors, the failure on part of the government to involve people being cited as a major one. The sewage pumps virtually failed in the flood season and during power disruptions, often for prolonged' periods, resulting in total sewage of the city draining into the river. The treated effluents from STP caused health hazards, ruined crops and polluted groundwater.

2.8.1 Civil Society's Campaign

In 1982 the citizens of Varanasi and teachers of Banaras Hindu University started Swachha (Pure) Ganga campaign, supported by the Sankat Mochan Temple establishment resolving to clean the Ganga, to start with at Varanasi. Sankat Mochan Foundation acted as "a catalytic agent to work with the people, to spread the message of 'clean Ganga', and the need for everybody's participation and assistance to achieve the objective" [39]. The campaign proved to be very effective.

Following only partial success of GAP, various alternatives were proposed from several quarters, but the one from Varanasi Nagar Nigam and Sankat Mochan Fund [39] which provided for people's participation in the operation, evinced interception and diversion of sewage flowing into the Ganga without using electricity and pumps, treatment plants using minimum electrical power, and a technology eliminating faecal coliform and harmful bacteria, using existing facilities. The cleaning operation involved Advanced Integrated Wastewater Oxidation Pond System (AIWPS) which allowed fermentation in anaerobic condition leaving no residue, release of maximum free molecular oxygen in dissolved state, algal removal, which have fertilizing value and disinfection of the residues. This is the basis of a pilot project sponsored by Government of India. However, it has now been realized that decentralized system is cheap, efficient and self regulating. Domestic waters contain biodegradable elements. Separation of black water (faecal matter) and grey water (remaining waste water) is possible by adopting two-pipe system. Primary treatment is feasible at plot-holding level. Treated and semi-treated waters may be recycled and reused. Existing sewage treatment plants (STPs) can provide additional treatment to water outflows. Use of chemicals and energy is minimal. STPs can provide secondary and tertiary treatment for the water to be recycled. Citizens and NGOs share responsibility.

Jal Biradari under the leadership of Rajendra Singh started its country-wide campaign: Ganga Revival Yatra in 2003, Yamuna Satyagraha and River Revival Satyagraha in 2007, Save Bhagirathi River in 2008, joined by various other organizations and even individuals in 'mass rallies', 'sit ins'and 'fast unto deaths' to stir the civil society and the government into actions to restore the Ganga and also conserve 144 other endangered water courses in the country. Jal Biradari has also launched 'Jal Kumbh' movement on these rivers to resolve and take collective actions on protection, purification and conservation of these rivers. The campaign is supported by religious heads, too.

2.8.2 Cleaning of International Rivers

The clean-up program of the Thames is internationally treated as a model [30]. The river was cleaned up in 20 years' time with an investment of Rs. 5,000 million from 1950 to 1980. Similar river restoration of Danube is an ongoing program, selecting the programs on cost benefit basis.

2.8.2.1 Thames

The pollution problems of the Thames dates back to Roman times in history. The targets of concerted actions since 1970 were to improve river water quality, permit fish to survive and migrate at all stages of the tide. During 1910–1950 the pollution became so intense that long stretches of the Thames became anaerobic and ecologically dead. The action policy in the second restoration plan setting location specific environmental quality objectives (EQO) based on detailed scientific monitoring and modeling under jurisdiction of one authority, has become the foundation of all modern aquatic pollution control (Fig. 2.5). Achieving the EQO requires adoption of a pollution budget which would restrict the pollution load applied to any stretch of the

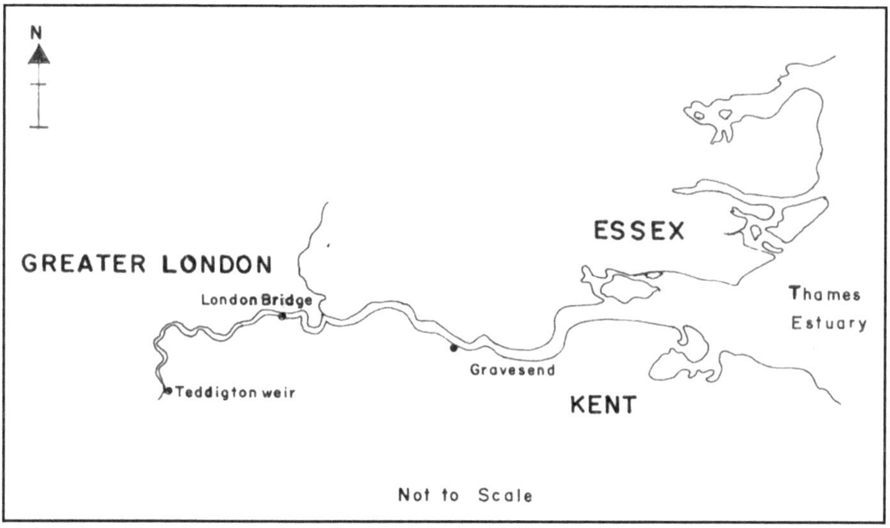

Fig. 2.5 Thematic Map of the Tidal Thames River

river to an amount which would not reduce DO levels below those set out in EQO. The amount was decided based on mathematical modeling. The river water quality objectives in different sections were set as: (i) Teddington to London bridge: the minimum percentage of air saturation with dissolved oxygen (A) should be at least 40 % and the water should be nontoxic to fish; (ii) London bridge to Canway island: 'A' should be atleast 10 % and quarterly value of 'A' should not be below 30 %; (iii) Canway Island to Seaward limit: 'A' should be atleast 60 % and the quality should be suitable for whole life cycle of marine organisms including fish.

2.8.2.2 Danube

Danube has a course of 2,857 km through 11 countries. It has great self-purification potential. The problem relates mainly to nitrate pollution, eutrophication and microbial contamination, and contamination from hazardous substances causing problems with fisheries in the Black Sea. Human population contributed 25 % of nutrient load as a result of absence or poor working condition of waste water treatment plants. Another 25–30 % nutrient input arises from industrial sources e.g., Fertilizer factories. As in 1991, 540,000 tons of nitrogen, 45,000–50,000 tons of phosphorous were carried into the Black Sea creating ecological threat. Concerted actions to control pollution started in 1985, divided into short, middle and long term Strategic Action Plans including identification of hotspots, setting water quality objectives for the tributaries and the main river Danube, as also wetland conservation and management programs.

2.8.2.3 Lessons Learnt

The cleanup programs in the Thames and Danube highlight the need for a multilateral approach and constitution of an autonomous central authority, established by law and armed with executive and financial powers, which will be responsible to execute the following policies to control environmental Problems.

1. Monitoring and control, assisted by analytical research on protection and conservation of river.
2. Preparing and executing river basin management plan.

The above objectives may be translated into several strategic actions programs

1. Collecting data. Establishing national and regional data bases, and national river and river basin authority information systems. Identification of 'hot spots' of pollution.
2. Based on monitoring data and water quality modeling, pollution budget may be drawn to estimate the maximum pollutant load compatible with quality objectives to be achieved in these identified sections.
3. Phased enhancement of municipal waste water treatment capacity. Construction of waste treatment plants.
4. Reduction in effluent discharges from industries and agriculture.
5. Prohibition of discharge of untreated sewage, and use of plastic bags.
6. Wetland inventory. Conservation, restoration and management of wetland, and flood plain areas through integrated water management'.
7. All real estate activities or construction or other related developmental activities around the River must have permission from the Authority to accord maximum protection against damage and pollution to the riverine regime.

2.8.3 Restoration of River Arvari: A Shining Example

Rainwater harvesting and watershed treatment using traditional knowledge may play a major role in river rejuvenation of the country. The dead river Arvari in Alwar district of Rajasthan in India came alive through traditional rainwater harvesting methods, – construction of thousands of johads along with water conservation measures, achieved entirely through community participation under the aegis of Tarun Bharat Sangh [41]. Arvari River Parliament is a showcase of community management of water and land based natural resources. Likewise through rainwater harvesting in various parts of the country dying rivers have been regenerated. The experiment of Arvari River is the finest example before the nation. Success of Arvari has triggered a country wide campaign of river rejuvenation. The community may take the lead in restoration of rivers, streams through rainwater harvesting, along with soil and water conservation measures using traditional knowledge with modern inputs. This will significantly contribute towards revival of the River Ganga and its tributaries as also sub tributaries.

Harvesting and storing the monsoon flood waters in the flood plain deposits of the rivers/streams of the Ganga basin by induced recharge or rainwater harvesting techniques may revive lean month flows in the rivers through groundwater discharge, thereby improving river water quality and ecology.

2.8.4 Ganga River Basin Regulatory Authority

Recognizing this problem as of national dimension, the river Ganga was declared as National River by the Prime Minister in 2008 with the constitution of a Ganga River Basin Regulatory Authority. It is a solemn pledge to ensure its perennial flow in all its natural quality, quantity and natural piety. It introduces a code of conduct and self discipline for the citizens towards this national symbol. It aims at maximum involvement and participation of the people or stake holders, and innovative strategies for its restoration. This has been enshrined in the profile of actions as envisaged in the Government Notification (February 2009) and briefly quoted below.

1. Ensuring effective abatement of pollution and conservation of River Ganga by adopting a river basin approach.
2. Maintaining minimum ecological flows in the river Ganga with the aim of ensuring water quality and environmentally sustainable development.
3. Development of river basin management plans and regulation of activities to maintain its quality and take such measures as relevant to river ecology and management.
4. Collection, analysis and dissemination of information, monitoring and evaluation of the management plans including social auditing.
5. Investigation and research regarding problems of environmental pollution and conservation of the river.
6. Promotion of water conservation practices including recycling and reuse, rainwater harvesting and decentralized sewage treatment systems.
7. Constitution of State River Conservation Authorities and comprehensive management of the river in the states through these authorities based on integrated basin management plans drawn by National Ganga River Basin Authority.

The Authority should have a Management Board drawing its Members from stake holders, including the public representatives and priests, NGOs, all related scientific and technical institutions or organizations, both government and non-government, engaged in management and research on conservation and protection of river regime. It should draw state wise plans in consultation with the members. It will operate through State River Conservation Authorities, which will come under its complete control. It will also introduce a system of social auditing and ensure transparent actions. The Ganga river should be in the concurrent list of the Constitution to facilitate implementation of its restoration projects.

The *'Mission Clean Ganga'* is an ambitious project of the Ganga River Basin Authority with a massive investment of 15,000 Crores of rupees for its revival and purification including installation of Purification Plants, by the targeted year of 2020 (*The Hindu*, October 6, 2009). In 10 years no municipal waste water will be discharged into the Ganga. No new project that endangers the ecosystem will be approved by the Ministry.

2.9 Conclusions

River regime is lifeline to civilization. Modern life style, technological developments and industrial revolution, together with Green revolution in the country have all accentuated demand and consumption of water stupendously over the last half a century or so. But uncontrolled and mindless development activities, have led to overuse and abuse of the River with adverse impacts on its environment and ecology. Consequently, the River Ganga, once a mighty river, has now paled to polluted pools, or reduced to a thin sluggish flow in places breeding germs and turning unfit for human or even animal use. Wetlands, in-stream resources like flora, fauna, – fish population, and dolphins, – are all endangered or facing extinction. This unequivocally calls for conservation measures through (1) stoppage of river flow diversion through construction of dams and barrages; (2) augmentation of the river flow by technological devices or means like rainwater harvesting and artificial recharge to groundwater; and (3) protection from pollution or overexploitation.

The core issue is integrated water management and water governance in the river basin. People have a big role to play in this gigantic river rejuvenation program. In this context Rajendra Singh, the Water Man, Magsaysay Awardee, in his "Ganga – People's Mandate" [40] has reminded the society of its bounden duties to cleanse and salvage the river.

1. Generating civic consciousness about this national problem, and inculcating this awareness from the school levels.
2. Prohibition of defecation in the river, throwing cremation articles into the river, and use of plastic bags.
5. A new set of rituals for Ganga worship to be decided by the Saints without involving excessive use of flowers, leaves, 'aarti' (obeisance) etc, which are harmful to the aquatic life.
6. Promotion of organic farming.
7. Encouraging rainwater harvesting and watershed management.
8. Demarcation of a Reserve Area around the Ganga to accord maximum protection against religious and public activities causing damage and pollution to the riverine regime.

The Government and the civil society have specific roles in the management of the river. The society should strive to unite man's consciousness with the River to reinstate Ganga in its pristine natural glory. Amalgamation of the civil society's role with science and technology is now a major issue.

> … if the river does not have water, then the river dies. And with it, the story of Ganga will be over
>
> (Veer Bhadra Mishra, Founder of the 'Clean Ganga Movement').

Acknowledgements The author places on record his deep sense of gratitude to Late Dr. B.P. Radhakrishna of Geological Society of India, for initiating him in the study with many valuable suggestions. Grateful thanks are also due to Dr. Rashmi Sanghi for her invitation to write the article and constant encouragement in this work. Lastly, the author also acknowledges the valuable support received from Dr.K. M. Najeeb, Regional Director, Central Ground Water Board and Shri Dwaipayan Das, Director (Engg), ARICENT, in this work.

References

1. Central Board For the Prevention and Control of Water Pollution (1984) The Ganga Basin, Part II. In: Dasgupta SP (ed) Centre for Study of Man and Environment, Department of Environment, Government of India. New Delhi, p 202
2. Rao KL (1979) Water wealth of India. Orient Longman Ltd., New Delhi, p 166
3. CWC (2010) Water and Related Statistics. Central Water Commission. http://www.cwc.nil.in/ISO_DATA_Bank/W&Relatedstatistics_2010/pdf
4. Central Pollution Control Board (CPCB) (2009) Ganga water quality trend. Ministry of Environment and Forests, New Delhi, p 63
5. Subramanian V (1979) Chemical and suspended sediment characteristics of rivers in India. J Hydrol 44:37–55
6. Bandyopadhyay J (2011) Prayers for a little. Down To Earth, 16–31 Aug 2011, p 26
7. Seth BL (2011) Prayers for a little. Down To Earth, 16–31 Aug, pp 24–26
8. Richter B, Baumgartner J, Powell J, Braun D (1996) A method for assessing hydrologic alteration within ecosystems. Conserv Biol 10(4):1163–1174
9. Katz D (2006) Going with the flow: preserving and restoring in stream water allocations. In: Gleick P (ed) The world's water. Island Press, Washington, DC, pp 29–45
10. Bhambri R, Tobias B, Chaujar RK (2012) Frontal recession of Gangotri Glacier, Garhwal Himalayas, from 1965 to 2006, measured through high resolution remote sensing data. Curr Sci 102(3):489–493, 10 February 2012
11. Shripad D (2008) Mountains of concrete. International rivers, 2008, p 32 (In: There is little hope. A civil society view. Himanshu Thakkar 2009)
12. Thakkar H (2009) India's National Action Plan on Climate Change: a civil society view etc. South Asia Network on Dams, Rivers & People, New Delhi, p 56
13. Moore E (2012) Ganga basin temperature to rise by 1-2°C by 2050. Down To Earth, 16–30 June, p 560
14. Bhargava DS, Agarwal GD (1977) Use of integrated quality indices in water pollution control. J Inst Public Health Eng 74:95
15. Bhargava DS (1982) Purifying power of Ganga unmatched. IST Bull 34–35. In: Ghosh NC, Sharma CB (eds) Pollution of Ganga, Ashish Publishing House, p 36

16. Sharma CB, Sinha SN, Ghosh NC (1982) Pollutions study of river Ganges at Patna. In: Proceedings of All India seminar on environmental engineering. Pollution control and water conservation, Patna, Institute of Engineering, India, pp 111–117
17. Sharma CB, Ghosh NC (1987) Pollution of the river Ganga by municipal waste: a case study from Patna. J Geol Soc India 30:369–385
18. Sarin MM, Krishnaswamy S (1984) Major ion chemistry of Ganga-Brahmaputra river system, India. Nature 312(5994):538–541
19. Ghose NC, Sharma CB (1989) Pollution of Ganga. Ashish Publishing House, New Delhi, p 38, 39, 42, 141
20. Das S (2011) Cleaning of the Ganga. J Geol Soc India 78(2):124–130
21. Pahwa DV, Mehrotra SN (1966) Observations on fluctuations in the abundance of planktonin relation to certain hydrobiological conditions of river Ganges. Proc Natl Acad Sci India 36 B(92):157–189
22. Saxena KL, Chakravarty AK, Khan AQ, Chattopadhyay RN, Chandra H (1966) Pollution studies of the River Ganges near Kanpur. Indian J Environ Health 8:270
23. Jayaraman S (1977) A study of water-mineral equilibria in the river Ganges. Unpublished M.Phil. thesis, Jawaharlal Nehru University, New Delhi
24. Nath AJ (1985) Hooghly: Bengal's Sewer. Sunday, 24, p 29
25. Handa BK (1972) Geochemistry of the Ganga River water. Indian Geohydrol 8:72–78
26. CPCB (2010) Water quality data of river Ganga and Yamuna, 2010. Central Pollution Control Board (A-14011, dt. 14.01.2010). In: J Geol Soc India 78(2):126
27. Centre for Science and Environment (2007) Sewage canal: how to clean the Yamuna. Centre for Science and environment, New Delhi, p 178
28. Mathur PP (1965) Pollution studies of river Jamuna at Delhi. In: Proceedings of the Symposium on Water Pollution Conference, CPHERI, Nagpur
29. Srikantia SV (2011) Yamuna river at Noida, Uttar Pradesh – Polluted and foamy (Photograph). In: Subhajyoti Das: Cleaning of the Ganga. J Geol Soc India 78(2):127
30. Markandya A, Murty MN (2000) Cleaning up the Ganges. A cost benefit analysis of Ganga Action Plan. Oxford University Press, New Delhi, p 1, 222
31. Agarwal A, Chopra R, Shukla K (1982) The state of India's environment. Centre for Science and Environment, New Delhi, p 189
32. Central Board for the Prevention and Control of Water Pollution (1982) The water: killer drink. Hindustan Times, 16th May. In: Ghosh NC, Sharma CB (eds) Pollution of Ganga, Ashish Publishing House, p 42
33. Mehrotra MN (1990) The role of sediments in environmental pollution. A case study of the Gangaat Varanasi. J Indian Assoc Sedimentologists 9:1–14
34. Mallikarjun Y (2003) Pollution levels in Ganga alarming. The Hindu, 15 Sept
35. Priyadarshi N (2009) Ganga river pollution in India – a brief report. American Chronicle, 8 July. http://www.americanchronicle.com/articles/view/109078
36. Ajmal M, Khan MA, Nomani AA (1989) Levels of trace metals in waters and sediments of Yamuna river. On status and impact of heavy metal pollution in India, Madras. In: Ghosh NC, Sharma CB (eds) Pollution of Ganga, Ashish Publishing House, p 134
37. Bashir T (2010) Ganges river dolphin (Platanista gangetica) seeks help. Curr Sci 98(3):287–288, 10 February, 2010
38. Anon (2011) Dynamic ground water resources of India, Central Ground Water Board, Faridabad, p 69
39. Mishra VB (2005) The Ganga at Varanasi and a travail to stop her abuse. Curr Sci 89(5): 755–763
40. Singh R (2009) The Ganga – people's mandate. Tarun Bharat Sangh, Bhikampura
41. UN-IAWG-WES (1998) Johad: Watershed in Alwar district, Rajasthan. UN Inter Agency Working Group on Water and Environmental Sanitation, Oct 1996, pp 1–24

A Ganga ghat scene (Courtesy Julian Crandall Hollick)

Part II
Rich Biodiversity of Ganga Basin

Birds spotted at Sangam in Allahabad (Courtesy Nitin Kaushal)

Chapter 3
Distribution of Major Floral and Faunal Diversity in the Mountain and Upper Gangetic Plains Zone of the Ganga: Diatoms, Macroinvertebrates and Fish

Prakash Nautiyal, Jyoti Verma, and Asheesh Shivam Mishra

Abstract Present study examines the distribution of biodiversity elements (diatom, macroinvertebrate and fish) in respect of their taxonomic richness, density, composition in the high gradient mountain zone (MZ) and low gradient plains zone (PZ) consisting PZ I (Hardwar to Allahabad) and PZ II (Allahabad to Varanasi). The richness of benthic diatom and algal flora declines with loss of substrate heterogeneity from MZ to PZ, but that of plankton, macroinvertebrate and fish, increases gradually. Diatom predominate the benthic flora in MZ but richness declines mildly in PZ. The plankton community of PZ is mildly richer in diatom flora, as share of green algae is substantial. The macroinvertebrate fauna is predominated by insects in the MZ and a combination of insects and molluscs in the PZ. Low Jaccard's similarity and high distances among MZ and PZ I, illustrate unique flora in these zones. Since, PZ II shows greater distance from both MZ and PZ I, this zone has different biodiversity in respect of these communities. In contrast the fauna is unique in MZ, while PZ I and PZ II have relatively more similarity, less in case of macroinvertebrate but high in fish.

Keywords Biodiversity • Taxonomic richness • Benthic biota • Plankton • Piscine fauna • Zonation • Biogeography

P. Nautiyal (✉) • A.S. Mishra
Aquatic Biodiversity Unit, Department of Zoology and Biotechnology, Hemwati Nandan Bahuguna Garhwal University (A Central University), Srinagar 246174, Uttarakhand, India
e-mail: lotic.biodiversity@gmail.com; pn.mahseer@yahoo.com

J. Verma
Department of Zoology, University of Allahabad, Allahabad 211001, India

R. Sanghi (ed.), *Our National River Ganga: Lifeline of Millions*,
DOI 10.1007/978-3-319-00530-0_3, © Springer International Publishing Switzerland 2014

Himalayan Mahseer picture (Courtesy Prof. Atul Borgohain)

3.1 Introduction

The course of River Ganga consists of a short mountain zone-MZ (Alaknanda-Ganga spans 266.7 km; Bhagirathi – Ganga spans 283.8 km from the source to Hardwar; [1]) and long Gangetic Plains (over 2,000 km long), referred as plains zone-PZ in the text. The MZ lies in the west Himalaya and the PZ in the Upper Gangetic Plains (UGP) biogeographic regions, respectively. The Ganga has a north and a south basin. Excluding its source river systems of the Bhagirathi and Alaknanda, the major tributaries of the north basin (Ramganga, Kali-Sarda-Ghaghara, Yamuna, Gomti, Gandak, Kosi, Mahananda) have their source in the Himalaya. The Vindhyan ranges extending from the Central to eastern India (north of the River Narmada) give rise to the tributaries of the south basin; the Chambal, Sind, Betwa and Ken that form the lower half of the Yamuna river system and the rest join the Ganga itself (Tons, Son). Ramganga is the only north basin tributary that joins the Ganga before Varanasi, other tributaries join the Ganga after Varanasi.

The length of the river, the diversity of the terrain, climate zones (mountain, tropical wet), geology, soil, vegetation (temperate, tropical; both wet and dry) physiographic conditions of the tributaries from diverse topography (mountain, plains, plateau) shows that the basin has all necessary ingredients for rich biodiversity. For instance, the Yamuna is unique as its source lies in the Himalaya but the lower half receives the Plateau Rivers from the Vindhyan ranges, as happens in the Ganga also. Along its basin there is variation because of altitude, longitude and latitude. The ice-cold fresh-waters not only at origin but considerable distance from the source and marine influence at its mouth in the Bay of Bengal also factor into it. From the biodiversity viewpoint it possesses elements from three wildlife biogeographic regions; the West Himalaya, the Upper and the Lower Gangetic Plains [2] along its main course and north basin while the Deccan Peninsula (Central Highlands, Chhota Nagpur) in the south basin. In view of the mosaic of biogeographic conditions along its course the present study examines biodiversity and its distribution in the main channel till Varanasi and a portion of the south basin in Bundelkhand region. Barring a few small rivers all major tributaries that form the northern basin join the source waters after Varanasi. Only south basin tributaries which belong to Deccan Peninsula biogeographic region join the Ganga through Yamuna before Varanasi.

The rationale for selecting this course of the Ganga lies in our opinion that the quality and quantum of biodiversity occurring in the Ganga till Allahabad is influenced by properties of the source waters from the Himalaya and after that by the Yamuna through its tributaries from the plateau with semi-arid climate. The stretch from Allahabad to Varanasi is ca. 144 km, sufficient to examine zonation even for fish. The essential feature of the present work was to document the present state of biodiversity and distribution of characteristic elements of benthic biota, plankton and fish fauna in the mountain and the Upper Gangetic Plains (UGP) zone of the Ganga. In order to develop a basin management plan for the National Ganga Authority of India, it remains to be analysed if river regulation has modified the continuum of the Ganga and if so to what extent, as the scope of this contribution is limited to knowing the broad patterns of biodiversity and its distribution. The knowledge of ecosystem and hence its ecological integrity is fundamental to concept of *nirmal* and *aviral dhara*.

3.2 Study Area: Physical and Chemical Features of Mountain and Plains Zone

The zone of interest for the present study lay in the upper half of the Ganga river from Gangotri to Varanasi (1,296.2 km, [1]). The Bhagirathi-Ganga from Gangotri and Badrinath to Hardwar constitutes the mountain zone (MZ), and the rest (till Varanasi) the Plains zone (PZ). For understanding the distribution of biodiversity elements the PZ is split into PZ I (Hardwar to Allahabad) and PZ II up to Varanasi. Geographical and physiographic details of the sampled locations that have been used to arrive at biodiversity scenario in large zones of the river based on the premise detailed above are described in Krishnamurti et al. [3]. The details of physiography, land use and physico-chemical features of MZ have been described by Singh et al. [4] and Nautiyal [5]. In general MZ flows through gorges and narrow valleys. In PZ, the river bed is wide with extensive flood plains and meandering channel. The habitats in MZ include rapids and riffles near source, short deeper pools alternate between these habitats in the middle mountain zones with moderate slopes, while longer pools, short runs and riffles in the lower mountain zone with gentle slopes (towards foothills). The river habitat in PZ consists of pools, riffles and runs. The river bed consists of hard sediments (large particle size) in the mountains in contrast to soft sediments (small particle size-silt, clay, sand, pebbles) in varied proportions in the plains. The vegetation consists of forests in the MZ and scrub in the floodplain (PZ). Cultivation is extensive and intensive in PZ.

Since the Ganga has glacier at source, the waters are ice-cold in headwaters and cold in the foothills (4.3–25 °C). The water is moderately cool even in Narora – Allahabad zone. In the Plains, water temperature ranges from 12(PZ I) to 31.5 °C (PZ II). In MZ the water current varies from torrents to fast flowing waters (3.3–0.38 m s^{-1}) due to high gradient of river bed compared with 20–30 cm s^{-1}in the PZ. Therefore, DO declines from MZ to PZ [1, 4, 6]. The pH ranges from neutral to moderately alkaline (7.0–7.6–8.0), conductivity low to high (20–571.6–796 mho cm^{-1}) from Gangotri to Hardwar to Varanasi. The maximum turbidity is high around 1,500 NTU in the MZ and declines to <300 NTU in the PZ. All other ions; alkalinity, hardness and nutrients increase gradually from MZ to PZ [3].

3.2.1 *River Regulation*

The Bhimgoda barrage at Hardwar is the first instance of water abstraction as a very large volume of water is diverted to Upper Ganga canal (10,500 cusecs), Eastern Ganga Canal (4,850 cusecs). At Bijnor the barrage diverts water to Madhya Ganga canal stage I (8,250 cusecs) and II (proposed for 4,300 cusecs). Upstream of Narora a canal from Ramganga augments the flow of the river at Brijghat (5,100 cusecs). Lower Ganga (LG) Canal and Parallel LG Canal are designed for abstraction of 8,500 and 4,200 cusecs respectively at Narora. The Narora barrage supplies water to the atomic power station and two parallel lower canals which run up to the Kanpur. Between Kanpur and Rae Bareilly there are 4 < 50 cusecs and 2 larger (Dalmau <1,000 cusecs) lift canals (http://www.rid.go.th/thaicid/_6_activity/Technical-Session/SubTheme2/2.15Ravindra_K.pdf accessed, 27 July 2012). In addition to the municipal wastes from the towns and cities of varying magnitude, industrial effluents are also discharged into the river between Narora to Kanpur. Agriculture is the major landuse along the river banks from Hardwar to Varanasi, as is evident from the barrages and canal systems.

3.3 General Procedures for Documenting Biodiversity

The bottom-dwelling (benthic) biota, plankton and nekton (fish) constitute the bulk of major aquatic communities and hence biodiversity in a river. The diatoms and algae (green, blue-green, others) among the benthic flora and phytoplankton, insects, molluscs and worms among the macroinvertebrates and fish have the highest number of species in the freshwater. The aim was to prepare a consolidated list of species or higher taxon level and see how these elements and the communities formed by them are distributed in the mountain (MZ) and the UGP zone of the Ganga (called as plains zone – PZ in the text). The species found in MZ are listed first, followed by PZ I and PZ II (species listing in alphabetic order). Inventories are based on unpublished and published sources specified below; **Benthic flora** [7–15]; **Net plankton** [3, 4, 16–20]; **Macroinvertebrates and Fish** [3, 4, 8, 9, 12, 13, 15, 21–32].

The observations are based on studies conducted during last three decades. The data of this period is considered comparable as the mountain zone from Gangotri to Hardwar was studied during the Ganga Action Plan I (prior to Tehri Dam). Except for some quick EIA surveys [33] that are site specific, there are no substantial studies from Gangotri to Devprayag and there is practically no river downstream Maneri to Tehri after the Dam was commissioned in 2005. Further, the locations for the post Dam studies below Devprayag are not immediately below the Dam between Tehri and Devprayag where the impact of regulation is visible as the discharges tend to be normal after confluence with the Alaknanda at Devprayag. All locations sampled recently (Narora 10 km u/s barrage, Kachla Ghat 1 km d/s, Bithoor u/s outskirts of Kanpur) are distant from the other aforesaid barrages on the Ganga. Distance from Hardwar to Narora is ca. 305 km, Kachla Ghat is 60 km d/s of Narora barrage and Narora to Bithoor is 330 km. The barrage at Chilla, Bhimgoda, Bijnor and Narora on the Ganga were commissioned prior to 1980.

The inventory of benthic flora, plankton, macroinvertebrates and fish for the MZ, PZ I and PZ II were processed to obtain valuable information on species/

taxonomic richness of these major groups of biota and their major components: e.g. Bacillariophyceae, Chlorophyceae etc. for benthic algae and phytoplankton; cladocera and copepods for zooplankton; Heptageniidae for macroinvertebrates; Cyprinidae, Bagridae for fish. These inventories were used to determine the species-rich genera in case of diatoms-algae, species-rich genera/family in case of fish and family-rich order in case of the macroinvertebrates. Each inventory was compared among the zones with the help of CAP 4 (version 4, [34]). This helped to determine the number of species in each zone MZ and those common to it. The software was also used to compute similarity (Binary-no double zeros, Jaccard's measure) among the zones. Cluster technique, Wards method and 1-Jaccard's distance was used to find the distances for each biodiversity component among these zones. The density estimates were also obtained from the existing published and unpublished literature.

3.4 Biodiversity of the Upper Ganga

The first comprehensive information on the aquatic life of the Ganga and various ecological aspects from source to mouth was generated during GAP I [3]. Despite such a huge effort there were many gaps as inventory of biota was highly fragmentary since all flora and fauna (fungi, bacteria, algae, protists to mammals) was not available for all trophic levels and for all zones. Even prior to GAP I, there was considerable fishery related information for sizeable part of the Ganga and Yamuna with respect to physical, chemical and some biological features (periphyton, plankton, macroinvertebrates) of direct relevance to fisheries. There was extensive information on the fish fauna of economic relevance owing to which a considerable number of species were known [35]. Besides the fish fauna, there is information on the fish-food items especially algae (primarily plankton) and invertebrates (zoobenthos). Fish fauna of the Ganga was however known much earlier [36] and the efforts continued for the Himalaya [37, 38] and Ganga, primarily Misra [39], Venkateswarlu and Menon [40] which are also documented by Talwar and Jhingran [41] and Jayaram [42, 43]. The interest in fish fauna continues not only due to its commercial value, but also because it is a source of livelihood to a sizeable zone of the population that resides in the villages along the banks of the Ganga and Yamuna. The notable efforts are from Payne and Temple [44], Payne et al. [45] and Lakra et al. [46].

It is also important to note that much more is known about the lower Ganga below Varanasi to Bay of Bengal [47, 48], than from Gangotri to Varanasi. Rao [29] also noted that the information about the biological community of the Ganga River between Rishikesh and Kanpur is very scanty. Only recently, the efforts to compute e-flows for the upper Ganga resulted in studies at certain locations [49]. Besides this, information on various components of biodiversity is available at local level, especially in the stretches close to the Universities. Therefore these fragments of the information have been put in place to draw a large-scale biodiversity profile. Since, river is a continuum [50], it is always interesting to know the longitudinal

Table 3.1 Composition of benthic flora (PB), periphyton (PP) and zooplankton (ZP) richness at coarse taxonomic level in the mountain and plains zone. Acronym s.s. = sensu stricto

	MZ		PZI		PZII		PZI	
Flora	PB	PP	PB	PP	PB	PP	Fauna	ZP
Bacillariophyceae	443	47	101	90	51	135	Protozoa	10
Chlorophyceae	29	26	52	5	0	73	Rotifera	27
Cyanophyceae	12	10	21	7	0	45	Cladocera	14
Euglenophyceae	0	1	4	0	0	4	Copepoda	5
Dinophyceae	0	0	0	0	0	8	Miscellaneous	5
Chrysophyceae	0	0	2	0	0	1		
Xanthophyceae	0	0	1	1	0	0		61
	484	84	181	103	51	266		
Species-rich genera								
Navicula s.s	57	4	8	10		17		
Achnanthes s.s	51							
Gomphonema	37							
Cymbella s.s	30	5	10	9		6		
Nitzschia	28	2	13	5		17		
Synedra	22	4	10	5		6		

distribution of all components of biodiversity, especially those with large number of species, such as the diatoms, insects, molluscs. This does not undermine the importance of higher taxonomic units with fewer species and families. Rao [39] reported a total of 40 species of zooplankton, 4 crustaceans, 15 molluscs, 51 insects, 83 fishes, 12 freshwater turtles, 2 crocodiles, 48 aquatic birds and two mammal species in the upper Ganga River (Rishikesh to Kanpur).

3.4.1 *Species/Taxonomic Richness*

3.4.1.1 Benthic Flora

The information on the benthic flora of the river Ganga was available largely for MZ compared with PZ I and PZ II, where benthic flora has not been investigated keenly. In the study area the benthic flora contains five taxonomic groups; Bacillariophyceae, Chlorophyceae, Cyanophyceae, Euglenophyceae and Xanthophyceae, amounting to 501, 132 and 67 species of Bacillariophyceae, Chlorophyceae, Cyanophyceae, respectively. The diatoms (D) dominate the benthic flora, being exclusive in MZ compared with PZ I and II (Table 3.1). In MZ the richness apparently increases from middle reaches of the Alaknanda (140 taxa; [15]) to its lower reach (199 taxa; [10]), but declines in the Ganga between Rishikesh and Hardwar (217 diatom taxa from 35 genera; (Appendix I)) The present inventory contains flora mainly from MZ of which 328 taxa have been reported till Hardwar [11] and thus very few appear to occur as benthic flora in PZ I. Lack of hard substrate and prevalence of softer sediments in the PZ explain the reason for low richness of benthic diatom flora. Very few taxa are common to MZ and PZ I. Taxa specific to PZ I are also few, *Mastogloia danseii* and *Anomoeoneis serians* (Table 3.2), the taxa not found in MZ.

In the Alaknanda-Ganga that constitutes MZ, the diatom flora comprised 16 % araphid, 17.5 % monoraphid and 63 % biraphid elements [10, 11]. In the PZ I the mono and araphid flora declines (<10 taxa), while the biraphids increase marginally. This decline in PZ I is linked to decrease in substrate heterogeneity. Consequently, the species-rich genera differ in the upper MZ (MZ-U) (*Navicula – Achnanthidium*), and lower MZ (MZ-L) *Cymbella – Navicula* and PZ I (*Navicula-Cymbella – Nitzschia*) compared with *Achnanthidium – Cymbella* in Alaknanda [15]. *Pediastrum, Scenedesmus, Oscillatoria* and *Merismopedia* are species-rich green algae [GA] and blue green algae [BGA].

3.4.1.2 Plankton

The phytoplankton has been widely studied till Varanasi (PZ II). This community was represented by six taxonomic groups; Bacillariophyceae (48–195), Chlorophyceae (54–126), Cyanophyceae (18–60), Euglenophyceae (4–10), Xanthophyceae (1) and Chrysophyceae (3) from MZ to PZ II, amounting to 397 species from 131 genera (Table 3.1). Thus, diatoms dominate qualitatively in plankton community also, followed by GA and BGA. Comparison of the benthic flora and phytoplankton shows that many benthic diatoms lead planktonic existence in the Plains. Zooplankton (61 taxa) is represented by rotifers (27 taxa), crustaceans (cladocera 14, copepod 5 taxa), protozoa (10 taxa) and miscellaneous (4 taxa), primarily in PZ I. The total plankton amounted to 464 taxa from Gangotri (MZ) to Varanasi (PZ II). Zooplankton richness increases from the MZ to PZ I.

Singh et al. [4] reported 42 genera of phytoplankton from the MZ (Gangnani to Rishikesh) of which diatoms were represented by 27 genera. Krishnamurti et al. [3] report an increase in phytoplankton flora from foothills (83 taxa) to Varanasi (237 taxa) in the Ganga, 100 taxa (43 BGA, 35 D, 22 GA) between Rishikesh to Garhmukteshwar. The richness of diatoms decreases in the phytoplankton community in the PZ II. The zooplankton rotifera are common downstream of Narora to Varanasi. About 18 taxa of phytoplankton and 11 taxa of zooplankton were recorded in the stretch between Kanpur and Allahabad [51]. Vass et al. [52] noted that in the upper Ganga (Tehri to Kannauj) rotifers and protozoa were recorded in the Anupsahar and Farrukhabad stretches. Anonymous [16] recorded 54 taxa (Bacillariophyceae 28, Chlorophyceae 20, Cyanophyceae 6) around Kanpur. Shukla and Asthana [53] reported 577 algal components from the Ganga around Kanpur; Euglenophyceae 5> Chlorophyceae 151> Cyanophyceae 198> Bacillariophyceae 223, from a very large number of habitats.

It augurs that plankton richness increases from MZ to PZ I. The richness of zooplankton increases enormously in PZ I (Table 3.1). The phytoplankton richness increases from the PZ I to PZ II. A wide bed span, rich nutrient load, open surface area for incoming radiations and moderate water current possibly contribute to this.

Table 3.2 Benthic flora of the Ganga: The floral elements common to the mountain (MZ) and the Upper Gangetic Plains – UGP (PZ I) and the flora specific to UGP reported only in the Plains [53]. These genera indicated by * may have some species in the MZ

Common to MZ and PZ	Exclusive to PZ
Achnanthes brevipes	*Anomoeoneis serians*
Achnanthidium biasolettianum	*Asterionella formosa*
A.chitrakotense	*Caloneis silicula var. lapponica*
A. minutissimum	*Cocconeis* sp.
Amphora montana	*Cyclotella meneghinana* fo. *unipunctata*
Amphora ovalis	*Cyclotella stelligera*
Cocconeis placentula	*Delicata delicatula*
C. placentula var. *lineata*	*Encyonema perpusillum*
Cyclotella glomerata	*Cymbella prostrata*
C. meneghinana	*Diatoma vulgare* var. *brevis*
Cymbella australica	*Fragillaria crotonensis*
C. parva	*Gomphonema helveticum*
Diatoma mesodon	*Gyrosigma acuminatum*
D. vulgare	*Gyrosigma distortum*
Encyonema minutum	*Luticola mutica* var. *geopertiana*
Gomphonema minutum	*Mastogloia danseii*
Gyrosigma scalproides	
Navicula cryptotenelloides	*N. hungarica*
N. lanceolata	*N. ignorata*
N. radiosa	*Navicula krasskei*
N. rostellata = *N. germainii*	*N. rhyncocephala* var. *amphiceros* = *N. vaneei*
N. stagnorum	
Pinnularia sp.*	*P. subcapitata* var. *lapponica*
Placoneis elegans	*Achnanthes* sp.*, *Amphora* sp.*, *Cyclotella* sp.*, *Cymbella* sp.*, *Diatoma* sp.* *Fragillaria* sp.*
Panothidium lanceolatum	*Frustulia* sp.*, *Gomphonema* sp.* *Gyrosigma* sp.* *Navicula* sp.* *Nitzschia* sp.* *Surirella* sp.* *Stauroneis* sp.*
Surirella apiculata	*Tabellaria* sp.*
S. delecatissima	*Chlorella* sp.
Synedra ulna	*Chlorogonium* sp.
S. ulna v. aequalis	*Closterium* sp.*
S. ulna var. *subaequalis*	*Cosmarim* sp.*
Tabellaria fenestrate	*Draparnalcia* sp.
Tetracyclus rupestris	*Hydrodictyon* sp.
Cladophora sp.	*Kirchneriella* sp.
Oedogonium sp.	*Merismopedia glauca*
Spirogyra sp.	*Pediastrum duplex*
Ulothrix sp.	*Pediastrum simplex var. bivaense*
Zygnema sp.	*Scenedesmus diamorplus**
Anabaena sp.	*Scenedesmus quadricauda*
Lyngbya sp.	*Scenedesmus* sp.*; *Schizogonium* sp.*; *Vaucheria* sp.*; *Nostoc* sp.*
Oscillatoria limosa	*Anacystis* sp.*; *Rivularia* sp.*
Oscillatoria sp.	*Euglena acus*; *Euglena viridis*; *Tribonema bombycinum*

The benthic floral community depicted almost reverse trend of that of phytoplankton. This is because of absence of hard substratum and the low current velocity of river that helps to the growth of planktonic forms in the river. The maximum abundance of plankton in terms of quality and quantity was observed in the stretch between Kanpur (lower PZ I) and Varanasi (PZ II). The species-rich plankton genera in PZ I were *Nitzschia, Cymbella, Synedra,* while in PZ II *Navicula, Nitzschia* in case of diatoms (Table 3.1), *Pediastrum, Scenedesmus, Spirogyra, Crucigenia, Cosmarium, Closterium* for GA and *Spirulina, Phormidium, Merismopedia, Oscillatoria, Anabaena* and *Euglena* for BGA, in order of species richness.

3.4.1.3 Benthic Macroinvertebrate Fauna

The benthic macroinvertebrate fauna in the MZ and UGP (PZ I, PZ II) belongs to the phyla; Annelida, Arthropoda and Mollusca. The Annelida is represented by one taxon each of Oligochaeta, Polychaeta and Hirudinea. The Arthropoda consisted of 10 orders, 52 families of Insecta, while one taxon each of Crustacea and Arachnida. The Mollusca consisted of Class Gastropoda (5 families) and Pelecypoda (2 families). They were present only in PZ. Thus, fauna consists of 65 taxa from the MZ to PZ II. Of these 32 taxa are present only in MZ and 55 in PZ. The Arachnida is present in MZ only, while Ostracoda (Crustacea) and Hymenoptera (Insecta), along with Annelida are present only in PZ (Appendix I). Earlier studies in MZ of the Bhagirathi have reported 29 taxa belonging to 7 insect orders, of which 20 genera belonged to 3 orders; Ephemeroptera-7, Diptera-8 and Trichoptera-5 [4]. Studies by various authors reveal 20 families from 8 Orders from Devprayag to Rishikesh, 9 families at former and 17 at latter [4, 6]. Anonymous [16] report 10 taxa from Devprayag to Kanpur.

3.4.1.4 Fish

The fish fauna consisted of 149 fish species from 25 families along the study stretch of the river Ganga. Fifty-eight species are present in MZ and 113+9 (Hardwar to Bijnor) = 122 species in PZ. Orders Tetradontiformes, Osteoglossiformes and Clupeiformes are restricted to PZ only. Many other fish families are also peculiar to PZ (Table 3.3).

Most fish species common to MZ and UGP belong to Order Cypriniformes and Siluriformes (Table 3.4). In recent past, 21 fish species (Carp: Catfish: Other = 18:2:1) were observed from Devprayag to Rishikesh zone, 22 fish species (Carp: Catfish: Other = 13:6:3) at Narora and 27 species at Kachla Ghat (Carp: Catfish: Other = 8:7:14) and 31 fish species at Bithoor (Carp: Catfish: Other = 12:07:12). Earlier, 30 fish species (Carp: Catfish: Other = 20:3:7) were recorded in Rishikesh-Hardwar zone [54]. Sharma and Rajput [55] recorded 26 fish species (Carp: Catfish: Other = 14:4:9) around the Bijnor District. Further down in the Ganga the 82 fish species (Carp: Catfish: Other = 45:17:20) have been reported between Brijghat to Narora [29, 56].

Table 3.3 Fish fauna: Species rich fish families and their distribution in the mountain (M) and the Upper Gangetic Plains (PZ I, II)

Family	No. of species	MZ	PZ	Family	No. of species	MZ	PZ
Amblyceptidae	1		+	Chandidae	2		+
Anabantidae	1		+	Clupeidae	3		+
Cichlidae	1		+	Mastacembelidae	3	+	+
Clariidae	1	+	+	Belontidae	4	+	+
Engraulidae	1		+	Channidae	5	+	+
Gobiidae	1		+	Cobitidae	5	+	+
Heteropneustidae	1		+	Schilbeidae	6	+	+
Mugilidae	1	+	+	Siluridae	6		+
Pangasiidae	1		+	Bagridae	9		+
Sciaaenidae	1		+	Balitoridae	9	+	+
Tetraodontidae	1		+	Sisoridae	17	+	+
Nandidae	2		+	Cyprinidae	65	+	+
Notopteridae	2		+	Total No. of Species	**149**		

Table 3.4 Fish fauna common among zones and to all zones. Nomenclature as in original documents

Common MZ-PZ I	*Contd.*	Common to all zones
Barilius vagra (Ham.)	*Chitala chitala* (Ham.)	*Barilius barila* (Ham.)
Glyptothorax pectinopterus (McClel)	*Cirrhinus mrigala* (Ham.)	*Barilius bendelisis* (Ham.)
Labeo dero (Ham.)	*Cirrhinus reba* (Ham.)	*Barilius bola* (Ham.)
Labeo dyocheilus (Day)	*Colisa fasciaotus* (Bloch & Sehn)	*Botia dario* (Ham.)
Nemacheilus beavani (Gunther)	*Glossogobius giuris* (Ham.)	*Channa gachua* (Ham.)
Nemacheilus montanus (McClelland)	*Gudusia chapra* (Ham.)	*Clarias batrachus* (Linn.)
Nemacheilus rupecola (McClelland)	*Heteropneustes fossilis* (Ham.)	*Clupisoma garua* (Ham.)
Nemachielus savona (Ham.)	*Labeo bata* (Ham.)	*Crossocheilus latius* (Ham.)
Rasbora daniconius (Ham.)	*Labeo gonius* (Ham.)	*Danio devario* (Ham.)
Schizothoracthys progastus (McClelland)	*Labeo pangusia* (Ham.)	*Danio rerio* (Ham.)
Schizothorax plagiostomus (Heckel)	*Labeo rohita* (Ham.)	*Esomus danricus* (Ham.)
Schizothorax richardsonii (Gray)	*Lepidocephalus guntea* (Ham.)	*Garra gotyla gotyla* (Gray)
Schizothorax sinuatus (Heckel)	*Macrognathus aculeatus* (Bloch)	*Labeo calbasu* (Ham.)
Tor putitora (Ham.)	*Macrognathus pancalus* (Ham.)	*Mystus tengara* (Ham.)
Common PZ I-PZ II	*Mastacembelus armatus* (Lacepede)	*Notopterus notopterus* (Pallas)
Ailia coila (Ham.)	*Mystus aor* (Ham.)	*Puntius sarana sarana* (Ham.)

(continued)

Table 3.4 (continued)

Common MZ-PZ I	*Contd.*	Common to all zones
Amblypharyngodon mola (Ham.)	*Mystus bleekeri* (Day)	*Puntius sophore* (Ham.)
Anabas testudineus (Bloch)	*Mystus cavasius* (Ham.)	*Puntius ticto* (Ham.)
Aspidoparia morar (Ham.)	*Mystus seenghala* (Sykes)	
Bagarius bagarius (Ham.)	*Mystus vittatus* (Bloch)	*Rita rita* (Ham.)
Catla catla (Ham.)	*Nandus nandus* (Ham.)	*Salmostoma bacaila* (Hamilton)
Chagunius chagunio (Ham.)	*Nemacheilus botia* (Ham.)	*Tor tor* (Ham.)
Chanda nama (Ham.)	*Ompok bimaculatus* (Bloch)	*Wallago attu* (Bloch)
Chanda ranga (Ham.)	*Ompok pabda* (Ham.)	*Xenentodon cancila* (Ham.)
Channa marulias (Ham.)	*Osteobrama cotio* (Ham.)	
Channa punctatus (Bloch)	*Oxygaster bacaila* (Ham.)	
Channa stewartii (Playfair)	*Pangasius pangasius* (Ham.)	
Channa striatus (Block)	*Cyprinus chagunio* (Ham.)	
Chela laubuca (Hamilton)	*Silonia silondia* (Ham.)	

No fish species were common to MZ and PZ II, except *C. carpio* (accidental introduction into Tehri reservoir on the Bhagirathi)

A recent information shows 58 fish species (Carp: Catfish: Other=26:18:14) at Narora (web page accessed March 2010). Still further, total 88 (Carp: Catfish: Other=36:23:29) have been recorded below Allahabad [15]. The earlier records are; 13 species [15], 38 species [15] and 75 species [15]. Shukla and Vandana [53] recorded 25 fish species around Kanpur (Carp: Catfish: Other=7:10:7).

Thus, a general increase in fish richness is evident from MZ to UGP (PZ). Stretch of the river from Hardwar to Bijnor (77 km. apart) is of special interest from the viewpoint of fish distribution since it is the junction of two biogeographic regions, the west Himalaya and the Upper Gangetic Plains. Sixty-eight species are listed in this zone of which 17 species are common with the upper mountain zone (MZ-U, above Devprayag), 37 species (including 17 from MZ-U), that are present in MZ-L (i.e. Devprayag to Hardwar), 9 species are specific to this junction and 21 species are common to PZ I. In the present analysis this junction has been included in PZ I (Table 3.5). Thus, this zone has a larger share of mountain element and nearly one-third from the plains. It is notable that some essentially coldwater species i.e. snow trout (3 species), *Garra* and *Glyptothorax* (1 sp. each) extend their range into the PZI but are few in junction zone. The coldwater character of these species is also obvious because they are not present after this junction zone, while the cool-water elements continue to be found in PZ I. The other elements include migratory *Tor* and *Labeo* and a mixture of loaches and barils that can be called as cool-water forms.

From fisheries perspective the ratio of Carp: Catfish: Other has changed in last few decades; 46:30:24 *1960s; 36:36:28 in 1970s, 14:45:40 in 1990s. Thus fisheries, has shifted from carp to catfish dominated to others dominated. However, earlier Hilsa was prominent among others, whereas now miscellaneous fish have become dominant component in 'others' category.

Table 3.5 Fish fauna of the junction zone (JZ) between the west Himalaya (WH) and Upper Gangetic Plains (UGP) biogeographic regions

		WH		UGP	
Fish fauna	Family	MZ-U	MZ-L	JZ	PZ I
1. Schizothorax plagiostomus (Heckel)	Cyprinidae	+	+	+	
2. Schizothorax richardsonii (Gray)	Cyprinidae	+	+	+	
3. Schizothorax sinuatus (Heckel)	Cyprinidae	+	+	+	
4. Schizothoraicthys progastus (McCl)	Cyprinidae	+	+	+	
5. Schistura rupecula (McCl)	Cyprinidae	+		+	
6. Glyptothorax pectinopterus (McCl)	Sisoridae	+	+	+	
7. Labeo (Bangana) dero (Ham.)	Cyprinidae	+	+	+	
8. Labeo dyocheilus (Day)	Cyprinidae	+	+	+	+
9. Schistura montanus (McCl)	Balitoridae	+	+	+	+
10. Barilius bendelisis (Ham.)	Cyprinidae	+	+	+	+
11. Barilius vagra (Ham.)	Cyprinidae	+	+	+	+
12. Tor putitora (Ham.)	Cyprinidae	+	+	+	+
13. Tor tor (Ham.)	Cyprinidae	+	+	+	+
14. Barilius barila (Ham.)	Cyprinidae	+	+	+	+
15. Crossocheilus latius (Ham.)	Cyprinidae	+	+	+	+
16. Clupisoma garua (Ham.)	Schilbeidae	+	+	+	+
17. Garra gotyla gotyla (Gray)	Cyprinidae	+	+	+	+
18. Acanthocobitis botia (Ham.)	Balitoridae		+	+	+
19. Raiamas bola (Ham.)	Cyprinidae		+	+	+
20. Salmostoma bacaila (Ham.)	Cyprinidae		+	+	+
21. Rasbora daniconius (Ham.)	Cyprinidae		+	+	+
22. Puntius sarana sarana (Ham.)	Cyprinidae		+	+	+
23. Puntius ticto (Ham.)	Cyprinidae		+	+	+
24. Xenentodon cancila (Ham.)	Belontidae		+	+	+
25. Danio devario (Ham.)	Cyprinidae		+	+	+
26. Danio rerio (Ham.)	Cyprinidae		+	+	+
27. Botia dario(Ham.)	Cobitidae		+	+	+
28. Puntius sophore (Ham.)	Cyprinidae		+	+	+
29. Colisa fasciatus (Bloch & Sehn)	Belontidae		+	+	+
30. Channa gachua (Ham.)	Channidae		+	+	+
31. Esomus danricus (Ham.)	Cyprinidae		+	+	+
32. Labeo calbasu (Ham.)	Cyprinidae		+	+	+
33. Labeo gonius (Ham.)	Cyprinidae		+	+	+
34. Bagarius bagarius (Ham.)	Sisoridae		+	+	+
35. Mystus tengra (Ham.)	Bagridae		+	+	+
36. Rita rita (Ham.)	Bagridae		+	+	+
37. Rhinomugil corsula (Ham.)	Mugilidae		+	+	+
38. Barilius dimorphicus (Tilak & Husain)	Cyprinidae			+	
39. Puntius conchonius (Ham.)	Cyprinidae			+	
40. Botia almorhae (Gray)	Cobitidae			+	
41. Botia lohachata (Chaudhari)	Cobitidae			+	
42. Amblyceps mangois (Ham.)	Amblyceptidae			+	
43. Clupisoma montana (Hora)	Schilbeidae			+	

(continued)

Table 3.5 (continued)

Fish fauna	Family	WH		UGP	
		MZ-U	MZ-L	JZ	PZ I
44. *Glyptothorax dakpathri* (Tilak & Husain)	Sisoridae			+	
45. *Glyptothorax indicus* Talwar	Sisoridae			+	
46. *Puntius chola* (Ham.)	Cyprinidae			+	+
47. *Labeo boga* (Bloch)	Cyprinidae			+	+
48. *Schistura corica* (Ham.)	Balitoridae			+	+
49. *Notopterus notopterus* (Pallas)	Notopteridae			+	+
50. *Labeo pangusia* (Ham.)	Cyprinidae			+	+
51. *Lepidocephalus guntea* (Ham.)	Cobitidae			+	+
52. *Macrognathus pancalus* (Ham.)	Mastacembelidae			+	+
53. *Glossogobius giuris* (Ham.)	Gobiidae			+	+
54. *Nandus nandus* (Ham.)	Nandidae			+	+
55. *Aspidoparia morar* (Ham.)	Cyprinidae			+	+
56. *Chagunius chagunio* (Ham.)	Cyprinidae			+	+
57. *Labeo bata* (Ham.)	Cyprinidae			+	+
58. *Chitala chitala* (Ham.)	Notopteridae			+	+
59. *Heteropneustes fossilis* (Ham.)	Heteropneustidae			+	+
60. *Cirrhinus mrigala* (Ham.)	Cyprinidae			+	+
61. *Cirrhinus reba* (Ham.)	Cyprinidae			+	+
62. *Labeo rohita* (Ham.)	Cyprinidae			+	+
63. *Mystus seenghala* (Sykes)	Bagridae			+	+
64. *Mastacembelus armatus* (Lacepede)	Mastacembelidae			+	+
65. *Channa marulius* (Ham.)	Channidae			+	+
66. *Channa striatus* (Bloch)	Channidae			+	+
67. *Channa punctatus* (Bloch)	Channidae			+	+

MZ-U, above Devprayag; MZ-L Devprayag to Hardwar. JZ is the uppermost part of the PZ I

Much like the macroinvertebrate fauna there is very distinct distribution of fish fauna in each zone, being least in the MZ and higher in PZ I compared with PZ II (Table 3.4). The most typical fish in MZ are Cyprinidae; snowtrouts (*Schizothorax* sp., *Schizothoraichthys* sp. *Naziritor chelynoides*) Balitoridae (*Schistura* sp.) and Sisoridae (*Pseudecheneis* sp., *Glyptothorax* sp.). In PZ I besides *Schistura* and *Glyptothorax* sp., a wide variety of other genera of these families are exclusive. Presence of diverse silurid (four) families (Order Siluriformes) is the unique feature of this zone. In PZ II cyprinid elements become fewer and among cat fish besides few Bagridae, Siluridae, Schilbidae, there is more representation of Sisoridae as in the MZ, but the genera are different. Some unique elements appear in PZ II, especially clupid, mugil and perch (Table 3.6).

Thus, within the UGP biogeographic zone, there are ichthyozones. This can possibly be explained by ichthyogeography of India, where clupid, mugil and perch may be peninsular elements and the carps, loaches and certain catfishes are extra Peninsular elements. This, however, is just an opinion based on the

Table 3.6 Fish fauna specific to zones MZ, PZ I and PZ II. The superscript letters for each species, indicate the Family of each species to emphasize the diversity at Family level in PZI and II

MZ	PZ I	PZ II
Barilius barna (Ham.)[c]	*Amblyceps mangois* (Ham.)[i]	*Cyprinus carpio* (Linn.)[c]
Garra lamta (Ham.)[c]	*Amblypharyngodon melettinus* (Valenciennes)[c]	*Barilius tileo* (Ham.)[c]
Glyptothorax cavia (Ham.)[j] = *G. lineatum*	*Amblypharyngodon microlepis* (Blk)[c]	*Botia dayi* (Hora)[e]
Glyptothorax conirostris (Steindachner)[j]	*Aspidoparia jaya* (Ham.)[c]	*Colisa chuna* (Ham.)[p]
Glyptothorax trilineatus (Blyth)[j]	*Badis badis* (Ham.)[n]	*Eutropiichthys murius* (Ham.)[h]
Glyptothorax madraspatanum (Day)[j]	*Barilius dimorphicus* (Tilak and Husain)[c]	*Eutropiichthys vacha* (Ham.)[h]
Glyptothorax prashadi (Mukerji)[j]	*Barilius modestus* (Day)[c]	*Gagata cenia* (Ham.)[j]
Labeo angra (Ham.)[c]	*Botia almorhae* (Gray)[e]	*Glyptothorax telchita* (Ham.)[j]
Pseudecheneis sulcatus (McCl.)[j]	*Botia lohachata* (Chaudhari)[e]	*Goniolosa manmina* (Ham.)[a]
Schizothorax curvifrons Heckel[c] = *Schizothorax micropogon* Heckel[c]	*Chaca chaca* (Ham.)[k]	*Hilsa ilisha* (Ham.)[a]
Schizothorax esocinus (Heckel)[c]	*Clupisoma montana* (Hora)[h]	*Hypophthalmichthys molitrix* (Val.)[c]
Schizothorax intermedius (McCl. and Griffith)[c]	*Colisa lalia* (Ham.)[p]	*Ilisha motius* (Ham.)[a]
Schizothorax niger (Heckel)[c]	*Garra prashadi* (Hora)[c]	*Johnius coitor* (Ham.)[m]
Naziritor chelynoides(McCl.)[c]	*Glyptothorax indicus*[j]Talwar	*Rama chandramara* (Ham.)[f]
	Labeo boga (Bloch)[c]	*Mystus menoda* (Ham.)[f]
	Chela atpar (Ham.)[c]	*Nangra punctata* (Day)[j]
	Nangra nangra (Ham.)[j]	*Nangra viridescens* (Ham.)[j]
	Acanthocobitis botia (Ham.)[d]	*Ompok pabo*(Ham.)[g]
	Schistura corica (Ham.)[d]	*Pseudotropius atherinoides* (Bloch)[h]
	Schistura scaturingina (McCl.)[d]	*Puntius* spp. (Ham.)[c]
	Ompok boopis (Ham.)[g]	*Sciamugil cascasia* (Ham.)[l]
	Oxygaster gora (Ham.)[c]	*Securicula gora* (Ham.)[c]
	Puntius chola (Ham.)[c]	*Setipinna phasa* (Ham.)[b]
	Puntius conchonius (Ham.)[c]	*Sisor rabdophorus* (Ham.)[j]
	Rasbora elanga (Ham.)[c]	
	Tetraodon cutcutia (Ham.)[q]	

Clupeiformes: [a]Clupeidae, [b]Engraulidae
Cypriniformes: [c]Cyprinidae, [d]Balitoridae, [e]Cobitidae
Siluriformes: [f]Bagridae, [g]Siluridae, [h]Schilbeidae, [i]Amblyciptidae, [j]Sisoridae, [k]Chacidae
Mugiliformes: [l]Mugilidae
Perciformes: [m]Scianidae, [n]Nandidae, [p]Belontidae
Tetraodontiformes: [q]Tetraodontidae

abundance of cyprinid and silurid families specific to the MZ and PZ I and the elements that are common in the MZ, PZ I and PZ II zones (Tables 3.4 and 3.6). They continue to dominate in the Lower Gangetic Plains also, but then they are not the only elements that relate to Himalayan orogeny and evolution of fish fauna in the Himalaya and immediate Plains, subsequent to glaciations in the recent history of India.

3.4.2 *Density*

3.4.2.1 Phytobenthos

The community density during summer decreased in MZ-L from Devprayag (1,026 cells mm^{-2}) to Rishikesh (797 cells mm^{-2}). Anonymous [16] recorded an annual density of 240 units cm^{-2} at Devprayag. The density of this community declines in the PZ I (Narora 288–311, Kachla Ghat 266, Bithoor 333–255–400 cells mm^{-2}) compared to the MZ. In the MZ, the periphyton density is low at Gangotri (131 cells mm^{-2}) and downstream nearby Harsil (187–373 cells mm^{-2}, [12]) compared to downstream stations where it is high (313 to 509 cells mm^{-2}). The periphyton density is little higher in the Alaknanda (440–1,045 cells mm^{-2}) in similar stretch (unpublished data). Further downstream in the Alaknanda, the peak densities were observed in winter that increased from 1,204 to 2,201 cells mm^{-2} with downward flow of the river from 980 to 550 m asl [24, 25]. In the foothills (MZ-L), the density ranged from 35.5 to 916.7 cells mm^{-2} [8, 57], which indicates a gentle decline from source to foothills. The phytobenthos density seems to declines from MZ to PZ I. However, density 1,395 units cm^{-2} was reported at Kanpur [15], which is shade higher than the foothills.

3.4.2.2 Plankton

The density of plankton increased in MZ-L from Devprayag 510 to Rishikesh 777 u l^{-1}. Singh et al. [4] registered a general increase in plankton density from Gangnani to Rishikesh (MZ). In respect of MZ, density shows a general decline in PZ I attributed to river regulation and influx of effluents but within the PZ I density increases mildly (Narora 255–366 u l^{-1}, Kachla Ghat 330–433 u l^{-1}, Bithoor 388–577 u l^{-1}). The density was reported to vary from 100 to 7,439 u l^{-1} from Rishikesh to Garhmukteshwar [3]. Vass et al. [52], reported an increase in the average total plankton density in the upper stretch of the Ganga, between Tehri and Kannauj, from 58 to 1,578 number per litre, 95 to 1,050 l^{-1} and 60 to 1,435 l^{-1} during summer, monsoon and winter months, respectively. The overall plankton density in the middle stretch varied from 24 to 782 l^{-1}, 146 to 3,649 l^{-1} and 14 to 8,049 l^{-1} respectively, during summer, monsoon and winter seasons respectively. The maximum density was found in the Kanpur stretch (8,049 u l^{-1}) during winter. In the lower stretch, between Sultanpur and Farakka, the plankton density ranged between 34 and 1,204 l^{-1}. A lower density of plankton in middle and lower freshwater stretches of the river was reported in the mid 1990's compared to the levels reported in early 1960's [51], but the composition showed little change. The minimum phytoplankton density (471.17 cells ml^{-1}) has been observed from Bithoorghat and maximum (14,294.10 cells ml^{-1}) from Jajmaughat [58].

3.4.2.3 Benthic Macroinvertebrate

The benthic macroinvertebrate density ranges from 327 to 1,053 indiv. m^{-2} in the Bhagirathi and 125–1357 indiv. m^{-2} in the Alaknanda and 169–612 indiv. m^{-2} in the Ganga. Recent investigations from Devprayag to Rishikesh during summer revealed low densities15 and 22 indiv. m^{-2}. At Narora density varied from 143 to 178 indiv. m^{-2}, 59–62 indiv. m^{-2} at Kachla Ghat and 11 to 27 indiv. m^{-2} at Bithoor. The density was relatively higher (ca. 100–150 indiv. m^{-2}) in shallow areas along the banks (15 cm depth). Anonymous [16] report density (**annual**) range from 44 to 286 units m^{-2}in **MZ-L**, 528 unit m^{-2} at Kanpur [15]; 1,122 unit m^{-2} at Kanpur [15].

3.4.3 *Community Composition*

3.4.3.1 Phytobenthos

Despite decline in the species richness, the quantitative share of diatom flora does not appear to decline. The composition in lower MZ (MZ-L) from Devprayag to Rishikesh was dominated by diatoms (D 51–84 %, GA 5–31 %, BGA 11–20 % during summer and D 96 % in winter). In contrast Anonymous [16] reported a dominance of green (GA) and blue-green algae (BGA) over diatoms (D) at Devprayag. The share of diatom fluctuates in PZ I (Narora – D 58–85 %, GA 11–36 %, BGA 4–6 %; Kachla Ghat – D 90–97 %, GA 3–10 %; Bithoor – D 54–70 % and GA 27–46 %) and is almost similar to MZ.

3.4.3.2 Plankton

Quantitative share (counts of number of individuals irrespective of species) of diatoms is relatively higher than GA and BGA. The plankton composition varies mildly in MZ-L from Devprayag to Rishikesh (D 44–61 %, GA 23–26 %, BGA 9–34 %, Z 3.5–5 %). In respect of MZ, the diatom share increases marginally, BGA declines and zooplankton enhances in PZ I (Narora – D 53–55 %, GA 24–32 %, BGA 5.4–9 %, Z 4 %; Kachla Ghat – D 50–56 %, GA 35–38 %, BGA 9.3–10 %, Z 0–3 %; Bithoor – D 60–70 %, GA 9–37 %, and BGA 5–7 % and Z 3–14 %).

Singh et al. [4] observed that phytoplankton accounts for 81.3 % of the total plankton density. The diatoms contributed (92.0 %) to the phytoplankton density. The other groups recorded were desmids (6.3 %), green algae (0.93 %), yellow-green algae (0.7 %), and blue green algae (0.1 %). The zooplankton consisted mainly of ciliates (97.0 % of the total zooplankton density). In the PZ, Vass et al. [52] noted that the phytoplankton accounted for a very large share while zooplankton formed only 16.6 %. They further state that zooplankton formed 7.9–34.8 % of the total plankton in the stretch between Haridwar and Kannauj. Bacillariophyceae accounts for 83.4 % of phytoplankton between Tehri and Kannauj. In Sultanpur and Farakka zone, phytoplankton was 70.9–89.2 % and rest was zooplankton.

3.4.3.3 Benthic Macroinvertebrate Fauna

In the MZ, Ephemeroptera, Diptera and Trichoptera account for 40.8, 35.5, and 13.1 % respectively of the total benthic density. Other orders, such as Plecoptera (1.4 %), Hemiptera (1.4 %), Coleoptera (4.8 %), Odonata (0.3 %), Molusca and Nematoda (2.8 %) and other miscellaneous forms, accounted for the remaining (10 %) of the density [4]. The benthic macroinvertebrates from Devprayag to Rishikesh in MZ, were represented by Trichoptera 100 % in MZ-L in summer (Diptera 90 %, Trichoptera 2 % and Ephemeroptera 8 % in winter) in contrast to diverse community at Narora (Diptera 11 %, Odonata 11 %, Coleoptera 2 %, Gastropoda 72 % and Pelecypoda 5 % right bank; Ephemeroptera 1 %, Diptera 57 %, Odonata1%, Hemiptera 7 %, Gastropoda 30 and Pelecypoda 4 %) and Kachla Ghat (Ephemeroptera 4 %, Diptera 50 %, Odonata 21 %, Gastropoda 11 % and Pelecypoda 14 %, Diptera 100 % leftbank). At Bithoor composition was restricted (Diptera (92 %), Odonata (8 %) – right bank; Pelecypoda (100 %) – left bank). Thus, prominent insect components of MZ (Ephemeroptera, Trichoptera, Colepotera) are either few or absent (Plecoptera) in the PZ.

3.4.4 Distribution of Flora and Fauna

3.4.4.1 Benthic Flora and Plankton

Distribution studies show that the benthic flora consists of 392 taxa exclusive to MZ, compared with 56 taxa only in the UGP (PZ I). Similarity is too low as only 41 taxa are common to these zones and distances are too high (Table 3.7, Fig. 3.1a). The richness decreases tremendously from the mountain to plain zone. In case of phytoplankton very few elements were exclusive to MZ (32 taxa) and common (53 taxa) with the UGP till PZ I (127 taxa). Further reduction in the number of common taxa and proportionate increase in the exclusive taxa occurs till PZ II. Within the lower part of UGP (PZ I to PZ II) there were notably high numbers of exclusive taxa in both zones and very few common elements (46 taxa). Fewer

Table 3.7 Species matching and similarity for major aquatic communities using ecological software, CAPS 4

Zones	Test sample	Common	Compare sample	Jaccards similarity
Benthic flora				
MZ-PZI (PB)	391	41	56	0.0838446
MZ-PZII				
Plankton				
MZ-PZI	32	53	208	0.180887
MZ-PZII	73	12	224	0.038835
PZI-PZII	215	46	190	0.101996
Phytoplankton				
MZ-PZI	32	53	127	0.25
MZ-PZII	73	12	224	0.038835
PZI-PZII	134	46	190	0.124324
Macroinvertebrate fauna				
MZ-PZI	11	21	35	0.313433
MZ-PZII	22	10	12	0.227273
PZI-PZII	34	22	0	0.39287
Fish fauna				
MZ-PZI	20	38	71	0.294574
MZ – PZII	32	26	67	0.208
PZ I-PZ II	42	67	26	0.496296

The first three columns of this Table are reproduced exactly as displayed by the software. When two samples are compared for instance MZ-PZ I, MZ is Test sample and PZ I is Compare sample while those common to them are called Common. The values in each column are the number of species in each category. The last column displays the measure of distance between the compared samples

common elements indicate low similarity in these zones, not only for phytoplankton but for plankton as well. This has also happened due to non-availability of zooplankton data in MZ and PZ II. The results of Cluster analysis support the observations on similarity because high distance (>0.5) is evident in dendrograms among these zones. Relatively lower distance among MZ and PZ I compared with PZ II supports the hypothesis that the influx from Yamuna affects the plankton community qualitatively in the PZ II. The distance is greater with zooplankton rather than with phytoplankton only (Fig. 3.1b, c). The level of similarity among these zones was least and distance was high for the benthic flora among MZ and PZ I compared to plankton (Fig. 3.1a–c, Table 3.7).

3.4.4.2 Macroinvertebrate Fauna

The comparison of macroinvertebrate fauna in concerned zones reveals that only 11 taxa are specific to the MZ, 21 taxa are common and 35 taxa are peculiar to PZ II. With respect to PZ II, the MZ is far more distinct in fauna as 22 taxa are

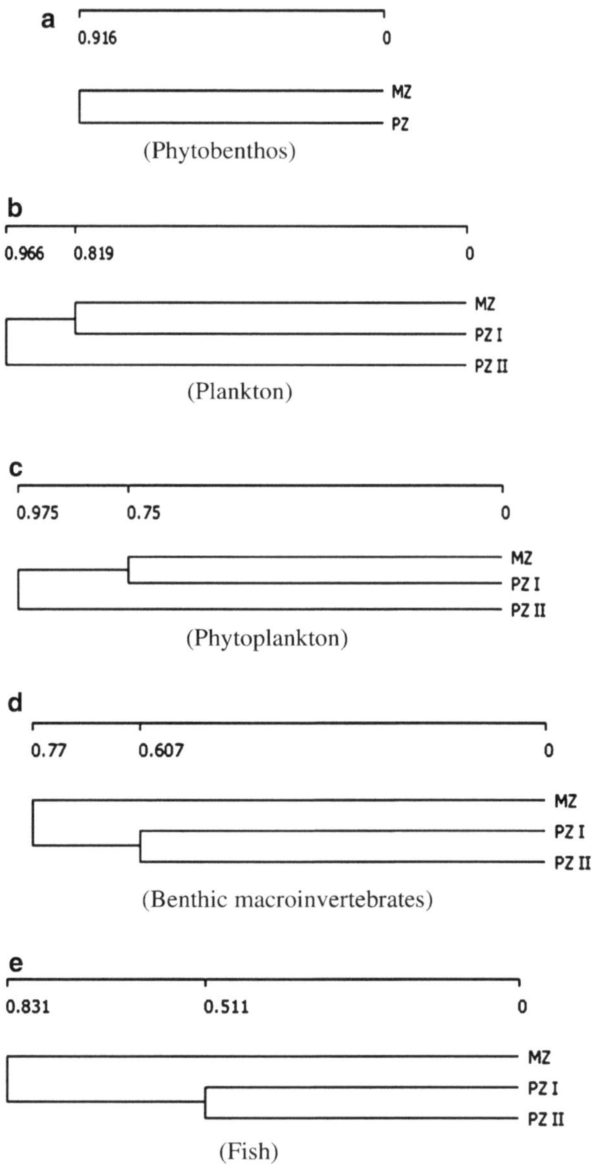

Fig. 3.1 Cluster (Agglomerative) Analysis: 1-Jaccards distance (Wards method) to demonstrate differences in the floral (benthic, phytoplankton) and faunal (macroinvertebrate, fish) composition in the mountain (MZ) and plains zone (PZ I, II). (**a**) Phytobenthos. (**b**) Plankton. (**c**) Phytoplankton. (**d**) Benthic macroinvertebrates. (**e**) Fish

specific to MZ, only 10 taxa are common and 12 restricted to PZ II. The macroinvertebrate fauna of PZ II is rather similar to PZ I because 34 taxa are peculiar to PZ I, and a very large share is common with PZ II. Notably, no taxon is peculiar to PZ II rather 34 elements are absent in this zone. Since the number of common taxa is higher in MZ-PZ I and PZ I – PZ II, they are relatively more similar than MZ-PZ II (Table 3.7), because the latter is farther than PZ I from the MZ.

The Cluster analysis (Fig. 3.1d) clearly shows lesser distance within PZ (among PZ I and PZ II) and greater distance from MZ. Thus, the MZ benthic fauna (at family level) differs sufficiently from the whole PZ, just like the benthic flora where the distance is high among the MZ and PZ (Fig. 3.1a). This is in slight contrast to the plankton where distance between MZ and PZ I is less than that of both with PZ II (Fig. 3.1b, c).

Fish fauna: Fewer fish species were common between MZ and PZ I (38) and still less among MZ and PZ II. However, there is a noticeable increase in the exclusive taxa among these zones. There is an immense increase in common taxa between PZ I and PZ II as expected and hence higher similarity and less distance (Fig. 3.1d). Of these 25 species are exclusive in the Kanpur-Varanasi zone while 57 species are common to Bijnor-Kanpur zone. There are more taxa peculiar to PZ I than to PZ II zone (Table 3.7). The few taxa those are unique to PZ II in respect of PZ I should be useful for zonation as already discussed. Probably motility and physiological requirements in respect of tolerance for temperature restrict the faunal elements in a different fashion from floral components. However, the similarity among MZ and PZ I is low is case of benthic flora and plankton while PZ I and PZ II in case of the fauna (Fig. 3.1d, e). Similarity of zone in respect of fauna was relatively higher than benthic flora and plankton. Hence, each zone is distinct from the other; MZ and PZ I in case of benthic flora and plankton while PZ I and PZ II in case of fauna.

The river is extensively regulated in the MZ and UGP till Kanpur compared with LGP, primarily for power generation in the MZ and irrigation in the UGP. This has modified the continuum of the Ganga in the examined stretch such that the similarity in the flora and fauna that would be expected to be high (at least 0.5 Jaccard's Similarity) in a natural river stretch, is as low as 0.03 (Table 3.7), and barely reaches 0.39 for macroinvertebrates (feeble motility) and 0.49 in case of fish (high motility) in PZ I and PZ II.

The continuum of flora and fauna depends on their dispersal ability so essential for population dynamics and since aquatic organisms can disperse only if there are no barriers, their dispersal is hindered, inhibited and impaired. This leads to decline in the similarity in the UGP, where no major river from different biogeographic zone is joining the Ganga.

There is a serious lacuna about the natural range (however small in size) of each species, as there have been no dedicated research programs for so important a river as the Ganga with economic implications for the country, especially irrigation and the variety of livelihood it provides to poverty ridden areas of north India. For instance we know that the snow trouts *S. richardsonii* and *S. plagiostomus* reside upto Hardwar, and *Tor putitora* resides in the foothills (even Bijnor) but migrate even beyond Srinagar (Alaknanda) and Tehri (not now because of Tehri Dam). If these species are not found in impounded areas then it is obvious that the continuum does not exist, but then the fish needs the food web and thus each component has an ecological function in the ecosystem and hence the emphasis on the knowledge of various components of biodiversity and their distribution to know the health of the ecosystem. It can inform us about disturbances in the continuum.

3.5 Conclusions

The river flows through mountain and plains. The ice-cold waters and shooting torrents in MZ transforms into a thick sheet of slow-moving nutrient-rich warm-waters. The voluminous flows prevent rapid warming of the water in UGP, especially PZ I. There are pockets of knowledge about the ecosystems along the Ganga river but the profile of biodiversity and many other structural and functional aspects of the ecosystems remain to be known. River regulation has modified the continuum of the Ganga and the knowledge of the state of biodiversity is now a necessity for developing a basin management plan, under the aegis of National Ganga Authority. The PZ II is qualitatively different from MZ and PZ I.

3.5.1 Species/Taxonomic Richness

The species richness of benthic flora declines from MZ to PZ I and PZ II. In contrast there is a proportionate increase in plankton richness (both phyto and zoo). This relates well with diminishing substrate size from MZ to PZ II and increasing width and depth (surface area). The macroinvertebrate and fish fauna also increases gradually in richness. Since, soft sediments prevail in the PZ free floating plankton, burrowing invertebrates and highly motile fish are important components of biodiversity. The macroinvertebrate community consists of insects in MZ and insects, mollusc and annelids in PZ.

3.5.2 Density

The density of all phytobenthos, plankton and macroinvertebrate communities is almost similar in the MZ and PZ. The decline at certain locations is attributable to river regulation and pollution. The densities are high during winter and declines substantially with snow-melt in summer and floods in monsoon. The life-histories of the organisms forming these communities are synchronised with seasonal changes in hydrology, substrate availability so important in the MZ. In the PZ, the densities are high in winter but the decline during summer and even monsoon is not conspicuous like MZ.

3.5.3 Community Composition

The community composition of plankton and phytobenthos is similar from MZ to PZ as diatoms dominate (>50 %) followed by GA, BGA and Z. The composition of macroinvertebrate community shows shift from dominance of insect flies (may, caddis and diptera) in the MZ to insect and molluscs dominance in PZ.

3.5.4 *Distribution*

The elements of phytobenthos and plankton in the MZ and PZ I have relatively more similarity than with PZ II attributed to the influx from Yamuna river system. However, the macroinvertebrate and fish fauna have higher similarity among PZ I and II compared with MZ. Thus floral and planktonic biodiversity is more distinct in the PZ II, while the faunal in MZ and to some extent in PZ II. Possibly motility and tolerance for temperature restrict the faunal elements in a different fashion from floral components. However, the similarity among zones is very low which shows that each of them is a distinct zone; MZ and PZ I in case of benthic flora including plankton and PZ I and PZ II in case of fauna.

Acknowledgements I am very grateful to Prof. H. R. Singh (Former Vice Chancellor) University of Allahabad, Former Head and Dean Department of Zoology H. N. B. Garhwal University, Srinagar, for inspiration, guidance and academic support, over last 3 decades. I sincerely thank and acknowledge the facilities provided by the University and Head, Prof. N. Singh and Former Head, Department of Zoology and Biotechnology Prof. J. P. Bhatt. The research funding by MoEN and F, New Delhi vide J-11013/8/92-GPD, 24.9.92 and 14/28/92 – MAB/RE/01.06.94, DCFR, Bhimtal under ICAR outreach program and the resources (including financial) under the "Living Ganga Programme (2008-12) of WWF-India under the HSBC Climate Partnership", are also thankfully acknowledged.

Appendices

Appendix I

Distribution of benthic flora (PB) and plankton (PL) in the mountain (M) and the Upper Gangetic Plains (PZ I, II). The diatom taxa marked by * are present in the mountain zone (MZ) of the Ganga river. The taxa mentioned as genus only e.g. *Ceratoneis* sp. (mostly from secondary sources) have not been counted for species richness estimate as some species of the genus are already listed elsewhere in this Appendix. The appendix has been arranged to show distribution of species in longitudinal fashion. Thus, the taxa occurring in the all the zones (as PB or PL) have been indicated first, preferable in alphabetical order, followed by those found only in each section from MZ to PZ II. *Achnanthes, Navicula* and *Cymbella* sensu stricto that have been further split into more genera in recent literature e.g. *Planothidium* from *Achnanthes; Placoneis, Sellaphora* and *Craticula* and from *Navicula*; *Cymbopleura, Encyonema, Encyonopsis, Reimeria* from *Cymbella*. This Appendix contains most species that belong to *Cymbella* and *Navicula* sensu lato, but that of *Achnanthes* are sensu stricto.

	MZ		PZ I		PZII	
	PB	PL	PB	PL	PB	PL
Bacillariophyceae						
*Achnanthes affinis**	+	+				+
Amphipleura sp.		+				
Aulacoseira sp.		+				
Bacillaria sp.		+				
Ceratonies sp.		+				
Cymatopleura sp.		+				
*Cymbella turgidula**		+			+	
*Cymbella ventricosa**	+	+				+
Denticula sp.		+				
Gomphoneis sp.		+				
Rhoicosphenia sp.		+				
*Achnanthes biasolettiana**	+	+	+	+	+	
*Achnanthes minutissima**	+	+	+	+	+	+
Epithemia sp.		+	+			
*Cymbella affinis**	+	+	+			+
Melosira sp.		+	+			
Meridion sp.		+	+			
Planothidium lanceolatum var. *lanceolatum**	+	+	+	+		+
Stephanodiscus sp.		+	+			
*Diploneis ovalis**	+	+				
*Fragilaria capucina**	+	+		+		+
*Encyonopsis leei**	+	+			+	
*Navicula exilis**	+	+			+	
*Nitzschia dissipata**	+	+			+	
*Nitzschia filiformis**	+	+				+
Pinnularia sp.	+	+		+		
*Cymbella excisa**	+	+	+		+	
*Cymbella turgidula**	+	+	+	+		
*Cymbella tumida**	+	+	+			+
*Cocconeis pediculus**	+	+	+	+		+
Cocconeis placentula var. *euglypta**	+	+	+	+	+	+
C. diminuta	+	+	+			
*Diatoma mesodon**	+	+	+	+		+
*Encyonema minutum**	+	+	+	+	+	+
*Diatoma monoiliformis**	+	+	+	+		
Frustulia sp.	+	+	+			
*Gomphonema minutum**	+	+	+	+	+	+
*Hantzschia amphioxys**	+	+	+		+	+
*Navicula caterva**	+	+	+			+
*Navicula cryptotenella**	+	+	+	+	+	
*Navicula rhyncocephala**	+	+	+	+		+
Stauroneis sp.	+	+	+			
*Synedra ulna**	+	+	+	+		+
Synedra ulna var. *aequalis**	+	+	+	+		
Amphora sp.	+	+	+			

(continued)

(continued)

	MZ		PZ I		PZII	
	PB	PL	PB	PL	PB	PL
Synedra ulna var. *danica**	+	+	+	+		
Synedra ulna var.*oxyrhynchus* fo.*mediocontracta**	+	+	+	+		
Amphipleura pellucida	+					
*Amphora montana**	+		+	+		+
*Amphora pediculus**	+		+	+	+	
*Caloneis bacillum**	+		+		+	
*Amphora veneta**	+		+	+	+	+
Asterionella formosa	+		+			
Cocconeis placentula var. *lineata**	+		+	+		+
*Cocconeis placentula**	+				+	+
*Cyclotella glomerata**	+		+	+	+	+
*Cyclotella meneghiniana**	+		+	+	+	+
Cyclotella stelligera	+		+			
Cymbella excisa var. *aungusta**	+		+	+	+	
*Cymbella kolbei**	+		+	+		
*Cymbella laevis**	+		+	+		+
*Cymbella parva**	+		+	+	+	
Diatoma sp.	+		+			
*Encyonema silesiacum**	+		+	+		
Cymbella sp.	+		+	+		
*Fallacia pygmaea**	+		+			+
*Gesslaria decussis**	+		+	+	+	+
Fragilaria crotonensis	+		+			+
*Fragilaria construens**	+				+	
Fragilaria sp.	+		+			
*Gomphonema lagenula**	+		+	+	+	
Gomphonema pumilum var. *rigidum**	+		+	+	+	
*Gomphonema parvulum**	+		+	+	+	+
*Gyrosigma acuminatum**	+		+	+		+
*Gyrosigma scalproides**	+		+	+		+
*Hannaea arcus**	+		+	+		+
Hannaea arcus var. *amphioxys**	+		+	+		+
Hannaea arcus var. *recta*	+		+	+		+
*Luticola mutica**	+		+			+
Gyrosigma sp.	+		+			
*Melosira varians**	+		+	+		+
*Navicula cari**	+		+	+		
*Navicula cryptofallax**	+		+	+	+	+
*Navicula cryptotenelloides**	+		+	+		
*Navicula radiosafallax**	+		+	+	+	+
Navicula sp.*	+		+	+		+
*Navicula rostellata**	+		+	+	+	+
*Nitzschia amphibia**	+		+	+		+
*Nitzschia communis**	+		+	+	+	+
*Nitzschia fonticola**	+		+	+	+	+

(continued)

(continued)

	MZ		PZ I		PZII	
	PB	PL	PB	PL	PB	PL
*Nitzschia linearis**	+		+	+		+
*Nitzschia microcephala**	+		+			+
*Nitzschia palea**	+		+	+	+	+
*Nitzschia hungarica**	+		+			+
*Nitzschia frustulum**	+		+	+	+	+
Nitzschia sp.	+		+			
*Nitzschia recta**	+		+	+		+
*Placoneis elegans**	+		+	+	+	+
Pinnularia gibba	+		+			+
*Reimeria sinuata**	+		+	+		
*Sellaphora pupula**	+		+	+		+
*Synedra acus**	+		+	+		+
Synedra acus var. *radians**	+		+	+		+
Synedra ulna var. *subaequalis**	+		+	+		
Tabellaria fenestrata	+		+			
*Achnanthes subhudsonis**	+		+	+		
Synedra acus var. *angustissima**	+		+			+
*Synedra amphirhynchus**	+		+			+
*Diatoma elongatum**	+					+
*Adlafia muscora**	+			+	+	
*Achnanthes austriaca**	+					
Achnanthes biasolettianum var. *subatomus*	+					
Achnanthes bicapitata	+					
Achnanthes boyei	+					
*Achnanthes brevipes**	+					
Achnanthes brevipes var. *intermedia**	+					
Achnanthes brevipes var. *parvula*	+					
Achnanthes cf. *pseudoswazi*	+					
*Achnanthes clevei**	+					+
Achnanthes clevei var. *rostrata*	+					
Achnanthes conspicua	+					
Achnanthes delicatula	+					
Achnanthes delicatula sp. *hauckiana*	+					
*Achnanthes exigua**	+				+	+
Achnanthes exigua var. *constricta*	+					
Achnanthes exigua var. *heterovalva*	+					
*Achnanthes exilis**	+					
Achnanthes flexella	+					
*Achnanthes fragilarioides**	+					
*Achnanthes grimmei**	+					
*Achnanthes hauckiana**	+					
Achnanthes helvetica	+					
Achnanthes holistica	+					
*Achnanthes kryophila**	+					
*Achnanthes lacunarum**	+					

(continued)

(continued)

	MZ		PZ I		PZII	
	PB	PL	PB	PL	PB	PL
Achnanthes lapponica	+					
Achnanthes laterostrata	+					
*Achnanthes lemmermannii**	+					
Achnanthes levanderi	+					
*Achnanthes linearis**	+					+
*Achnanthes marginulata**	+					
*Achnanthes microcephala**	+					+
Achnanthes microcephala var. *gracillima**	+					
Achnanthes minutissima var. *cryptocephala**	+					
Achnanthes minutissima var. *robusta*	+					
Achnanthes nodosa	+					
Achnanthes orientalis	+					
Achnanthes oestrupi	+			+		
Achnanthes petersenii	+					
Achnanthes pusilla	+					
Achnanthes rupestris	+					
*Achnanthes saxonica**	+					
Achnanthes sp.	+					
Achnanthes sphacelata	+					
*Achnanthes subsalsa**	+					
*Achnanthes suchlandti**	+					
*Achnanthes taeniata**	+					
*Achnanthes trigibba**	+					
*Amphora delicatissima**	+					
Amphora nagpurense	+					
Amphora normanii	+					
*Amphora ovalis**	+					+
*Amphora perpusilla**	+					
Karayevia ploenensis	+					
Anomoeoneis serians	+					
Brachysira vitrea	+					
Caloneis beccariana	+			+		
Caloneis s. var. *elliptica*	+					
*Caloneis obtusa**	+					
Caloneis silicula	+					+
Caloneis silicula fo. *truncatula**	+					
Caloneis silicula var. *lapponica*	+					
Cymbella subtruncata var. *subtruncata*	+					
Cocconeis sp.	+					
Craticula ambigua	+					
Craticula buderi	+			+		
*Craticula citrus**	+				+	
*Craticula halophila**	+					+
Craticula submolesta	+					
Cyclotella meneghiniana fo. *unipunctata**	+					

(continued)

(continued)

	MZ		PZ I		PZII	
	PB	PL	PB	PL	PB	PL
Cymatopleura solea	+					
*Cymbella amphicephala**	+					
Cymbella aspera	+					
*Cymbella australica**	+			+		
*Cymbella austriaca**	+					
*Cymbella brehmi**	+					
*Cymbella cistula**	+					
*Cymbella cymbiformis**	+					+
*Cymbella delicatula**	+					
Cocconeis disculus	+					
Cymbella excisa var. *procera*	+					
*Cymbella gonzalvesii**	+					
*Cymbella helvetica**	+					
*Cymbella hustedtii**	+					
Cymbella kappi	+					
Cymbella lanceolata	+					
Cymbella metzeltinii	+					
Cymbella neocistula	+					
Cymbella novazeelandiana	+			+		
*Cymbella nagpurensis**	+					
*Cymbella perpusilla**	+					
Cymbella pervarians	+					
Cymbella prostrata	+					+
Reimeria sinuta fo. *ovata*	+					
Cymbella superparva	+					
*Cymbella tropica**	+				+	
*Cymbella tumidula**	+					
Cymbella turgidula var. *bengalensis*	+					
Cymbella turgidula var. *venezolana*	+					
Cymbella vulgata	+					
Cymbopleura angustata	+					
Cymbopleura diminuta	+					
Cymbopleura naviculiformis	+			+		
Cymbopleura reinhardtii	+					
*Diatoma anceps**	+			+		
*Diatoma hyemale**	+			+	+	+
*Denticula kuetzingii**	+					+
Diatoma minus	+					
*Achnanthes oblongella**	+					
*Diatoma tenuis**	+			+		
Diatoma vulgare var. *brevis**	+					
Diatoma vulgare	+					+
Diatoma vulgare var. *ovalis**	+					
Diatoma vulgare var. *producta**	+					
Diatomenella balfouriama	+					

(continued)

(continued)

	MZ		PZ I		PZII	
	PB	PL	PB	PL	PB	PL
Diploneis minuta	+					
Diploneis oculata	+					
Diploneis puella	+					
Diploneis subovalis	+					
Encyonema bipartitum	+					
Encyonema gracilis	+					
Encyonema gracilis var. *aurangabadensis*	+					
Encyonema hebridicum	+					
Encyonema jemtlandicum var. *venezolanum*	+					
*Encyonema lacustre**	+					
Encyonema latens	+					
Encyonema neomesianum	+					
Encyonema prostrata	+					
Encyonopsis leei var. *leei*	+					
Encyonopsis leei var. *sinensis*	+					
Encyonopsis microcephala	+			+		
Eunotia arcus	+					
Eunotia bigibba	+					
Eunotia crista var. *galli*	+					
Eunotia exigua	+					
Eunotia kocheliensis	+			+		
Eunotia lapponica	+					
Eunotia pectinalis. var. *bidens*	+					
Eunotia pectinalis var. *minor*	+					
Eunotia polydentula	+					
Eunotia praerupta	+					
Eunotia tenella	+					
Eunotia valida	+					
Eunotia veneris	+					
*Epithemia sorex**	+					
Epithemia sorex var. *gracilis*	+					
Epithemia zebra	+					
Epithemia zebra var. *saxonica*	+					
Fragilaria bicapitata	+					
Fragilaria bidens	+					
*Fragilaria inflata**	+					
Fragilaria intermedia var. *chandrapurensis**	+					
Fragilaria vaucheriae	+					
Fragilariforma virescens	+					
Fragilaria capucina var. *lanceolata*	+					+
*Fragilaria intermedia**	+					+
Frustulia rhomboides fo. *undulata*	+					
Frustulia vulgaris var. *elliptica*	+					
Frustulia weinholdii	+					
Gomphoneis herculeanum	+					

(continued)

(continued)

	MZ		PZ I		PZII	
	PB	PL	PB	PL	PB	PL
Gomphonema abbreviatum	+					
Gomphonema abbreviatum var. *linearis*	+					
Gomphonema acuminatum var. *brebissonii*	+					
Gomphonema acuminatum var. *pusillum*	+					
Gomphonema acuminatum var. *turris*	+					
*Gomphonema angustatum**	+					+
Gomphonema augur	+			+		+
*Gomphonema sphaerophorum**	+					+
Gomphonema bohemicum	+					
Gomphonema brasiliense sp. *pacificum*	+					
*Gomphonema clevei**	+					+
Gomphonema constrictum	+					+
Gomphonema constrictum var. *capitata*	+					
*Gyrosigma distortum**	+					
*Gomphonema gracile**	+					+
*Gomphonema helveticum**	+					
*Gomphonema intricatum var. pumilla**	+					
*Gomphonema lanceolatum**	+					
Gomphonema lanceolatum var. *insignis**	+					
Gomphonema longiceps var. *subclavata**	+					
Gomphonema nagpurense	+					
*Gomphonema olivaceum**	+					+
Gomphonema olivaceum var. *balticum*	+					
Gomphonema olivaceum var. *calcarea**	+					
Gomphonema olivaceum var. *minutissima**	+					
Gomphonema parvulum var. *micropus**	+					
Gomphonema parvulum var. *subelliptica*	+					
Gomphonema parvulum var. *exilissima**	+					
Gomphonema productum	+					
Gomphonema pseudoaugur	+					
*Gomphonema pseudosphaeophorum**	+					
Gomphonema pumilum var. *pumilum*	+					
Gomphonema rhombicum var. *rhombicum**	+					
Gomphonema sp.	+					
Gyrosigma distortum	+					
Gyrosigma kutzingii	+					+
Gyrosigma spencerii	+					
Gyrosigma spencerii var. *nodifera*	+					
Gomphonema subtile	+					
Gomphonema tergestinum	+					
Gomphonema ventricosum	+					
Gyrosigma vidarbhense	+					
Hantzschia amphioxys var. *capitata*	+					
Hantzschia amphioxys var. *major*	+					
Hantzschia virgata	+					

(continued)

(continued)

	MZ		PZ I		PZII	
	PB	PL	PB	PL	PB	PL
Hippodonta ruthnielseniae	+					
Luticola mutica var. *geopertiana**	+					
*Meridion circulare**	+					
Meridion circulare var. *constrictum*	+					
Navicula alineae	+					
Navicula antonii	+					
Navicula broetzii	+					
Adlafia bryophila	+					
Navicula capitatoradiata	+					
*Nitzschia capitellata**	+					
Navicula cari var. *angustata*	+					
Navicula cataracta-rheni	+					
Navicula cincta	+					+
Navicula cincta var. *heufleri**	+					
Navicula constans var. *symmetrica**	+					
Navicula cryptocephala	+					+
Navicula cryptocephala var. *intermedia*	+					
Navicula cryptocephala var. *sabsalina**	+					
Navicula digitulus	+					
*Navicula grimmei**	+					
*Navicula hustedti**	+					
Navicula hustedti var. *obtusa.*	+					
Nitzschia hybrida	+					
Navicula krammerae	+					
Navicula krasskei	+					+
*Navicula lanceolata**	+					
*Nitzschia lancettula**	+					
Navicula minima	+					+
Navicula minima var. *atmoides*	+					
Navicula notha	+					
*Navicula radiosa**	+					
Navicula radiosa var. *acuta**	+					
Navicula radiosa var. *minutissima**	+					
Navicula radiosa var. *tenella**	+					
Navicula reichardtiana	+					
Navicula rhyncocephala var. *amphiceros**	+					
Navicula rhyncocephala var. *grunowii**	+					
Navicula rhyncocephala var. *tenua**	+					
*Navicula salinarum**	+					+
Navicula salinarum var. *intermedia*	+					
Navicula seminulum	+					+
Navicula subrhyncocephala	+					
Navicula subtilissima	+					
*Navicula symmetrica**	+					
Navicula tripunctata	+					

(continued)

(continued)

	MZ		PZ I		PZII	
	PB	PL	PB	PL	PB	PL
Navicula veneta	+					
*Navicula venezuelensis**	+			+		
Navicula vermicularis	+					
Navicula viridula fo. *capitata*	+					
Navicula viridulacalcis	+					
Navicula viridula var. *lanceolata**	+					
Navicula viriduloides var. *lanceolata**	+					
Navicula vitabunda	+					
Navicymbula pusilla	+					
Nedium denticula var. *curta*	+					
Neidium halophila fo. *subcapitata*	+					
Neidium viruduloides	+					
Neidium affine var. *longiceps*	+					
Nitzschia acuta Hantzsch	+					
Nitzschia clausii	+			+		
Nitzschia commutata	+					+
*Nitzschia denticula**	+					
*Nitzschia gracilis**	+				+	+
*Nitzschia hantzschiana**	+				+	
*Nitzschia ignorata**	+				+	
Navicula intergracilis	+					
Nitzschia kuetzingiana	+					
Navicula leistikowii	+					
*Nitzschia romana**	+					
*Nitzschia paleacea**	+					+
*Navicula schorteri**	+					
*Nitzschia sigmoidea**	+					
Nitzschia sinuata var. *tabellaria**	+					
*Nitzschia stagnorum**	+					
*Nitzschia sublinearis**	+					+
Nitzschia tenuis	+					
Nitzschia thermalis	+					+
Nitzschia thermalis var. *minor**	+					
Nitzschia vitrea	+					
*Nitzschia flexa**	+				+	
*Nitzschia gandersheimiensis**	+				+	
*Nitzschia intermedia**	+				+	
*Nitzschia woltereckoides**	+				+	
Nupela lapidosa	+					
Peronia fibula	+					
Planothidium lanceolatum fo. *capitata*	+					
Planothidium lanceolatum var. *elliptica**	+		+	+		
Planothidium lanceolatum var. *lanceolatum**	+					
Planothidium lanceolatum fo. *rostrata**	+					
Planothidium lanceolatum fo. *ventricosa*	+					

(continued)

(continued)

	MZ		PZ I		PZII	
	PB	PL	PB	PL	PB	PL
Planothidium lanceolatum var. *frequentissimum**	+				+	
Pinnularia appendiculata	+					
Pinnularia appendiculata var. *budensis*	+					
Pinnularia borealis	+					
Pinnularia brauni	+			+		
Pinnularia lata	+					
Pinnularia microstauron	+					
Pinnularia microstauron var. *diminuta*	+					
Pinnularia pusilla	+					
Pinnularia subcapitata	+					
Pinnularia subcapitata var. *lapponica**	+					
Pinnularia sublanceolata	+					
Placoneis constans var. *symmetrica*	+					
Planothidium spec. cf. *biporomum*	+					
Rhopalodia gibba	+					+
Rhopalodia gibberula	+					
Rhopalodia rupestris	+					
*Rhoicosphenia vanheurcki**	+					
Rhoicosphenia abbreviata	+					
*Rhoicosphenia curvata**	+					
*Sellaphora bacillum**	+					
Sellaphora bacillum var. *minor*	+					
*Sellaphora pupula**	+					+
Sellaphora pupula var. *rectangularis*	+			+		
Sellaphora pupula var. *rostrata*	+					
Stauroneis smithii	+					
Stauroneis anceps	+					+
Stauroneis phoenicenteron	+					
Stauroneis smithii var. *borgei*	+					
Stauroneis smithii var. *sagitta*	+					
*Staurosira construens**	+					
Staurosira martyi	+					
*Staurosirella leptostauron**	+					
Staurosirella leptostauron var. *dubia**	+					
*Staurosira mutabilis**	+					
*Staurosirella pinnata**	+					
Staurosirella pinnata var. *lancettula*	+					
Staurosirella lapponica	+					
*Surirella angusta**	+					
Surirella elegans	+					+
Surirella islandica	+					
*Surirella ovata**	+			+		
Surirella ovata var. *pinnata*	+					
Surirella vidarbhensis	+					
*Synedra amphicephala**	+					

(continued)

(continued)

	MZ		PZ I		PZII	
	PB	PL	PB	PL	PB	PL
Synedra amphicephala var. *austriaca*	+					
*Surirella apiculata**	+					
*Surirella delicatissima**	+					
Synedra dorsiventralis	+					
Synedra miniscula	+					
*Synedra rumpens**	+					+
Synedra rumpens var. *familiaris**	+					
Synedra rumpens var. *fragilarioides**	+					
Synedra rumpens var. *scotica*	+					
Synedra tabulata	+					
Synedra tabulata var. *acuminata*	+					
Surirella tenera	+					
Synedra ulna var. *contracta**	+			+		
Synedra ulna var. *impressa*	+					
Synedra ulna var. *oxyrhynchus**	+					
Synedra utermohli	+					
Synedra vidharbensis	+					
Tabellaria flocculosa	+			+		+
Tabellaria sp.	+			+		
Cymbella lancettuliformis			+	+		
Cymbopleura sp.			+	+		
Craticula cuspidata			+			+
Nitzschia acicularis			+		+	+
Nitzschia capitellata			+			+
Nitzschia obtusa			+	+		+
Synedra sp.			+			
Craticula molestiformis			+		+	+
Achnanthes crenulata			+	+		+
Cyclotella operculata			+			+
Asterionella sp.			+			
Melosira granulata			+	+	+	+
Bacillaria paradoxa			+	+		+
Cymbella turgid						+
Diploneis intermedia						+
Fragilaria brevistriata						+
Gomphonema macropuctatum						+
Gomphonema subclavatum						+
Melosira ambigua						+
Melosira distans						+
Melosira exigua						+
Melosira granulata var. *angustatum*						+
Navicula cocconiformis						+
Navicula exigua						+
Navicula gracilis						+
Navicula gregaria						+

(continued)

(continued)

	MZ		PZ I		PZII	
	PB	PL	PB	PL	PB	PL
Navicula microcephala						+
Navicula miniscuIus						+
Navicula simplex						+
Nitzschia kittonii						+
Nitzschia punctata						+
Nitzschia sigma						+
Nitzschia subtilis						+
Nitzschia tryblionella						+
Pinnularia major						+
Pinnularia viridis						+
Pinnularia viridis var. *intermedia*						+
Adlafia minuscula				+	+	
Amphora inariensis				+	+	
Synedra fasciculata						+
Synedra ulna var. *biceps*						+
Cyclotella antigua						+
Cyclotella kutezingiana						+
Anomoeneis sphaerophora						+
Caloneis amphisbaena						+
Cyclotella comta						+
Chlorophyceae						
Chaetophora sp.		+				
Characium sp.		+				
Chroococcus sp.		+				
Phormidium sp.		+				
Sphaeroplea sp.		+				
Pediastrum sp.		+	+			
Protococcus sp.		+	+			
Eudorina sp.		+	+			
Coelastrum sp.		+	+			
Gonatozygon sp.		+	+			
Cladophora sp.	+	+	+			
Cosmarium sp.	+	+	+			
Desmidium sp.	+	+	+			
Hydrodictyon sp.	+	+	+			
Microspora sp.	+	+	+			
Oedogonium sp.	+	+	+			
Scenedesmus sp.	+	+	+			
Stigeoclonium sp.	+	+	+			
Spirogyra sp.	+	+	+			
Ulothrix sp.	+	+	+			
Vaucheria sp.	+	+				
Genicularia sp.		+				
Pandorina sp.		+				
Desmidium aptogonum		+				+

(continued)

(continued)

	MZ		PZ I		PZII	
	PB	PL	PB	PL	PB	PL
Pediastrum boryanum		+				+
Zygnema sp.	+	+	+	+		+
Chlorella sp.	+					
Chlorogonium sp.	+					
Closterium sp.	+					
Cosmarium granatum	+					+
Draparnalcia sp.	+					
Kirchneriella sp.	+					
Merismopedia glauca	+					
Scenedesmus diamorplus	+					+
Tetracyclus rupestris	+					
Ulothrix zonata	+					+
Closterium leibleinii	+			+		
Schizogonium sp.	+			+		
Chlorococcum humicola	+		+			
Pediastrum simplex v. *bivaense*	+		+	+		
Scenedesmus quadricauda	+		+	+		+
Hydrodictyon reticulatum	+		+			+
Pediastrum duplex	+		+			+
Actinastrum sp.			+			
Ankistrodesmus sp.			+			
Botryococcus sp.			+			
Bumillaria exilis			+			
Chlorobottrys			+			
Chlorogonium elongatum			+			
Chlorella vulgaris			+			
Closterium nitzsch			+			
Crucigenia lanterbornei			+			
Crucigenia sp.			+			
Gonotozygon kinahani			+			
Phormidium sp.			+			
Phormidium suletile			+			
Microspora amoena			+			
Netrium sp.			+			
Oedogonium macrandrous			+			
Palmella sp.			+			
Pleurodermus sp.			+			
Scenedesmus quadricuda v. *westii*			+			
Selenastrum sp.			+			
Sphaerocystis schroeteri			+			
Tetraspora sp.			+			
Treubaria varia			+			
Volvox globatum			+			
Chlorella vulgaris			+			+
Ankistrodesmus falcatus			+			+

(continued)

(continued)

	MZ		PZ I		PZII	
	PB	PL	PB	PL	PB	PL
Pediastrum simplex			+			+
Pediastrum tetras			+			+
Mougeotia sp.			+			+
Eudriona elegans			+			+
Actinastrum hantzschii						+
Ankistrodesmus acicularis						+
Ankistrodesmus angustus						+
Chlamydomonas derenbenji						+
Chlamydomonas laginula						+
Chlamydomonas mirabilis						+
Chlamydomonas truncata						+
Chlorella subsala						+
Chlorococcum infusionum						+
Cladophora glomerata						+
Closterium calosporum						+
Closterium incurvatum						+
Closterium maxima						+
Closterium parvulum						+
Closterium peracerosum						+
Closterium rostratum						+
Coelastrum cambricum						+
Coelastrum microporum						+
Coelastrum reticulum						+
Coelastrum sphaericum						+
Cosmarium anceps						+
Cosmarium denatum						+
Cosmarium dentiferum						+
Cosmarium undulatum						+
Tribonema sp.			+			
Crucigenia rectangularis						+
Crucigenia tetrapeda						+
Crucigenia truncata						+
Crucigenia apiculata						+
Dictyosphaerium pulchellum						+
Diaspora cuneiformis						+
Kirchneriella contorta						+
Kirchneriella obesa						+
Micractinium radiatum						+
Oocystis crassa						+
Oocystis marsoni						+
Oocystis parva						+
Oocystis solitaria						+
Pandorina morum						+
Pediastrum clathratum						+
Pediastrum consortium						+

(continued)

(continued)

	MZ		PZ I		PZII	
	PB	PL	PB	PL	PB	PL
Pediastrum simplex v. *duodenrium*						+
Scenedesmus acuminatus						+
Scenedesmus arcuatus						+
Scenedesmus bicaudatus						+
Scenedesmus bijugatus						+
Scenedesmus denticulatus						+
Scenedesmus falcatus						+
Scenedesmus longus						+
Ulothrix subtellisma						+
Selenastrum gracile						+
Spirogyra affinis						+
Spirogyra decimina						+
Spirogyra spingularis						+
Spirogyra subsala						+
Tetraedron bifidum						+
Tetraedron constrictum						+
Tetraedron minimum						+
Tetraedron muticum						+
Cyanophyceae						
Agmenellum sp.			+			
Anabaena sp.	+	+	+	+		
Anabaena circularis						+
Anabaena cylindrica						+
Anabaena laxa						+
Anabaena sphaerica						+
Anabaena torulosa						+
Anabaena variabilis						+
Anacystis sp.	+		+			
Aphanocapsa pulchra						+
Chroococcus dispersus						+
Chroococcus turgidus						+
Coccochloris stagnina	+			+		
Cylindrospermum sp.			+			
Genicularia sp.		+				
Gloeocapsa sp.		+				
Lyngbya gracilis						+
Lyngbya heironymusii						+
Lyngbya limnetica						+
Lyngbya magnifica						+
Lyngbya sp.	+	+				
Merismopedia convulata						+
Merismopedia glauca						+
Merismopedia marsinii						+
Merismopedia minima						+
Merismopedia punctata						+

(continued)

(continued)

	MZ		PZ I		PZII	
	PB	PL	PB	PL	PB	PL
Merismopoedia sp.			+			
Microcystis aeruginosa			+			+
Microcystis flosaquae						+
Microcystis protocystis						+
Microcystis sp.		+	+			
Nostoc calcicola						+
Nostoc sp.	+	+	+	+		
Oscillatoria aghardhii						+
Oscillatoria amphibia						+
Oscillatoria formosa						+
Oscillatoria irrigua						+
Oscillatoria limnetica						+
Oscillatoria limosa	+		+	+		+
Oscillatoria planktonica						+
Oscillatoria princeps	+		+			+
Oscillatoria raciboraki						+
Oscillatoria sp.	+	+	+	+		
Oscillatoria subbrevis						+
Oscillatoria subsalsa						+
Oscillatoria tenuis			+			+
Phacus sp.			+			
Phacus accuminatus			+			
Phacus pusillus			+			
Phormidium calcicola						+
Phormidium inundatum						+
Phormidium mucicola						+
Phormidium sp.	+	+	+	+		
Raphidiopsis curvata						+
Raphidiopsis indica						+
Raphidiopsis mediterranea						+
Rivularia sp.	+	+				
Spirulina laxissima						+
Spirulina maior						+
Spirulina princeps			+			+
Spirulina sp.	+	+	+	+		
Spirulina subsala						+
Spirulina subtillisma						+
Teterpedia sp.			+			
Trachelomonas granulata			+			
Trachylomonas planktonica v. *oblonga*			+			
Tribonema bombycinum	+					
Euglenophyceae						
Euglena acus			+			+
Euglena gracilis						+
Euglena proxima						+

(continued)

(continued)

	MZ		PZ I		PZII	
	PB	PL	PB	PL	PB	PL
Euglena sp.		+	+			
Euglena viridis			+			+
Dinophyceae						
Ceratium digitatum						+
Ceratium variegatum						+
Ceratium falcatum						+
Peridiriium brevipes						+
Peridiriium brochii						+
Peridiriium cinatum						+
Gymnodinicum album						+
Gymnodinium variabile						+
Chrysophyceae						
Botryococcus sp.						+
Dinophyceae sp.			+			
Ceratium sp.			+			
Xanthophyceae						
Tribonema bombycinum			+			

Sources: Nautiyal (1996, 1997), Nautiyal and Nautiyal (1999), Nautiyal et al. (2004), Unpublished list generated by the authors for the interim reports under the Living Ganga Program, WWF, India

Appendix II

Distribution of benthic macroinvertebrate fauna in the mountain (MZ) and the Plains (PZ I, II)

Category	Taxon	Family/class	M	PSI	PSII
Order	Ephemeroptera	Ephemeridae	+		
Order	Ephemeroptera	Leptophlebiidae	+		
Order	Trichoptera	Psychomiidae	+		
Order	Trichoptera	Brachycentridae	+		
Order	Diptera	Blepharoceridae	+		
Order	Diptera	Psychodidae	+		
Order	Odonata	Argionidae	+		
Order	Plecoptera	Peltoperlidae	+		
Order	Plecoptera	Isoperllidae	+		
Class-	Arachnida	Water mites	+		
Phylum-	Nematoda	Worms	+		
Order	Trichoptera	Hydrophiliae	+	+	
Order	Trichoptera	Limnephilidae	+	+	
Order	Trichoptera	Leptoceridae	+	+	
Order	Trichoptera	Glossosomatidae	+	+	
Order	Diptera	Tipulide	+	+	
Order	Diptera	Simuliidae	+	+	
Order	Diptera	Limoniidae	+	+	
Order	Diptera	Athericidae	+	+	

(continued)

(continued)

Category	Taxon	Family/class	M	PSI	PSII
Order	Diptera	Dixidae	+	+	
Order	Plecoptera	Chloroperlidae	+	+	
Order	Plecoptera	Perlodidae	+	+	
Order	Ephemeroptera	Siphlonuridae		+	
Order	Trichoptera	Polycentropodidae		+	
Order	Trichoptera	Helicopsychidae		+	
Order	Coleoptera	Elmidae		+	
Order	Coleoptera	Dytiscidae		+	
Order	Coleoptera	Amphizoidae		+	
Order	Coleoptera	Dytiscidae		+	
Order	Coleoptera	Gyrinidae		+	
Order	Coleoptera	Chrysomelidae		+	
Order	Coleoptera	Noteridae		+	
Order	Coleoptera	Psephenidae		+	
Order	Odonata	Corduliidae		+	
Order	Odonata	Aeshnidae		+	
Order	Odonata	Lestidae		+	
Order	Odonata	Macromiidae		+	
Order	Hemiptera	Belostomatidae		+	
Order	Hemiptera	Notonectidae		+	
Order	Hemiptera	Nepidae		+	
Order	Hemiptera	Corixidae		+	
Class	Gastropoda	Pleuroceridae		+	
Class	Gastropoda	Planorbidae		+	
Class	Gastropoda	Lymnaeidae		+	
Class	Pelyceopoda	Bythitidae		+	
Order	Diptera	Muscidae		+	+
Order	Diptera	Syrphidae		+	+
Order	Diptera	Culicidae		+	+
Order	Plecoptera	Perlidae		+	+
Order	Hymenoptera	Agaonidae		+	+
Subphylum	Crustacea	Ostracoda		+	+
Class	Oligochaeta	Glossoscolecidae		+	+
Class	Polychaeta	Nephthydae		+	+
Class	Hirudinea	Leech		+	+
Class	Gastropoda	Viviparidae		+	+
Class	Pelyceopoda	Corbiculidae		+	+
Order	Ephemeroptera	Baetidae	+	+	+
Order	Ephemeroptera	Caenidae	+	+	+
Order	Ephemeroptera	Ephemerellidae	+	+	+
Order	Ephemeroptera	Heptageniidae	+	+	+
Order	Trichoptera	Hydropsychidae	+	+	+
Order	Diptera	Chironomidae	+	+	+
Order	Diptera	Heleidae	+	+	+
Order	Diptera	Tabanidae	+	+	+
Order	Odonata	Gomphidae	+	+	+
Class	Gastropoda	Thiaridae	+	+	+

References

1. Mathur RP (1991 a–b) a) Stretches of the Ganga covered in the study; b) Trends of physico-chemical characteristics of the Ganga water. In: Krishnamoorti CR, Bilgrami KS, Das TM, Mathur RP (eds) The Ganga – a scientific study. Northern Book Centre, New Delhi, for the Ganga Project Directorate, New Delhi, pp 19–20, 27–38
2. Rodgers WA, Panwar HS, Mathur VB (2002) Wildlife protected area network in India: a review (executive summary). Wildlife Institute of India, Dehradun
3. Krishnamoorti CR, Bilgrami KS, Das TM, Mathur RP (eds) (1991) The Ganga – a scientific study, Northern Book Centre, New Delhi, for the Ganga Project Directorate, New Delhi, p 335
4. Singh HR, Nautiyal P, Dobriyal AK, Pokhriyal RC, Negi M, Baduni V, Nautiyal R, Agrawal NK, Gautam A (1994) Water quality of the river Ganga (Garhwal Himalayas). Acta Hydrobiol 36:3–15
5. Nautiyal P (2010) Food chains of Ganga River ecosystems in the Himalayas. Aquat Ecosyst Health Manag 13(4):362–373. doi:10.1080/14634988.2010.528998
6. Joshi BD (2005) Studies on the ecobiology of tributaries of river Ganga between Devprayag and Rishikesh. In: Nautiyal P, Bhatt JP, Gusain OP, Dobriyal AK (eds) Biological diversity in freshwater environments. Transmedia, Srinagar-Garhwal, pp 19–35
7. Khanna DR, Bhutiani R (2003) Limnological characteristics of the river Ganga at Hardwar (Uttaranchal). Uttar Pradesh J Zool 23(3):179–183
8. Nautiyal P (1996) The river valley projects in Garhwal region impact on the population dynamics of endangered Himalayan Mahseer with emphasis on bioconservation. Final technical report project no. J-1103/8/92-GPD, submitted to Ganga Project Directorate, Ministry of Environment and Forests Government of India, New Delhi, pp 1–93
9. Nautiyal P (1997) Migratory phenomenon of the endangered Himalayan mahseer *Tor putitora* in relation to the ecology of the river Ganga. Final technical report of research project no. 14/28/92, Ministry of Environment Forests, New Delhi, pp 1–85
10. Nautiyal R, Nautiyal P (1999) Altitudinal variations in the pennate diatom flora of the Alaknanda-Ganga river system in the Himalayan stretch of Garhwal region. In: Mayama S, Idei M, Koizumi I (eds) Proceedings of fourteenth international diatom symposium, Koeltz Scientific Books, Koenigstein, Tokyo, 2–8 September 1996, pp 85–100
11. Nautiyal P, Nautiyal R, Kala K, Verma J (2004) Taxonomic richness in the diatom flora of Himalayan streams (Garhwal, India). Diatom 20:123–132
12. Nautiyal P, Shivam A, Verma J, Semwal VP (2007) Bhagirathi river – an endangered ecosystem. In: Venkatramani B, Puranik VD, Apte SK, Gour HN, Sharma SK, Sharma LL, Durve YS, Gupta HCL, Verma PC, Sharma BK (eds) Proceedings of DAE-BRNS national symposium on limnology, Paliwal Printers, Udaipur, pp 164–166
13. Sinha M, Laal AK, Singh B, Tyagi RK, Pathak V, Katiha PK (1997) Likely impact on aquatic ecology in the context of barrage construction across the River Ganga at Kanpur. CIFRI report, Barrackpore, West Bengal
14. Verma J, Nautiyal P (2011) Lotic diatom diversity patterns in mountain chains (North India). Lambert Academic Publishing, Amazon Distribution GmbH, Leipzig, pp 274
15. Nautiyal R (2005) Altitudinal variations in the community structure of benthic diatoms in a mountain river the Alaknanda. In: Nautiyal P, Bhatt JP, Gusain OP, Dobriyal AK (eds) Biological diversity of freshwater environments as proceedings of the national symposium on aquatic biodiversity and emerging trend in freshwater biology, Transmedia, Srinagar-Garhwal, pp 224–241
16. Anonymous (1975–1985, 1985–1995, 1995–2005, 1975–2005, 1975–2006, 2000–2006, 2001–2002, 2001–2003, 2001–2006, 2002–2003). Annual reports. CIFRI, Barrackpore, West Bengal
17. Gopal K, Agarwal AK (2003) Studies on the distribution and abundance of zooplankton in Ganga River. In: Gopal K, Agarwal AK (eds) River pollution in India and its management. APH Publishing Corporation, New Delhi, pp 147–167

18. Khan AA, Alam A, Gaur RK (1998) Seasonal variation in the abundance and composition of phytoplankton in the river Ganga at Narora, U.P. J Inland Fish Soc India 30(1):79–86
19. Khanna DR, Badola SP, Dobriyal AK (1993) Plankton ecology of the river Ganga at Chandighat, Hardwar. In: Singh HR (ed) Advances in limnology. Narendra Publishing House, Delhi, pp 171–174
20. Zafar A, Sultana N (2005) Zooplankton and micro invertebrates of River Ganga at Kanpur. Uttar Pradesh J Zool 25(1):63–66
21. Bahuguna BK, Nautiyal R, Nautiyal P, Singh HR (2004) Stream regulation: variations in density, composition and diversity of benthic macroinvertebrates occurring in the up and downstream zone of the impounded zone of the river Ganga in the foothills. Trop Ecol 45(2):251–261
22. De DK (1999) Environment and fisheries. In: Sinha M, Khan MA, Jha BC (eds) Ecology fisheries and stock assessment of Indian rivers, CICFRI bulletin no 90. CICFRI, Barrackpore, pp 129–140
23. Jhingran AG, Pathak V (1988) Impact on man-induced environmental modifications on productivity potential and energy dynamics of river Ganga. J Inland Fish Soc India 20:43–53
24. Mishra AS, Nautiyal P (2011) Distribution of benthic macroinvertebrates in Central Indian rivers. Lambert Academic Publishing, Amazon Distribution GmbH, Leipzig, p 202
25. Nautiyal P (2005) Spatial and microdistribution of benthic macroinvertebrate community in tropical mountain streams of Garhwal region. Final Technical Report submitted to Department of Science and Technology, New Delhi (Ref No. SP/SO/C-42/98 dated 22 July 1999 [18.01.2002 to 17.01.2005]
26. Nautiyal R, Nautiyal P, Singh HR (2000) Species diversity of epilithic diatom community. Trop Ecol 41(2):255–258
27. Negi KS, Malik DS (2005) Fish fauna of Ganga river at Rishikesh. Himal J Environ Zool 19(2):145–148
28. Pushpa (2005) Studies on ichthyofauna of Haridwar district (Uttaranchal) with notes on systematic, ecology and zoogeography. D.Phil thesis, HNB Garhwal University, Srinagar
29. Rao RJ (2001) Studies on biological restoration of Ganga River in Uttar Pradesh: an indicator species approach. Hydrobiologia 458:159–168
30. Sahgal JK (1973) Fish and fisheries of Aligarh with notes on the fish catches. J Inland Fish Soc India 5:23–30
31. Singh HR, Badola SP, Dobriyal AK (1987) Geographical distributional list of ichthyofauna of the Garhwal Himalaya with some new records. J Bombay Nat Hist Soc 84:126–132
32. Singh HR, Nautiyal P (1990) Altitudinal changes and the impact of municipal sewage of the community structure of macrobenthic insects in the torrential reaches of the River Ganges in the Garhwal Himalaya (India). Acta Hydrobiol 32(3–4):407–421
33. Singh HR (2008) Ecological parameters of the Bhagirathi. Final technical report, submitted to National Thermal Power Corporation (NTPC), India, pp 1–78
34. Seaby RMH, Henderson PA (2007) Community analysis package v. 4. Pisces Conservation Ltd, Lymington, UK
35. Jhingran VG (1975) Fish and fisheries of India, 1st edn. Revised and enlarged 3rd edn. 1991. Hindustan Publishing Co., New Delhi, p 727
36. Hamilton F (1822) An account of the fishes found in the river Ganges and its branches. Archibald Constable and Company, Edinburgh, London, I-VII+1-405+1–39 pls
37. Menon AGK (1962) A distributional list of fishes of the Himalayas. J Zool Soc India 14(1 and 2):23–32
38. Menon AGK (1974) A check-list of fishes of the Himalayan and the Indo-Gangetic plains. J Inland Fish Soc India, Special publication, 1:1–136
39. Misra KS (1962) An aid to identification of commercial fishes of India and Pakistan. Rec Indian Mus 57:1–320
40. Venkateswarlu T, Menon AGK (1979) A list of fishes of the river Ganges and its branches. Acta Ichthyologica et Piscatoria 9:45–68
41. Talwar PK, Jhingran A (1991) Inland fishes of India and adjacent countries, vol I-II. Oxford and IBH Publishing Co Pvt. Ltd , New Delhi, pp 1158

42. Jayaram KC (1981) In: Director Zoological Survey of India (ed) The freshwater fishes of the India, Pakistan, Bangladesh, Burma and Sri Lanka. Zoological Survey of India, Kolkata, xxii+475
43. Jayaram KC (1999) The freshwater fishes of the Indian region. Narendra Publishing House, New Delhi, xxvii+551, Pls. xviii
44. Payne AI, Temple SA (1996) River and floodplain fisheries in the Ganges Basin. Final report. DFID Fisheries Management Science Programme R 5485, London, MRAG Ltd, pp 177
45. Payne AI, Sinha R, Singh HR, Huq S (2003) A review of the Ganges Basin: its fish and fisheries. In: Welcomme RL, Petr T (eds) Proceedings of the second international symposium of management of large rivers for fisheries, vol 1. FAO & The Mekong River Commission, 2004, RAP Publication 2004/16, 229–251, Bangkok.
46. Lakra WS, Sarkar UK, Kumar RS, Pandey A, Dubey VK, Gusain OP (2010) Fish diversity, habitat ecology and their conservation and management issues of a tropical River in Ganga basin, India. Environmentalist 30:306–319. doi:10.1007/s10669-010-9277-6
47. Bilgrami KS (1991) Biological profile of the Ganga: zooplankton, fish, birds and other minor fauna. In: Krishnamurti CR, Bilgrami KS, Das TM, Mathur RP (eds) The Ganga scientific study. Northern Book Centre, New Delhi, pp 81–94
48. Bilgrami KS, Datta Munshi JS, (1985) Ecology of River Ganges (Patna-Farakka): impact of human activities and conservation of aquatic biota. Bhagalpur University, Bhagalpur, pp 1–97
49. O'Keeffe J, Kaushal N, Bharti L, Smathkin V (2012) Assessment of Environmental Flows for the Upper Ganga Basin. WWF – India, pp 162
50. Vannote RL, Minshall GW, Cummins KW, Sedell JR, Cushing CE (1980) The river continuum concept. Can J Fish Aquat Sci 37:130–137
51. Sinha M, De DK, Jha BC (1998) The Ganga-environment and fishery. CIFRI, Barrackpore, pp 142
52. Vass KK, Mondal SK, Samanta S, Suresh VR, Katiha PK (2010) The environment and fishery status of the River Ganges. Aquat Ecosyst Health Manag 13:385–394
53. Shukla AC, Asthana V (1995) Ganga, a water marvel. APH Publishing, New Delhi, p 315
54. Khanna DR, Badola SP (1994) Foot hill zone of river Ganga – habitat of adult Mahseer. In: Nautiyal P (ed) Mahseer the game fish. Jagdmba Prakashan, Dehradun, pp 98–101
55. Sharma MK, Rajput DB (1986) Ichthyofauna of Bijnor district (Uttar Pradesh). J Bombay Nat Hist Soc 83:562–569
56. WWF-India (2004) Information sheet on Ramsar Wetlands (RIS), pp 1–14. http://jncc.defra.gov.uk/pdf/RIS/UK61002.pdf
57. Badoni K, Nautiyal R, Bhatt JP, Kishor B, Nautiyal P (1997) Variations in epilithic diatom community structure due to river valley projects on the Ganga between Rishikesh and Hardwar. Proc Indian Natl Sci Acad B 63:527–540
58. Rishi V, Awasthi AK (2012) Pollution indicator algae of river Ganga at Kanpur. Gold Res Thought 1(7):1–4

Fishing activity (Courtesy Sandeep Behera)

Chapter 4
Indicator Species (Gharial and Dolphin) of Riverine Ecosystem: An Exploratory of River Ganga

Sandeep Kumar Behera, Hari Singh, and Viveksheel Sagar

Abstract The Ganga River sustains diverse group of flora and fauna including the endangered Ganges river dolphin (*Platanista gangetica gangetica*) and critically endangered gharial (*Gavialis gangeticus*). Because, of the highest productivity and most populous regions on earth there is a strong demand and competition for natural resources, which threatened the survival of indicator species such as gharial and dolphin in river Ganga. There are several riverine indicator species which are threatened by human activities in the Ganga basin. However, the Ganges river dolphin and gharial are reliable indicator species to understand the health of the Ganga river ecosystem. Studying indicator species could create the basis for a sustained research programme to see how the changes of the said species can be related to the health of the river system. This would help to implement various programme for restoration of the river system. Human perturbations and anthropogenic disturbances have led to drastic declines in dolphin and gharial populations over much of their distribution ranges during the last several decades.

However, due to active involvement of WWF-India in dolphin conservation in certain areas the recent survey reported interesting trends in terms of a stable Dolphin population. Stretches that reported stable Dolphin population were areas where there had been some interventions taken by local communities, forest departments, and other non government organizations.

Keywords Gharial • Dolphin • Biodiversity • Ecosystem • Species

---

S.K. Behera (✉) • H. Singh • V. Sagar
River Basins and Biodiversity, WWF-India, Secretariat, 172-Lodhi Estate, New Delhi 110003, India
e-mail: sbehera@wwfindia.net; hsingh@wwfindia.net; vsagar@wwfindia.net

R. Sanghi (ed.), *Our National River Ganga: Lifeline of Millions*,
DOI 10.1007/978-3-319-00530-0_4, © Springer International Publishing Switzerland 2014

Dolphin 3: the national aquatic animal Dolphin (Courtesy Sandeep Behera)

4.1 Introduction

The Ganga river basin is one of the most populous regions on earth, home to 450 million people at an average density of over 550 individuals per square kilometer. As a result, there is a strong demand and competition for natural resources, especially water for domestic use and irrigation, and most of the basin tributaries are regulated by barrages. Ganga river system comprise of numerous torrents and tributaries. Most of the north India tributaries like Ram Ganga River, Sarda, Gomti, Ghagra, Gandok and Kosi arise from the lower Himalayas passes through the Terai region before joining the Ganga. The major southern tributaries are the Chambal, the Yamuna, the Son and the Subarnarekha rivers [1]. The land around the river is used largely for agriculture and fisheries. The fishery resources of the Ganga river system are of tremendous economic and nutritional sustenance to the people of riparian states. The system alone contributes as much as 89.5 % to the countries total carp seed production [1, 2]. This has lead people to live around the river and exploit natural resources for their livelihood. This exploitation is so great that some species in the river are likely to become extinct in the near future. The discharges from different factories along the Ganga River have affected water constituents to a great extent. By far scientists and experts have identified reduced river flows as one of the primary threats to the populations of dolphins, Mahseer, crocodiles, turtles and fishes, in this stretch [1]. The Ganges River has been constantly used for dumping of garbage, sewage disposals, tanneries disposal etc. An increased level of water quality deterioration has been observed year by year. The habitat loss has caused concern for the welfare of the aquatic animals that live in different areas along the Ganges River [3]. The living resources in the Ganga are threatened due to alarming pollution level, developmental activities, unsustainable collection and poaching. Over the past three decades or so, the pollution of the Ganga has reached serious levels leading to local and national campaigns and actions to restore the health of the sacred river [4].

4.2 Freshwater Diversity of the Ganga River

The Ganga River sustains diverse group of flora and fauna. Record of 268 fishes from the Ganges was the first-ever scientific documentation of the fauna of the river [5]. Jayaram [6] gave an account on distribution of freshwater fishes, amphibians and reptiles of the Ganges. Zoological Survey of India has documented 27 species of reptiles from the river [7]. Gharial (*Gavialis gangeticus*), a fish eating crocodile, *Aspideretes gangeticus,* a soft shell turtle, are some endemic reptiles of the river along with one species of monitor lizard (*Varanus bengalensis*). One of the most rare, endemic and endangered mammals of the Ganga is the Ganga river dolphin, *Platanista gangetica gangetica.* Beside the Ganga River dolphin a total of 40 species of zooplanktons, four crustaceans, 15 molluscs, 51 insects, 83 fishes,

12 freshwater turtles, 2 crocodiles and 48 aquatic birds species have been identified in the upper Ganges River [8]. These species have been heavily exploited in the last few decades, which have pushed them near to extinction.

4.3 An Overview of Gharial (*Gavialis gangeticus*)

The Indian Gharial (*Gavialis gangeticus*), is the oldest crocodilian and the sole survivor of family Gavialidae. The species is surviving from the great reptilian age and are recognized as Keystone species in their environment due to the role they play in maintaining the ecosystem and function by their activities. Gharial, a mythical creature, is revered as the vehicle (Vahana) of Ganga (River Deity) and Varuna (God of winds). The early records reveal that these aquatic reptiles at one time were very abundant throughout their distribution range. The late 1970s saw a drastic decline in the Gharial population and distribution due to commercial exploitation and habitat destruction populations of gharials was reduced to near extinction. In many of the habitats the populations of gharial was totally wiped out. Considering their vulnerability, the Government of India enforced protective legislation through the Indian Wildlife (Protection) Act, 1972 which prohibits killing [9]. Gharial was listed as 'Endangered' in the IUCN Red Data Book of 1975. The 2007 IUCN Red Data list of threatened species places the Indian Gharial in the 'Critically Endangered' Category [10].

4.3.1 Status and Distribution

The Gharials (*Gavialis gangeticus*, Gmelin, 1789), endemic to the Indian subcontinent, was once common in the river systems of Pakistan, Northern India, Bangladesh, Myanmar, Bhutan. However, they are now restricted to a few, scattered locations in India and Nepal (Fig. 4.1). In India many rivers, lakes and marshes offer a variety of habitats for three species of crocodiles the gharial (*Gavialis gangeticus*), mugger or marsh crocodile (*Crocodylus palustris*), and the saltwater or estuarine crocodile (*Crocodylus porosus*). Of these Gharials are present in the Ganga River and its tributaries particularly in northern India. The Ganga River is a major habitat for both the species of freshwater crocodile (Gharial and Mugger). Old records indicate that the crocodile abounded in all the great rivers of northern India including the Ganga River [11].

By 1976, the estimated adult population of wild Gharial had declined from what is thought to have been 5,000–10,000 in the 1940s to less than 200. In 2006, the mature gharial population in India stands at a similar figure, less than 200 [12]. In 1975, the Crocodile Conservation Project was initiated by the Government of India/UNDP, FAO in Uttar Pradesh, and Madhya Pradesh for the conservation of the Gharial. Under the crocodile project few important gharial habitats were identified in India and protected by declaring it as crocodile sanctuaries. In these sanctuaries

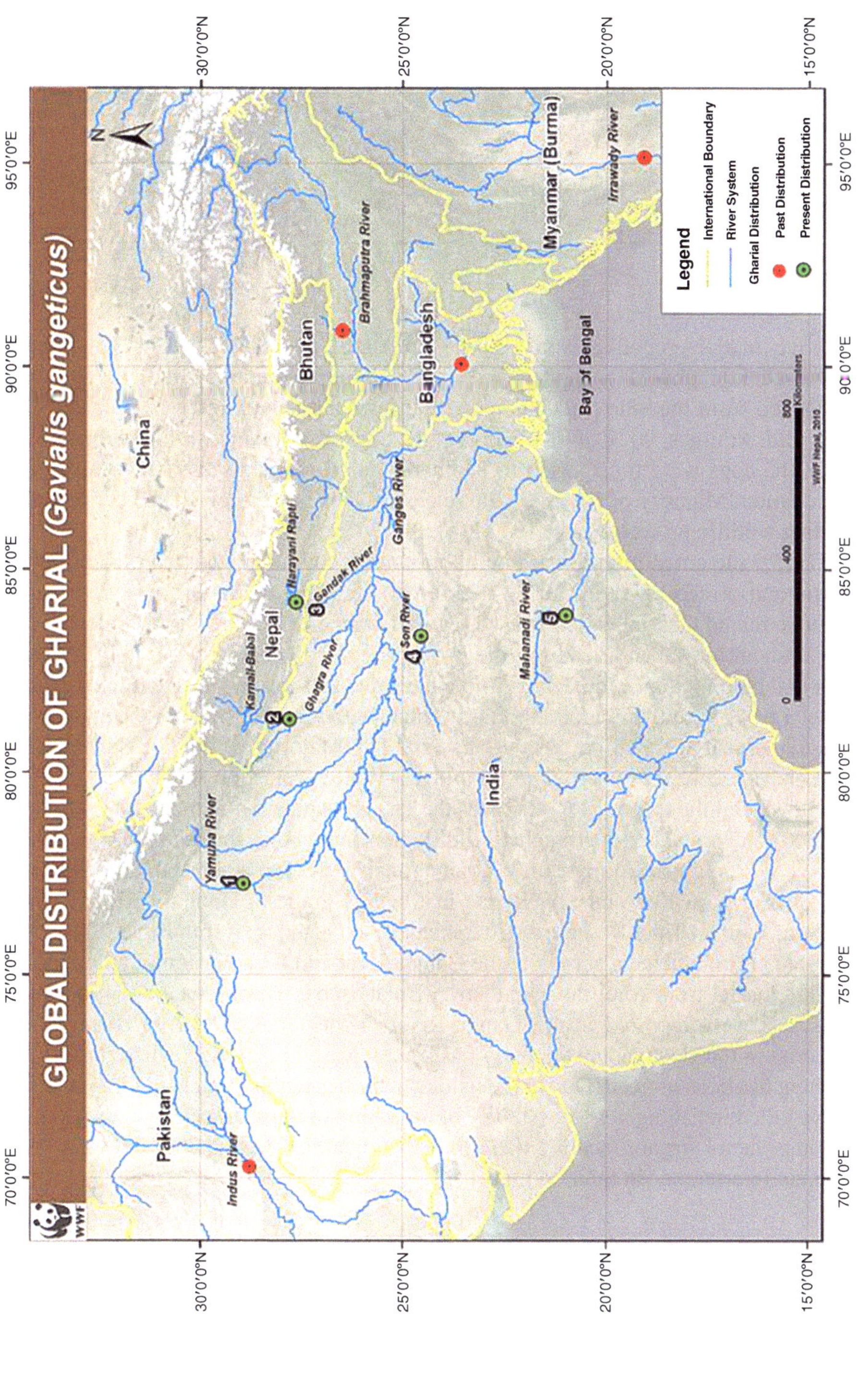

Fig. 4.1 Current distribution of the Gharial (Source: IUCN)

Table 4.1 Recent declines in the number of adult Gharial by subpopulation

Subpopulation	Past	Present	Estimated % reduction within one generation
Chambal	226 (1997)	78 (2006) [68 f +10 m]	13
Katerniaghat	30 (1997)	26 (2006) [20 f+6 m]	66
Others (India)	50 (1997)	40 (2006)	20
Chitawan (Nepal)	20 (1999)	8 (2006)	80
Others (Nepal)	100 (1994)	27 (2006)	73
Overall	**426**	**179**	**58**

Source: IUCN
f females, *m* males

captive reared crocodile were released regularly since 1977 [13]. The project involved two phases – captive breeding and rehabilitation. Gharial eggs were collected from the wild, incubated and the resultant hatchlings were released into the wild. At times, 90 % of incubation of the eggs was done *in situ* in wild nests and then the eggs were transported to hatcheries at rehabilitation centre for complete incubation. Majority of the crocodile releasing sites have received protection under Indian wildlife protection Act 1972 [14].

Under the crocodile project few important crocodile habitats were identified in India and protected by declaring it as crocodile sanctuaries. In these sanctuaries captive reared Gharial were released regularly since 1977. The Uttar Pradesh forest department has taken a leading role in releasing captive reared Gharial in various rivers including the Ganga River. The Uttar Pradesh forest department had released a total of 225 captive reared Gharial in the Ganga River upstream of Bijnor in the Histnapur sanctuary in the year 1991–1992 [1]. Majority of the Gharial releasing sites have received protection under Indian wildlife protection Act 1972.

In general the Chambal River holds the largest population with an upper estimate of 306 adult animals. Katerniaghat Wildlife Sanctuary holds the second largest population with an upper estimate of 68 adult animals [15]. The other smaller populations of Gharial is in Ken and Son Rivers in Madhya Pradesh, Hooghly River in West Bengal, Corbett Tiger Reserve in Uttarakhand and Ghandak river in Bihar (Table 4.1).

Rao [1] conducted a survey in the Ganga River and found a significant record of adult Gharial from Anupsahar in district Bulandsahar. During October 1994, three Gharials were reported in the Ganga River downstream of Narora barrage. The Gharial in the Hastinapur sanctuary has been released in an area, where large scale fishing has been noticed. Due to the fishing activities in this stretch all Gharial might have killed in fishing nets. Possibility of migration of these released Gharial may be another factor for not locating them in the study area. These animals always avoid human interference in their habitats [1].

4.3.2 Species Biology

The name 'Gharial' originated from a 'ghara' or earthen pot on their head [16]. Gharial are arguably the most thoroughly aquatic of the extant crocodilians, and adults apparently do not have the ability to walk in a semi-upright stance as other crocodilians do. They are typically residents of flowing rivers with deep pools that have high sand banks and good fish stocks. Exposed sand banks are used for nesting [17]. Young gharials eat insects, larvae, and small frogs. Mature adults feed almost solely on fish, although some individuals have been known to scavenge dead animals. Their snout morphology is ideally suited for preying on fish. Their long, narrow snouts offer very little resistance to water in swiping motions to snap up fish in the water. Their numerous needle-like teeth are ideal for holding on to struggling, slippery fish. The mating season is during November through December and well into January. The nesting and laying of eggs takes place in the dry season of March, April, and May. This is because during the dry season the rivers shrink a bit and the sandy river banks are available for nesting. Between 30 and 50 eggs are deposited into the hole that the female digs up before it is covered over carefully. After about 90 days, the juveniles emerge, although there is no record of the female assisting the juveniles into the water after they hatch (probably because their jaws are not suited for carrying the young due to the needle like teeth). However, the mother does protect the young in the water for a few days until they learn to fend for themselves.

4.3.3 Major Threats to the Gharial Population

Many factors, individually or in combination, were responsible for the drastic decline of the Gharial population in the Indian subcontinent. The Gharial, is becoming increasingly rare due to land-use changes, reduction in water flow, lack of stringent laws, modification of river morphology, loss of nesting sites, increased mortality in fishing nets and egg collection for consumption and is especially at risk from flow regulation because it prefers fast flowing river habitats, which are prime sites for dams [17–19]. Superstitions are part of the South Asian tradition, but some of them harm wildlife. A common belief of a local tribe called 'Tharu' of Nepal associated with the Gharial is that the 'ghara' kept under the pillow of pregnant women relieved her of her labour pain [20]. Another belief is that incense sticks made of 'ghara' when burnt in fields acted as a pesticide. In addition, the 'Tharu' community believes that Gharial eggs have medicinal value. All these superstitions have led to their decline in Nepal [21].

4.4 An Overview of Ganges River Dolphin (*Platanista gangetica gangetica*)

River Dolphins swim in some of the world's mightiest rivers, including the Ganges, Indus, Yangtze and the Amazon. But these river basins are also home to over 15 % of the world's population and include some of the most densely populated and poorest areas on the earth. Human perturbations and anthropogenic disturbances have led to drastic declines in dolphin populations over much of their distribution ranges during the last several decades. Several Asian species are now amongst the most endangered of all mammals.

Of the six species of freshwater dolphins present worldwide, three species (*Lipotes vexillifer, Inia geoffrensis and Platanista gangetica*) are recognized as true freshwater dolphins, which inhabit rivers and estuaries in Asia and South America. *Lipotes vexillifer* (the Chinese river dolphin), was declared functionally extinct [22]. The remaining species are *Inia geoffrensis* (the Amazon River dolphin) and *Platanista gangetica* which has two sub species *Platanista gangetica gangetica* (the Ganges River dolphin) and *Platanista gangetica minor* (the Indus River dolphin). Out of the other three, two river dolphins and one river porpoise that are found in both marine and freshwater, the tucuxi, *Sotalia fluviates* in South America and the Irrawaddy dolphin (*Orcaella brevirostris*) and finless porpoise, *Neophocaena phocaenoides* in Asia. Although several marine dolphin species are commonly found in other rivers quite far upstream, the true freshwater dolphins are morphologically and phylogenetically distinct from the marine dolphins and are restricted to only the freshwater ecosystems.

The Indus and Ganges populations were long regarded as identical until Pilleri and Gihr [23] divided them into two species based on differences in skull structure, but Kasuya and Haque [24] reduced the two taxa to sub-species of a single species, supported by the results of Yang and Zhou [25], who found that the difference between cytochrome-b sequences of Ganges and Indus river dolphins was very small. Thus one species is recognized in the genus *Platanista* and currently the Ganges River dolphins are *Platanista gangetica gangetica* and the Indus River dolphins of Pakistan are *Platanista gangetica minor.*

The Indians of course knew about this splendid creature since time immemorial but it was officially and scientifically discovered by Roxburgh in 1801 [27]. The animal abounds in our ancient literature and was also mentioned as "*Khuk-abi*" in *Babarnama* during the *Mughal* period, when the rural folk used to bum dolphin oil for lighting their homes. However, it was John Anderson who first published a scientific report on the dolphin in 1879 after which there is no scientific information available on this species for the next 100 years [26].

However, under few research works under the Ganga Action Plans project in 1980s provided the baseline information about the species population and distribution. Then it was estimated that around 5,000–6,000 dolphins are present in its distribution range and are distributed in the Ganga-Brahmaputra-Meghna and Karnaphuli river systems and its tributaries of Nepal, India, Bhutan and Bangladesh between the foothills of the Himalayas and the tidal zones [28]. In India, you may sight these beautiful dolphins along deep river reaches in Assam, Bihar, Jharkhand, Madhya Pradesh, Rajasthan, Uttar Pradesh and West Bengal.

Still, there is no complete scientific estimate of range-wise abundance of the species, it is assumed that presently around 2,500 individuals are surviving across their entire range (Table 4.2), out of which around 1,800 are within the Indian territory [36]. Its population is declining fast. It has already become extinct from most of its earlier distribution ranges and even in its present day distribution ranges the density of this animal is getting thinner.

Table 4.2 Current status of Ganges river dolphins in Ganga River and its tributaries

Name of the river	Length of the river surveyed (km)	Dolphin number	Source
The Ganga Mainstem			
The Ganga (Haridwar to Bijnor Barrage)	100	Nil	Sinha et al. [4]
The Ganga (Bijnor Barrage to Narora Barrage)	165	56	S. Behera, WWF-India personal communication
The Ganga (Narora to Kanpur)	300	3	S. Behera, WWF-India, personal communication
Kanpur to Allahabad	272	108	
The Ganga (Allahabad to Buxar)	425	172 (d/s survey)	Sinha et al. [4]
The Ganga (Buxar to Maniharighat)	500	808 (u/s survey)	Sinha et al. [4]
Vikramshila dolphin Sanctuary	65	179 (u/s)	Kelkar et al. [29]
The Ganga (Maniharighat to Farakka)	100	24 (d/s survey)	Sinha, personal communication
The Farakka Feeder canal	38	21 (d/s survey)	Sinha et al. [4]
The Bhagirathi (Jangipur Barrage to Triveni)	320	119 (d/s survey)	Sinha et al. [4]
The Hooghli (Triveni to Ganga Sagar)	190	32 (d/s survey)	Sinha [30]
Tributaries of the Ganga			
The Yamuna (Chillla to Saraswati ghat near Allahabad Sangam)	183	48 (d/s survey)	S. Behera, WWF-India, personal communication
The Kosi (Kosi Barrage to Kursela)	200	85 (discrete survey)	Sinha and Sharma [31]
The Gandak (Gandak Barrage to confluence with Ganga at Patna)	332	257 (d/s survey)	Choudhary et al. [32]
The Girwa river	18	30 (d/s survey)	S. Behera, WWF-India, personal communication
Ghagra River	505	295 (d/survey)	S. Behera, WWF-India, personal communication
The Sarda (Sarda Barrage to Palya)	100	Nil	Sinha and Sharma [31]
The Chambal (Pali to Pachhnada)	425	86	Singh and Rao [33]
The Ken(from confluence of Yamuna at Chilla to Sindhan Kala village)	30	08 (d/s survey)	Sinha et al. [4]
The Kumari (from confluence of Sind River)	100	Nil	Sinha et al. [4]

(continued)

Table 4.2 (continued)

Name of the river	Length of the river surveyed (km)	Dolphin number	Source
The Betwa (from confluence of the Yamuna at Hamirpur to Orai)	84	06 (d/s survey)	Sinha et al. [4]
The Sind (from confluence with the Yamuna)	110	05 (d/s survey)	Sinha et al. [4]
The Son	130	10 (d/s survey)	Sinha et al. [4]
The Brahmaputra river	600	400 (1996)	Mohan et al. [34]
	856	179 (2004–2005)	Wakid [35]
The Barak river	17	6 (2006)	Paulan Singh, personal communication
The Subhansiri River	99	26	Wakid [35]
The River Kulsi	76	27	Wakid [35]

Source: Reeves et al. [39]

4.4.1 Conservation Status

This species has been included in Schedule I of the Indian Wildlife (Protection) Act 1972, in Appendix I of the Convention on International Trade in Endangered Species (CITES), in Appendix II of the Convention on Migratory Species (CMS) and categorized as Endangered on the International Union for the Conservation of Nature's (IUCN) Red List [37].

4.4.2 Species Biology

The Ganges River Dolphin, often called '*Susu*' or 'Soons' locally, differs from the well known marine species. It is blind as its eyes have no lens, its snout is long as it grovels in mud for food and its dorsal fin is very small. Interestingly, to compensate the loss of vision the animal has a very efficient echolocation. Such an adaptation is indeed unique since the poor creature has to live in muddy swirling waters all the time. It is a thrilling sight to see the dolphin leap out of the Ganga. It certainly looks like a huge fish although it is in fact a mammal. The dolphin respires through lungs and emerges above the water surface to inhale air through a blow-hole on its melon-like head. Inhaling and exhaling produces a typical 'soooooosssss', which is the basis for its vernacular names.

The female dolphin is larger than the male and can attain a maximum size of 2.67 m. The male is about 2.12 m long. At the time of birth the calf is about 60–70 cm long. A female of 2.5 m had a weight of 114 kg. A mature female has a long up curved snout while a male has a straight snout. The calves and young ones are light pink in colour but as the animal grows in size the colour fades out and the adult and old individuals become darker in colour. The female attains sexual maturity probably at an age of 10–12 years whereas the male at 8–10 years. The gestation period is about 9–11 months and a female gives birth to only one calf once every 2–3 years. Maximum breeding takes place during the dry season i.e. between January to June. The calf feeds on mother's milk for about 2–3 months and then starts feeding on soft food like insect larvae and small fishes.

4.4.3 Dolphin Habitat

Although the Ganges dolphin is fluviatile in habit, it may also be found in brackish water, though it never enters the sea. It is generally assumed that salinity defines the downstream limits of its distribution, while physical barriers and low prey densities at high elevations define the upstream limits. The dolphins prefer to stay in all types of habitats which include deep zones of the rivers, around the confluence of two or more rivers, at the meandering and below sand bars where two channels reunite and create eddy-counter current system. Normally they chase surface dweller fishes and grovel mud dwellers in shallow water. The dolphins were also seen negotiating high rapids in River Karnali at the foothill of the Himalayas in Nepal. Seasonal habits and dispersal of the river dolphin are not well studied but there is some evidence of seasonal changes in distribution, with animals dispersing upstream as the water level rises. During the dry season many dolphins leave the tributaries and congregate in the main channels only to return to the tributaries the following rainy season.

Water levels in the Ganga are seasonal in nature. Peak flows occur between July and September when the river is fed by monsoon run-off and Himalayan melt-water, while leanest flow occurs from February to March. The flow in the Ganga and its tributaries is regulated, and the natural flow regime has been disrupted, by the construction of dams and barrages. While there were occasional reports of dolphin sightings in some of the smaller tributaries of the Ganga during the 1980s, it is now likely that dolphins have been completely extirpated from these rivers due to insufficient and inconsistent water supplies. The large-scale diversion of river water for irrigation in the dry season causes water flow to diminish especially between Haridwar and Allahabad (Fig. 4.2).

Being a mammal, the Ganges River dolphin can survive a wide range of temperature fluctuations. It can tolerate temperatures as low as 5 °C in the River Karnali in the winter in Nepal, and as high as 35 °C in the summer in the Gangetic plains. They have also been found in highly turbid water in monsoon and it is thus assumed that the water temperature and turbidity are not significant factors in determining the distribution of this species.

4.4.4 Threats to Dolphin Population

The complex geomorphology of freshwater and estuarine systems tends to concentrate the distribution of cetaceans in counter currents associated with confluences, meanders and mid-channel islands. The Ganges River dolphin population was fragmented by construction of the dams and barrages on the main stem of the Ganga and its tributaries.

Barrages are low, gated diversion dams comprised of a series of gates used to control the elevation of an upstream 'head pond'. Barrages also restrict the movement of river dolphins and other aquatic mega-fauna, thereby separating them into sub-populations.

Dolphins in the main channel of the Ganga were split into two sub-population in 1975 when the Farakka Barrage was commissioned. The Lower Ganga Barrage at Narora (1966) and the Middle Ganga Barrage at Bijnor (1984) further fragmented the Ganga main stem population into four subpopulations. Dolphins have now been extirpated above the Middle Ganga Barrage at Bijnor. Today they occur in three subpopulations bounded by the Bijnor, Narora and Farakka Barrages.

A few dolphins occasionally enter the irrigation canals and small tributaries during flood. Rarely can any such individuals successfully return to the main channel of the river; most of them get stranded and are subsequently killed, either by the locals or because of the water-regulating gates in these canals and rivulets.

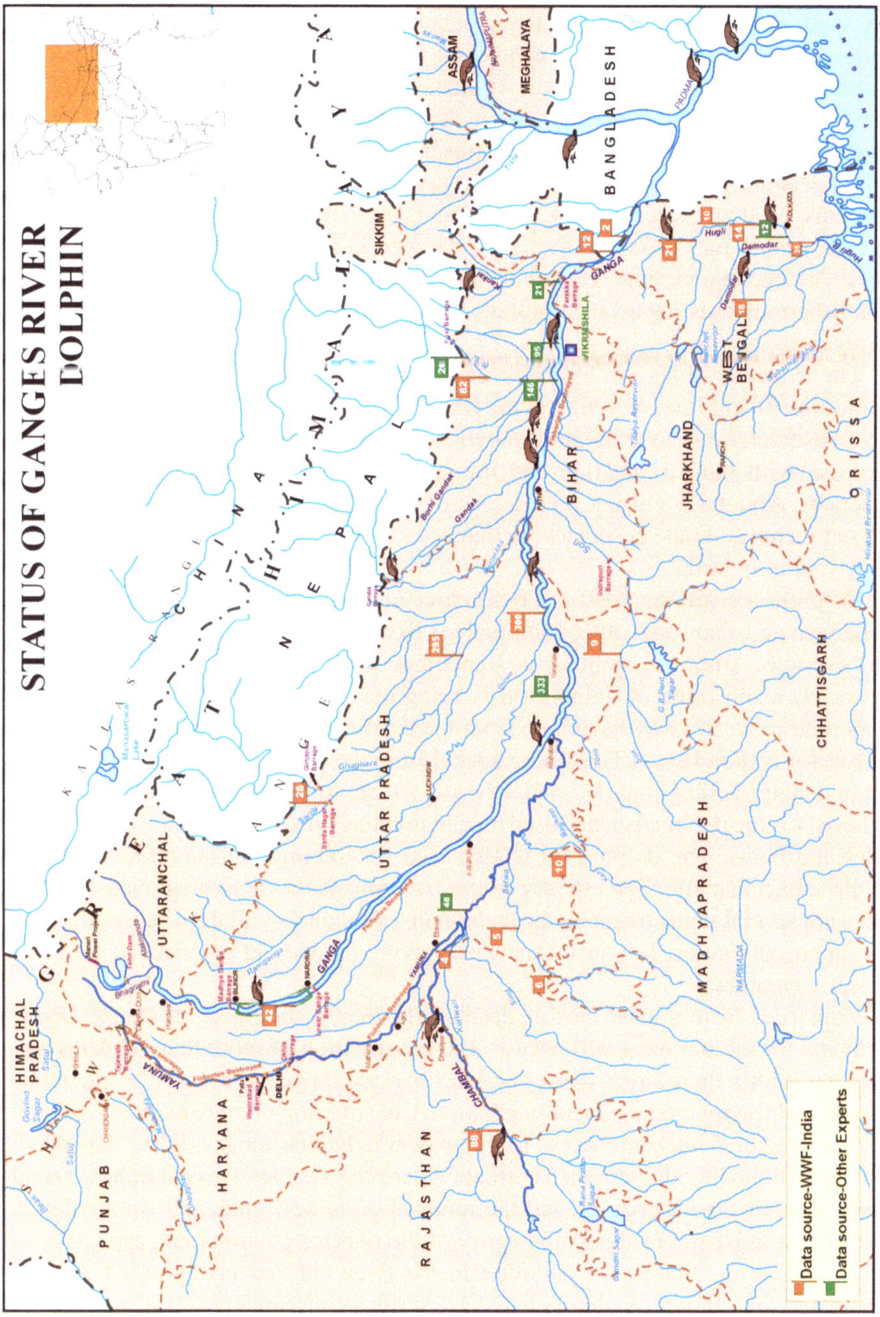

Fig. 4.2 Map of Ganges river basin depicting dolphin habitat

Ganges River Dolphin used to be and still is hunted commonly. Although hunting of dolphins is at present mainly limited to Bihar it is an enormous threat to the small surviving population. These nets are made of very fine nylon and many a times, the poor dolphins fail to echolocate the nets. While trying to get the fish the dolphins get entangled and drown. At times the dolphin also gets entangled in the line hooks laid on the river. The use of 'Kapdajal' (mosquito net cloth) is playing havoc not only with the riverine ecosystems but competing directly with the dolphins. These nets are used to catch-fingerlings, juvenile and other small fishes, which is dolphins' food too. Occasionally, the dolphins try to prey upon the fishes trapped in the nets and in turn drown as they get entangled in the net. It is usually the young ones who become victims. The adults are mostly able to tear the net and get away. The fishermen thus regard the dolphins as a nuisance and sometimes kill them out of revenge.

The concentrations of fertilizer and pesticide residues and industrial and domestic effluents are very high in the Ganges River. The effects of such pollutants may be deleterious to dolphin populations, and pollutant loads are expected to increase with industrialization and the spread of intensive modern agricultural practices. About 1.5 million metric tons of chemical fertilizers and about 21,000 tons of technical grade pesticides are dumped annually to the Ganges-Brahamaputra river system in India.

Dolphins are still regarded as a rich source of protein by poorest among the poor fishermen and also some other communities like Dheemars of the Chambal ravines. Oil extracted from the dolphin is sometimes used by fishermen to allure fish. However, a substitute of dolphin oil is being encouraged in riverine fishery. It is important to wean away fishermen from the existing practice of use of dolphin oil as bait for certain fishes. This oil, extracted from the dolphins blubber, is also used for massaging aching muscles, and serves as a tonic for general disability. The oil is believed to cure rheumatism and used for ameliorating bone fracture, cure old wounds, piles and bums. The dolphin oil is also used for treating Galsua (*Haemorrhagic septicaemia*) in cattle. The efficacy of these treatments is yet to be proved. Besides there are several mainstream medical alternatives for almost all the diseases that this oil supposedly cures. It is just a matter of giving up age-old practices to save this graceful creature.

Numerous factors exist for this decline; the deliberate killing of Ganges River dolphins for oil and meat still occurs. The construction of more than 50 dams and barrages within the Ganges River dolphin's historic range has drastically altered its habitat and fragmented the meta-population. Construction of embankments as flood control measures interrupt access to the spawning habitat for floodplain dependent fish and eliminate eddy-counter currents where the Ganges River dolphins spend much of their time. Dredging and the removal of stones, sand, and woody debris also compromise the ecological integrity of the riverine environment, especially in small tributaries. Increasing pollution in the river may adversely affect dolphin health and their bioaccumulation may have serious consequences.

No single strategy will facilitate recovery of depleted populations, reverse trends of population decline and habitat deterioration, and ensure that robust populations with high-quality habitat are secure. Approaches to conservation need to be multi-faceted, adaptable, and often tailored to particular local or regional conditions. The solutions must address the problems of unsustainable exploitation and habitat loss/degradation. In addition, some cross-cutting initiatives related to capacity building and governance is vital to achieve effective conservation.

4.4.5 *Major Gap Area*

Few are the major gap area in the knowledge, potential, utilization and conservation of Ganges river dolphin and Gharial:

- A lot of researches have been done on different aspects of the species in general but the information are discrete. Lack of proper documentation and compilation of information on the species as a ready reference is one of the biggest gap areas.
- Flood plains are an integral part of the river ecosystem, specially the food of the dolphins and gharial and also have direct influence on different components of the river, but these have been neglected in the policies for conservation, pollution abatement, and water quality improvement of the river. Even researches on this aspect provide scanty information.
- Lack of proper co-ordination among technocrats, ecological scientists, policy makers at local and national level is also a major gap area which restricts the information flow from one group to another ultimately affecting the formulation of proper plans of local interest.
- For making decisions over issues of resource utilization and conservation, the local community were neglected, especially the fisher folk who thrive on the resources of the Ganga, which often results into local conflicts.
- For proper conservation of freshwater resources in the Ganges River an optimum exploitation of the resources associated with the River Ganga demands a multi-pronged approach involving different government, non-government organizations, industrial houses, civic bodies, professional organization, academicians, researchers, co-operative societies and local people.

4.4.6 *Role of WWF in Dolphin Conservation*

WWF-India had set-up a Dolphin Action Group in early 1996 and initiated some activities for the conservation of the Ganges River Dolphin. Status surveys have been conducted in the selected having dolphin distribution in seven states of India covering around 6,000 km of the sites Ganges. The major rivers covered in the surveys are Ramganga, Ganga, Gomti, Ghagra, Girwa, Son, Yamuna, Betwa, Sindh, Ken, Chambal, Kosi, Brahmaputra, Hugli, Bhagirathi and Rupnarayan. A base line status of river dolphin was recorded as a result of the surveys.

WWF-India has been working comprehensively in the upper stretch of Ganga River in the state of Uttar Pradesh for last one decade. Research related to the habitat utilization and behavioral ecology were carried out to understand the basic needs of the dolphin habitat as well as special emphasis was given to involve the local community in the conservation of the species. Children's Awareness Programme, (CAP) was established by WWF-India, which includes an age appropriate learning package on river dolphins, their habitat and conservation. Also a Fishing Community Awareness Programme (FCEAP) has been launched within the target areas. Local language is used to train and educate the fishermen to prevent poaching, and adopt sustainable fishing practices. Moreover a programme has been undertaken to motivate the religious leaders (*Sadhus*) towards dolphin conservation. Workshops conducted along river banks involving religious leaders would have long term effects on the local population especially among followers. Besides, several other related programmes have been launched. All these effort resulted in increase in the dolphin population in the river stretch. The population of the river dolphin was doubled from 22 to round 56 in last 15 years [38].

Fifteen years of hard work under this WWF – India's programme has laid the foundation for challenging tasks ahead. Serious information gaps in the dolphin population status have been filled and identification of river stretches for priority conservation action recognized. State Forest Departments and other stakeholders have been motivated. It is now imperative that we look at the bigger picture, the river basin, and address the issues therein to ensure long- term conservation. There is still a long way to go and we need your support to fit the pieces of this bigger picture. The Ganges dolphin is a unique charismatic mega-fauna and is an indicator species for the river ecosystem. It is an endemic and rare aquatic mammal found only in the Indian sub-continents and is part of our natural aquatic heritage.

Developing a comprehensive program to conserve river dolphins in Ganga is required as India is the last strong hold with extant populations in the Ganges-Brahmaputra River Systems. The species is in peril in Nepal and Bangladesh. As a signatory to numerous international conventions, India must fulfill its treaty obligations to aid in the conservation of this species. As a flagship species, its conservation requires greater efforts along the lines of that provided to the Project Tiger, Project Elephant, Project Snow Leopard and Project Rhino.

4.4.6.1 Celebrating the Ganges River Dolphin

Commemorating the 3rd anniversary of Government of India's declaration of the Ganges River Dolphin as the 'National Aquatic Animal' in October 5, 2009, WWF-India, Uttar Pradesh Forest Department and HSBC organized "My Ganga, My Dolphin" campaign, 'beginning from 5th to 7th October, 2012' (Fig. 4.3). The campaign was a 3 day long awareness cum census of river dolphins across 2,500 KM stretch of Ganga & its tributaries (Geruwa, Ghaghra, Saryu, Rapti, Chambal, Yamuna, Ken and Betwa). It was targeted at generating awareness about the

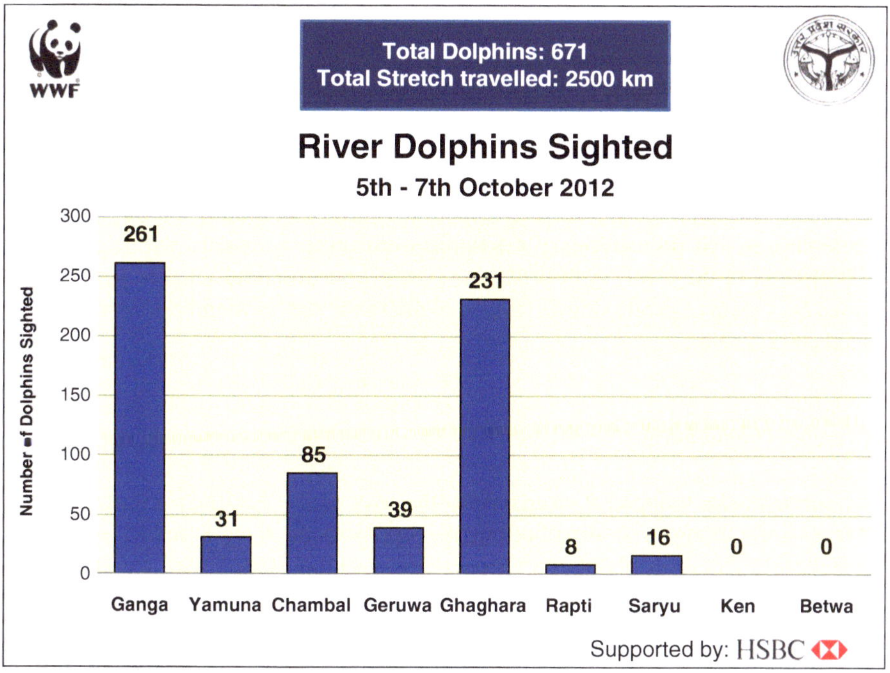

Fig. 4.3 Dolphins sighted during "My Ganga, My Dolphin" campaign

endangered species and to collect baseline information of the species distribution and population status, in the states of Uttar Pradesh, Rajasthan and Madhya Pradesh.

"My Ganga, My Dolphin" campaign led to successful partnership between WWF-India, Uttar Pradesh Forest Department and HSBC Bank, while garnering support of the people of Uttar Pradesh and of the Hon'ble Chief Minister of the State, Shri. Akhilesh Yadav. WWF-India adopted the Ganges River Dolphin as a species of special concern and initiated Ganges River Dolphin Conservation programme in 1997. Since then, it has been working closely with the communities for awareness generation about the Ganges River Dolphin.

As part of the preparation for the campaign launch, all the participating team members and volunteers were provided with field training. All observers/members of the survey team were trained to identify dolphins by colour, shape and size of the animal while conducting the survey. Apart from extensive study of the designated area and information sharing of endangered species conservation, the participating teams in the campaign were given training for adopting effective data collection skills, leading to a successful completion of the survey. One hundred and fifty people, comprising frontline staff of the UP state Forest Department, MP state Forest Department, local NGOs, researchers, scientists, WWF staff and the local community members, were trained through four training sessions on scientific ways of dolphin counting, and to build an understanding of the geographical area.

Field Survey

The survey, which was pivotal to the campaign activity, was carried out by 150 participants in 18 teams. The 2,500 km river expansion was divided into smaller stretches, enabling each team to cover 60–70 km of the river stretch each day. It was ensured that each team consisted of eight dolphin observers including an expert in dolphin counting. The boat speed was maintained at 10 km/h to reduce the duplication of the sightings. All precautionary measures were taken to ensure the safety of the participants in the river, with provision of a proper medical kit, life jackets supported by a team of professionals on land to guard the survey members from unanticipated emergencies.

Awareness Generation

The survey teams stopped at villages near the river banks, leading to fruitful interactive sessions along with distribution of pamphlets on river Dolphin, leading to awareness generation about the species conservation issues. The campaign was hugely successful in involving a number of young volunteers in the Dolphin survey, inspiring them to become active ambassadors of change.

The 3 day long dolphin counting concluded on October 7, 2012 with Uttar Pradesh Chief Minister, Shri Akhilesh Yadav declaring the presence of **671** dolphins in the state of UP.

The survey reported interesting trends in terms of Dolphin population. Stretches that reported stable Dolphin population were in areas where there had been some interventions taken by local communities, forest departments, and other non government organizations.

References

1. Rao RJ (1995) Studies on biological restoration of Ganga River in Uttar Pradesh; an indicator species approach. Final technical report, Project No. J-11013/10/92 GPD
2. Behera SK (2002) Ganges River dolphin—Wanted alive! (Status report submitted to WWF-Sweden). Ganges River Dolphin Conservation Project, Research Report, World Wide Fund for Nature-India, New Delhi
3. Krishnamurti CR, Bilgrami KS, Das TM, Mathur RP (1991) The Ganga. A scientific study. Northern Book Centre Publications, New Delhi
4. Sinha RK, Smith BD, Sharma G, Prasad K, Choudhury BC, Sapkota K, Sharma RK, Behera SK (2000) Status and distribution of the Ganges susu. Platanista Gangetica, in the Ganges river system of India and Nepal. In: Reeves RR, Smith BD, Kasuya T (eds) Biology and conservation of freshwater cetaceans in Asia, Occasional papers of the IUCN Species Survival Commission no. 23. IUCN, Gland, pp 54–61
5. Hamilton B (1822) An account of fishes found in the River Ganges and its branches. A. Constable and Co., Edinburgh, viii + 405
6. Jayaram KC (1974) Ecology and distribution of freshwater fishes, amphibian and reptiles. In: Mani MS (ed) Ecology and biogeography in India. W. Junk, The Hague, pp 517–584

7. Zoological Survey of India (1991) Faunal resources of Ganga. Part I. Zoological Survey of India, Calcutta, pp 145
8. Rao RJ (2001) Biological resources of the Ganga River, India. Hydrobiologia 458:159–168
9. FAO (1974) A preliminary survey of the prospects for crocodile farming (Based on the work of Bustard HR).F.O. Ind/71/033, pp 1–124
10. Choudhury BC, Singh LAK, Rao RJ, Basu D, Sharma RK, Hussain SA, Andrews HV, Whitaker N, Whitaker R, Lenin J, Maskey T, Cadi A, Rashid SMA, Choudhury AA, Dahal B, Win Ko Ko U, Thorbjarnarson J, Ross JP (2007) Gavialis gangeticus. In: IUCN 2011. IUCN Red List of Threatened Species. Version 2011.1. http://www.iucnredlist.org/. Downloaded on 29 July 2011
11. Shortt WHO (1921) A few hints on crocodile shooting. J Bombay Nat Hist Soc 29:77
12. Whitaker R, Rajamani V, Basu D, Balakrishnan V (1974) Preliminary survey of the gharial, Gavialis gangeticus. Madras Snake Park Trust Report, pp 1–16
13. Rao RJ (1994) Ecological studies of Indian crocodile. An overview proceedings of the 12th working meeting of the IUCN/SSC/CSG. Thailand, vol 1,259–273
14. Gad SD (2008) Indian Gharial (Gavialis gangeticus) on the verge of extinction. Curr Sci 12(94):1549
15. Converse L (2009) Katerniaghat Gharial Project 2008–2009. Report of Preliminary Findings. A report to GCA and James Cook University
16. Smith MA (1931) The Fauna of British India including Ceylon and Burma. Reptiles and Amphibians, 1. Taylor & Francis Ltd, London, p 185
17. Whitaker R, Basu D (1983) The gharial (Gavialis gangeticus): a review. J Bombay Nat Hist Soc 79(3):531–548
18. Dudgeon D (2000) The ecology of tropical Asian rivers and streams in relation to biodiversity conservation. Annu Rev Ecol Systematics 31:239–263
19. Hussain SA (1999) Reproductive success, hatchling survival and growth of a managed population of Ganges gharial. Biol Conserv 87:261–268
20. Mishra HR, Maskey TM (1981) Die Inter. Zeitschrift fur Tier, Mench undue, Nat, pp 15–18
21. Suman D (2008) Indian Gharial (Gavialis gangeticus) on the verge of extinction. Curr Sci 94(12):1549
22. Turvey ST, Pitman RL, Taylor BL, Barlow J, Akamatsu T, Barrett LA, Zhao X, Reeves RR, Stewar BS, Wang K, Wei Z, Zhang X, Pusser LT, Richlen M, Brandon JR, Wang D (2007) First human-caused extinction of a cetacean species? Biol Lett 3:537–540
23. Pilleri G, Gihr M (1971) Differences observed on the skull of Platanista indi and gangetica. Invest Cetac 4:44–70
24. Kasuya T, Haque AKM (1972) Some information on the distribution and seasonable movement of the Ganges dolphin. Sci Rep Whales Res Inst (Tokyo) 24:109–115
25. Yang G, Zhou K (1999) A study on the molecular phylogeny of river dolphins. Acta Theriol Sin 19:1–9
26. Roxburgh W (1801) An account of a new species of Delphinus, an inhabitant of the Ganges. Asiatick Res (Calcutta) 7:170–174
27. Anderson J (1879) Anatomical and Zoological Researches: Comparing an account of Zoological results of the two expeditions to western Yamuna in 1868 and 1875; Platanista and Orcella. Berward Quaritch, London
28. Smith BD, Ahmed B, Ali ME, Braulik GT (2001) Status of the Ganges river dolphin or shushuk Platanista gangetica in Kaptai Lake and the southern rivers of Bangladesh. Oryx 35:61–72
29. Kelkar N, Krishnaswamy J, Choudhary S, Sutaria D (2010) Coexistence of fisheries with river dolphin conservation. Conserv Biol 10:1523–1739
30. Sinha RK (1997) Status and conservation of Ganges River dolphin in Bhagirathi – Hooghly River systems in India. Int J Ecol Environ Sci 23:343–355
31. Sinha RK, Sharma G (2003) Current status of the Ganges river dolphin, Platanista gangetica in the rivers Kosi and Son, Bihar, India. J Bombay Nat Hist Soc 100(1):27–37
32. Choudhary S, Sushant D, Subhashish D, Viveksheel S, Nair T, Kelkar N (2012) River dolphin distribution in regulated river systems: implications for dry-season flow regimes in the Gangetic basin. Aquat Conserv: Mar Freshw Ecosyst 22:11–25

33. Singh H, Rao RJ (2012) Sighting frequency and Group composition of Ganges river dolphin (Platanista gangetica gangetica) in National Chambal Sanctuary, India. Zoo's Print 8:18–23
34. Mohan RSL, Dey SC, Bairagi SP, Roy S (1997) On a survey of the Ganges River dolphin Platanista gangetica of Brahmaputra River. Assam J Bombay Nat Hist Soc 94:483–495
35. Wakid A (2009) Status and distribution of the endangered Gangetic dolphin (Platanista gangetica gangetica) in the Brahmaputra River within India in 2005. Curr Sci 97:1143–1151
36. Behera SK, Sagar V, Nawab A (2008) Environmental flow requirements vis-à-vis Habitat use pattern of Freshwater Dolphins. In: Proceedings of the 11th international river symposium, Brisbane, 1–4 September 2008, pp 4
37. Smith BD, Braulik GT (2008) Platanista gangetica. In: IUCN 2010. IUCN Red List of Threatened Species. Version 2010.3. http://www.iucnredlist.org/
38. Bashir T, Khan A, Gautum P, Behera SK (2010) Abundance and prey availability assessment of Ganges River dolphin (Platanista gangetica) in the upper Ganges River, India. J Aquat Mammals 36:19–26
39. Reeves RR, Smith BD, Kasyua T (eds.) (2000). Biology and conservation of freshwater cetacean in Asia. IUCN,Gland,Switzerland and Cambridge, UK. VIII+152pp.

Gharial: A gharial resting at the river banks (Courtesy Sandeep Behera)

Chapter 5
Sundarban Mangroves: Impact of Water Management in the Ganga River Basin

Malavika Chauhan and Brij Gopal

Abstract Sundarban, the single largest contiguous area of mangroves in the delta of Ganga, Brahmaputra and Meghna, is shared between India and Bangladesh. The landscape consists of a network of tidal creeks, waterways, marshes and pools, low and high mangroves, strands of grasses and islands which disappear and reappear twice a day with the tides. Like other mangroves, the Sundarban has enormous ecological value as it stabilizes the coastline and protects against cyclonic storms, hosts very rich biodiversity, supplies forest products, hosts and supports coastal and near-shore fisheries, sequesters carbon disproportionate to its areal extent and serves the ecotourism. Sundarban is the only mangrove habitat of tigers, an endangered species.

Sundarban was declared as a reserve forest by the British in the mid 1800s for the sake of resource exploitation. However, after independence, India established three wildlife sanctuaries that were later merged into a national park. In late 1970s, Bangladesh also set up wildlife sanctuaries in its part of Sundarban. The entire Indian part of Sundarban has now been designated as a biosphere reserve whereas some parts in both India and Bangladesh were also inscribed on the World Heritage list in 1987 and 1997 respectively.

Despite the conservation status, the Sundarban mangroves are increasingly impacted by various land use and water management practices throughout the GBM basin through a complex interaction between hydrology, salinity and sediments. Hydrological alterations caused by the Farakka barrage affect the salinity gradients, transport of sediments, nutrients and pollutants, and accretion-erosion dynamics. Consequently,

M. Chauhan (✉)
Himmotthan Society, Vasant Vihar, Dehradun, India
e-mail: malavikachauhan@gmail.com

B. Gopal
Centre for Inland Waters in South Asia, Jaipur, India
e-mail: brij44@gmail.com

R. Sanghi (ed.), *Our National River Ganga: Lifeline of Millions*,
DOI 10.1007/978-3-319-00530-0_5, © Springer International Publishing Switzerland 2014

A mangrove forest with Nypa palm in Indian Sundarban (Credited to Dr Malavika Chauhan)

these changes impact the biodiversity, productivity, agriculture and people and their livelihoods. The projected climate change-induced sea level rise is another major threat looming large over the Sundarban.

While our understanding of the functioning of the Sundarban mangrove ecosystems remains poor and influenced by terrestrial approaches, their future depends upon the management of land and water resources throughout the Ganga-Brahmaputra basin, that in turn requires an integrated river-basin-wide appropriate strategy backed by political will.

Keywords Biodiversity • Mangroves • Climate change • Hydrological alteration • Land use • Salinity • Ganga-Brahmaputra-Meghna basin

5.1 Introduction

Mangroves are forested wetlands found extensively along estuaries and tidal backwaters across tropical and subtropical coastal areas. Tropical Asia accounts for about 42 % of the world's total mangrove area [1, 2], about half of which is in Indonesia. The Indo-Malayan mangroves are also the richest in species composition [3]. The world's largest contiguous mangrove complex however, is the Sundarban, which covers approximately one million ha over the Ganga-Brahmaputra-Meghna delta, across India and Bangladesh [4]. Divided into two by the India-Bangladesh border, the Sundarban has, over several hundred years, seen extensive human interventions which have caused ecological changes encompassing every aspect of the ecosystem.

More than 2000 studies have been published over the past century covering geological, biophysical, ecological, socio-cultural, economic, and practically all other aspects of the Sundarban. Several reviews summarise topical information and research. In recent years, the Sundarban has attracted especially greater attention in the context of likely impacts of sea level rise in the wake of global climate change [5–13]. Cyclone Aila, which struck the Sundarban in 2009, caused havoc unprecedented in the history of the Sundarban mangrove system, and changed considerably perceptions of its stability and resilience [14].

In this paper, we highlight the uniqueness of the combined Sundarban (in India and Bangladesh) and its importance in the lives of the people of the deltaic regions. We further discuss how changes have occurred over time in the mangrove system and how management of water resources of the Ganga basin are impacting the health and survival of the mangrove ecosystem and the people who depend upon it. For an overview of the climate, hydrology, geology and soils, variables which drive the Sundarban mangrove system, refer to Gopal and Chauhan [4].

5.2 The Sundarbans

5.2.1 A Short History

The history of the growth and development of the Sundarban dates back to the second century AD. The earliest traces of a developed civilization within the Sundarban itself date back to about 200–300 AD, the ruins of a city in the Baghmara Forest Block, India, built by the Chaand Saudagar line of merchants, who traded extensively in the region [15]. Studies show that possibly the earliest residents of the forest would have been religious and political refugees, driven into the forests for political reasons [16]. By the eleventh century systematic villages had developed which practiced shifting agriculture. Although the difficult terrain and dense, dangerous jungles kept humans at bay, it was also an enormous untapped mine of valuable resources including wood, honey, salt and fish. The region became a major centre of trade and commerce from the fifteenth century onwards when the Mugals began to lease out the fringe forests. In fact the Sundarbans hold the dubious distinction of being the world's first forest to be brought under known, large scale, formal management. After the Mugals, the East India Company took over proprietary rights to the forests. The area was first mapped in Persian in the mid 1700s. Much later the line of forest from the river Jamuna to the Hoogli was surveyed, between 1822 and 1823 by a Mr. Princep, who divided the forest land into numbered blocks. This was the beginning of the Sundarban "lots" [17]. The first comprehensive set of rules for the grant of leases in the Sundarbans were issued in 1853 following which the first Forest Management Division with formal jurisdiction over the Sundarbans was established in 1869. In 1875 a large portion of the mangrove forests was declared as reserve forests, under the Forest Act, 1865 and the area was placed under the control of the Forest Department. A Forest Division was created in 1879 with headquarters in Khulna, now in Bangladesh. Through the century the state licensed the large scale harvest of timber, bamboo and other forest projects. Only now, in recent years has the value of the forests once again increased relative to the now common rice lands [18]. By 1878, with the final demarcation of the reserve forest limits, the Bengal government put in place Wasteland Rules for the area, to prevent sporadic reclamation, and to push cultivation into the deeper deltaic areas. Blocks of up to 200 acres or more were leased for 40 years, while the government reserved rights to mineral resources. This opened up the Sundarban to farmers and speculators. Between 1870 and 1904 the recognized area of the Sundarban decreased by 13.3 %, reflecting increasing cultivation and settlement [18]. In 1904 the government introduced the ryotwari system of settlements, in which the government assisted by clearing land and constructing tanks. Revenue was received directly by the government, rather than by landlords, but the initial system failed. Over time however, reclamation became faster, and villages, rice paddies, schools and post offices came up in areas which had once been forbidden jungle. By 1910 large parts of the Sundarban had been successfully reclaimed. The ryotwari system, brought in again by 1915 became the normal means of settlement in the region. Between 1880

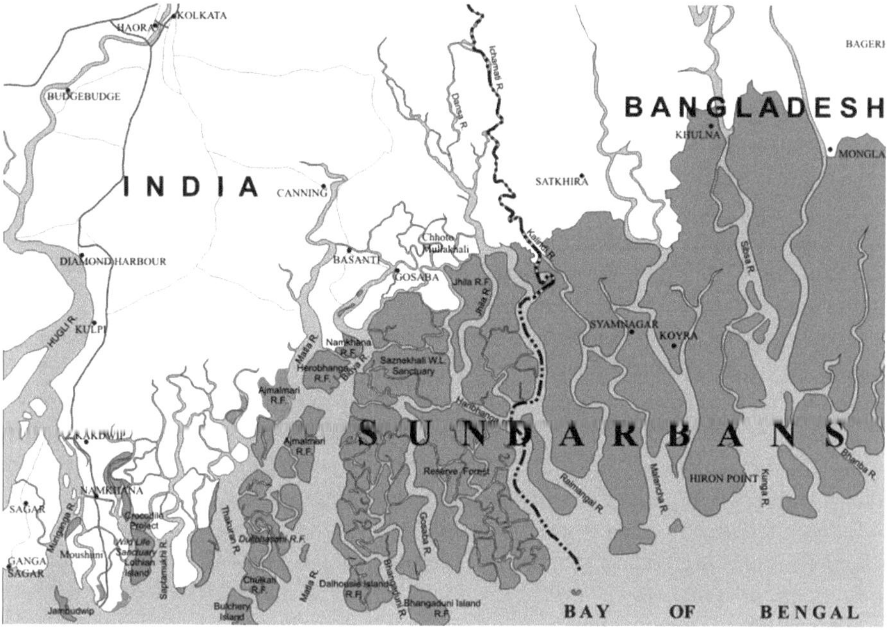

Fig. 5.1 The Sundarban mangroves are shared between India and Bangladesh. Mangroves once extended up to the Dampier-Hodges line – the inward limit of tidal influence. Dark hatched parts are reclaimed areas and the stippled areas are reserved forests (From Gopal and Chauhan [4])

and 1950, cultivated land in the area expanded by 45 %, while the human population went from 5.6 to 12.9 million [18]. Villages began to come up more rapidly, with small farmers, share croppers and landless labourers coming in from across the neighbouring regions, into the Sundarban. Refugees of drought and famine from the eastern plateau region, from Jhargram, western Medinipur, Bankura, Singhbhum and Santal parganas. Many were tribals including the Santals, Mundas, Oraons, Kurmis and Koras [19]. Then, post 1947, following the partition of India and creation of East Pakistan, the region underwent a sudden influx of migrant population, with hundreds of displaced families forced to occupy the Sundarban.

5.2.2 The Landscape

The Sundarban (21° 30′ to 22° 40′ N, 88° 05′ to 89° 55′ E) comprises of numerous deltaic islands formed by sediments deposited by three major rivers, the Ganga, Brahmaputra and Meghna, and the dense network of distributaries – rivers, channels and creeks which they form (Fig. 5.1). The landscape consists of a network of tidal creeks, waterways, marshes and pools, low and high mangroves, stands of grasses and islands which disappear and reappear twice a day with the tides. The delta in itself is formed by the alluvium brought down by the rivers which spread it across the coastal reaches, forming land. Most of the Ganga–Brahmaputra delta is located within Bangladesh, which hosts over 140 million inhabitants with a population density approaching 1,000 persons per km^2. The southwest sector of the delta extends into India, covering an area of approximately 25,000 km^2, or about 22 % of the subaerial surface. About 60 % of the mangrove forests lie in the Khushtia, Jessore, Khulna, Rajshahi, Pabna and Dhaka districts of Bangladesh [20], and the rest in the 24-Paragnas District of West Bengal, India, where the Dampier and Hodges line demarcates the inland margins of the mangroves.

A description of the Sundarban in the Bengal Gazetteer of 1914 [17] defines the landscape of the region much as it looks now – "*the district* (24 Parganas) *naturally falls into two divisions........the northern inland tract which is fairly well raised delta land of old formation, and the low lying Sundarbans towards the seaboard on the south. The northern tract is a land of sluggish or stagnant rivers, whose beds are out of reach of the scour of the tides, and of inland depressions which will never now be filled, because the rivers which should perform this office are locked into their channels by the high banks of silt which they have deposited. The Sundarbans on the other hand are a network of tidal channels, rivers, creeks and islands. Some of these islands are mere swampy morasses, covered with low forest and scrubwood jungle, but those to the north, which are embanked, grow rich crops of rice. As one approaches the coast, the land gradually declines to an elevation which throughout many hundred square miles is scarcely raised above the high water mark. This seaboard area is a typical specimen of new deltaic formation..... it exhibits the process of land-making in an unfinished state, and presents the last stage in the life of a great river – the stage in which it emerges through a region of half land, half water, almost imperceptibly, into the sea. It has been well described as* 'a sort of drowned land, broken up by swamps, intersected by a thousand river channels and maritime backwaters, but gradually dotted, as the traveller recedes from the seaboard, with clearings and patches of rice land' [21]".

The Sundarban is in essence a growing delta. However, at the same time, extensive deforestation over the last two centuries, and now rising sea levels have caused the physical losses of several hundred hectares of forest, and over the past few years, several islands to the sea. In 1911, the Sundarban stretched for about 270 km from the mouth of the Hoogli river to the mouth of the Meghna, and was bordered inland by the three settled districts of the 24 Parganas, Khulna and Bakerganj. The total area was estimated at 16,902 km^2. Today the width of the Sundarban is about 350 km, with a North-South length of 80 km from the coast to the maximum inland spread. In area it is less than a total of 10,000 km^2. In Bangladesh it covers about 5993.3 km^2 (1978 Landsat data, [22, 23]) and in India it is 4263.0 km^2 [24] at last count. The eastern and western limits of the Sundarban are defined by the course of the River Baleshwar in Bangladesh and the River Hooghly, a distributary of the Ganga, in India, respectively. The river Harinbhanga in India and the Ichamati or Raimongal in Bangladesh cuts through the mangroves, north to South, demarcating the border between the two countries.

The elevation of the delta is a maximum of 15 m in the extreme north and dips gently seaward from the upper deltaic plain. Average elevation at Khulna and Barisal in Bangladesh, is less than 2 m, while the average gradient of the delta is 0.16 m/km [20]. The landscape therefore is dominated by meandering creeks and streams, temporary islands and pools and the constantly moving tides. The thickness of deltaic sediments overlying the oceanic crust in the central and eastern basin may reach up to 20 km. Long-term tectonic subsidence in the eastern and southern basin of 1–2 mm per year has been balanced by the high sediment supply from Himalayan erosion [25]. Patterns of sediment deposition in the Bengal Basin reveal three main parameters which focus floodplain development including the channel process, overbank flooding and surface runoff. Accretion occurs mostly in the river braid belt and adjacent floodplain, decreasing rapidly with distance from the main channel. Accumulation rates increase again in low-lying distal basins. The extent and magnitude of this process indicate its comparable importance with overbank deposition for floodplain accumulation. A sediment budget for the study area reveals that at least 15 % of the estimated one billion tonnes of fluvial sediment discharge is stored annually and does not reach the oceans [26, 27]. On the Indian west side, new land formation occurs continuously because of heavy silt deposition by the Hooghly and its distributary, the Muriganga at their confluence with the Bay of Bengal. In the continuously moving landscape, the processes of bank erosion, the formation of new islands, and ever-changing soil texture and ever-varying salinity of water occur simultaneously.

5.2.3 *Biodiversity*

Communities of plants, animal, insects, plankton, fish and other living organisms in mangroves are defined by the degree and gradient of salinity in the waters and soils. Early studies [28, 29] showed that mangroves are salt tolerant, while the seeds of mangrove trees show retarded germination and retarded growth in excessively saline conditions. For healthy mangrove growth therefore, the right balance of fresh and saline waters is required, particularly at the right seasons across the year.

5.2.3.1 Plants

"*Mangroves are a taxonomically diverse group of salt-tolerant, mainly arboreal, flowering plants that grow primarily in tropical and sub-tropical regions*" [30]. This simple statement summarizes the basic characteristics of mangrove vegetation, and extrapolates to mangrove biodiversity. This biodiversity in the Sundarban includes about 350 species of vascular plants, along with numerous species of phytoplankton, fungi and microorganisms. Sundarban has a highly diverse algal flora comprised of both benthic and planktonic forms ranging from the freshwater to marine environments [4]. Seasonal and spatial variation follow the changes in water quality parameters, like pH, salinity, TSS and nutrients and DO. The phytoplankton community is dominated by diatoms (Biacillariophyceae), followed by Pyrrophyceae (Dinoflagellates) and Chlorophyceae. The highest abundance is noticed in the post-monsoon season [31, 32].

Ceriops roxburghiana, Avicennia alba, Excoecaria agallocha, Sonneratia apetala, Aegialitis rotundifolia and *Phoenix paludosa,* are dominant plant species in the western parts of the Sundarbans. Changes in salinity alter the distribution and density of various species in the area, while the topography of the land mass also exerts a major influence. Various studies show that mangrove trees like *Heritiera fomes* and *Nypa fruticans* are declining in both density and extent in the area near the Matla River [33, 34].

5.2.3.2 Animals

The Sundarban hosts the only known mangrove tiger in the world, the Royal Bengal Tiger (*Panthera tigris tigris*) which is a powerful swimmer. It is also known for an abundance of other fauna including 250 species of fish and 300 species of birds. Zooplankton, benthic invertebrates, molluscs, reptiles, amphibians and mammals have been reported from the region [4]. It is natural habitat of many rare and endangered species including the Royal Bengal tiger, the Estuarine Crocodile (*Crocodilus porosus*), Gangetic Dolphin (*Platinista gangetica*), Snubfin dolphin (*Orcella brevirostris*), River Terrapin (*Batagur baska*), *Batagur baska, Pelochelys bibroni, Chelonia mydas*, marine turtles like Olive Ridley (*Lepidochelys olivacea*), Green Sea Turtle (*Chelonia mydas*), Hawksbill Turtle (*Eritmochelys imbricata*). These numerous species contrive to make the Sundarban a natural biodiversity hot spot, as a result of which it is now protected in several sections, both in India and Bangladesh. It is also a World Heritage Site. However, continued degradation of the overall system is pushing an increasing number of species rapidly into extinction [4, 35].

The Sundarban is also famous for its marine and brackish water fisheries, as a very large human population depends upon them as a main livelihood activity. Jhingran [36] recorded a total of 172 species and mentioned that the diversity of the Hooghly-Matlah estuary increases along an increasing salinity gradient. Other sources record over 250 species of fish; 20 species of prawn and 44 species of crabs [37]. Commercially important marine and estuarine fishes include fin fish (*Lates calcarifer, Hilsha ilisha, Liza parsia, Liza taede, Harpodon hehereus, Plotosus canius, Pompus argenteus, Rhinobatus, Pangasius pangasius, Polydactylens, Chanos chanos, Eleutheronema tetradactylum, Polynemous indicum, Polynemous paradesious and Pama pama*); shellfish (*Penaeus monodon, Penaeus penicillatus* and *Metapenaeus monoceros*) and Crustaceans (edible crabs mainly *Scylla serrata* and *Neptunus pelagiens*).

Numerous species of wetland and terrestrial birds populate the mangroves, with over 170 species identified. Unique species include the endemic Brown-winged

Kingfishers (*Pelargopsis amauroptera*), the threatened Lesser Adjutants (*Leptoptilos javanicus*), Masked Finfoots (*Heliopais personata*), ospreys (*Pandion haliaetus*), White-bellied Sea Eagles (*Haliaeetus leucogaster*) and Grey-headed Fish-eagles (*Ichthyophaga ichthyaetu*). Others include the Open Billed Storks, Black-headed Ibis, Water Hens, Coots, Pheasant-tailed Jacanas, Pariah Kites, Brahminy Kites, Marsh Harriers, Swamp Partridges, Red Jungle fowl, Spotted Doves, Common Mynah, Jungle Crows, Jungle Babblers, Cotton Teals, Herring Gulls, Caspian Terns, Gray Herons, Brahminy Ducks, Spot-billed Pelicans, great Egrets, Night Herons, Common Snipes, Wood Sandpipers, Green Pigeons, Rose Ringed Parakeets, Paradise Flycatchers, Cormorants, Fishing Eagles, White-bellied Sea Eagles, Seagulls, Common Kingfishers, Peregrine falcons, Woodpeckers, Whimprels, Black-tailed Godwits, Little Stints, Eastern Knots, Curlews, Golden Plovers, Pintails, White Eyed Pochards and Whistling Teals.

5.2.3.3 Breeding Ground and Nurseries

The Sundarban ecosystem is one of the largest detritus-based ecosystems of the world [38]. Litterfall from the mangroves supplies the detritus and nutrients which regulate productivity, and this in turn creates the environment for a rich and productive nursery area. Numerous species of commercially important fish found in the Bay of Bengal and the Indian Ocean depend upon the mangroves to sustain their populations. The Sundarban mangroves and mudflats are the nursery for nearly 90 % of the aquatic species on eastern coasts, with the coastal fishery of eastern India and Bangladesh entirely dependent upon the system. Intertidal mudflats of the Sundarban occupy a significant component of the total estuarine habitat available to fishes, and play important roles as nursery and foraging grounds [39, 40], including for many important commercial species of the continental shelf that are harvested in India and neighboring countries. The high nutrient availability in the mangroves and large quantity of organic detritus present is an important source of energy for the fry. Studies show that species richness, evenness and diversity vary significantly between adjoining niche and habitats within the Sundarban, with clear temporal gradients noticed in fish diversity in both riverine and mudflat systems (*Tenualosa ilisa* in riverine and *Periophthalmus novemradiatus* in mudflats). Dominating vegetation appeared to have the greatest influence on fish distribution and density. Chaudhuri et al. [41] conclude that the disparity in fish populations and composition reflect the degree of hydrological connectivity of the habitat, and that the influence of spatial coupling of adjacent ecosystems affects the diversity-productivity pattern of local fish populations, leading to regional persistence of species. More than just a nursery, many commercial estuarine fish grow to maturity here and make up a large part of the near-shore fishery of the northern Bay of Bengal. Other fishes and prawns that spend most of their lives in freshwater descend annually to the estuary for spawning. Many more marine and freshwater prawn and fish require this environment to complete their lifecycle.

5.2.4 Nutrients

Nutrients are universally important as they are the raw materials for primary producers, and the Sundarban is rich in nutrients with a high influx rate. Runoff from the land during the monsoons and wastes from shrimp culture farms and villages and cities contribute to the vast nutrient load. Main estuarine nutrients include nitrogen, phosphorus and silica which usually limit phytoplankton growth in natural waters. Overall, in the Sundarban estuaries the phosphate concentration varies from 0.4 to 1.0 μmol/L, nitrates from 2.4 to 39.9 μmol/L, nitrite between 0.6 and 1.6 μmol/L and silicate from 4.8 to 49.1 μmol/L. Silicate is a primary growth limiting nutrient for diatoms. N-limitation is a wide spread phenomenon in tropical lakes rivers and estuaries, e.g., Mandovi-Zuari [42], Cochin estuary [43] and Hoogly estuary [44]. The Sundarban estuary was observed to be phosphorus limited during the post-monsoon and nitrogen-limited during the pre-monsoon and monsoon seasons. The seasonal phosphorous limitation characteristic is common to several estuaries (e.g., [45, 46]).

5.3 The Value of the Sundarban

Mangroves have enormous ecological value. They protect and stabilize coastlines, enrich coastal and estuarine waters, supply forest products and host and support coastal and deep sea fisheries. They produce large amounts of organic carbon and contribute significantly to the global carbon sequestration, disproportionate to their areal extent. They are sources of chemicals and medicine, and increasingly host the ecotourism industry.

The value of the Sundarban was recognized early on beyond being a source of wood, forest products and cultivated land. "*Along the sea face the timber is large, but generally with an undergrowth of low brushwood. This belt of forest serves as an admirable breakwater against the ocean; and in the recent cyclones of 1869, which were accompanied by storm waves, it broke the force of the tidal wave before the inundation reached the cultivated tracts, and thus prevented a great destruction of life and property. On the outer islands, and on parts of Rabnabad Island, where the forest once ran down without a single clearing to the water's edge, a belt of trees has been carefully preserved by the cultivators as a breakwater, varying in depth according to the exposure of the situation*" [21]. Storms and cyclones often move across the Bay of Bengal to break onto Sundarban and neighbouring regions, and the frequency of these storms appears to be changing with time. The Bay of Bengal is particularly prone to cyclones with exceptionally high storm surges, amongst the highest in the world, which occur mainly in the pre and post monsoon periods, and changes expected over time are well studied [47–49]. Between 1877 and 1995 Bangladesh was hit by 154 cyclones (including 43 severe cyclonic storms, 43 cyclonic storms, 68 tropical depressions). Since 1995, five severe cyclones hit the Bangladesh coast in May 1997, September 1997, May 1998, November 2007 and May 2009. On average, a severe cyclone strikes Bangladesh every 3 years [50]. Cyclones cause major hydrological disturbances along with major alterations in temperatures, surface water pH, dissolved oxygen, salinity and erosion rates [51–55]. Mangroves break the impact of storms, and reduce losses and impact.

The Sundarban plays an important role, particularly in the national economy of Bangladesh. It is the single largest source of forest produce in the country. In Bangladesh, it is estimated that over four million people depend upon the Sundarban for their livelihood. 45 % of the country's timber and wood related products come from the Sundarban. Further, the forest produces fuelwood, pulpwood, thatching material, honey, bees-wax, fish, and crustacean and mollusks for food. Shah and Dutta [56] report that about 49 % of the households in Southkhali near the Sundarban Reserve Forest (SRG) in Bangladesh depend upon the forest for livelihood support, and each household extracts fish, fuelwood and golpatta valued at an average of Tk 62,000 (US$1,000) per year. In India, people are granted permits to enter the Reserved Forests during summer for collecting honey which is purchased back by West Bengal Forest Development Corporation Limited, India.

For the poor and very poor, the Sundarban holds survival. Nearly 95 % of the population of the Sundarban depend primarily on agriculture which is basically rainfall-based mono-cropping. About 50 % of the agriculturists are landless labor. During the non-agriculture season people survive by fishing and collecting and selling prawn seed (*Penaeus monodon*). Prawn cultivation provides ready cash, and can be more paying than agriculture. Agricultural land which is inundated with brackish water, is used for prawn cultivation following the winter paddy harvest. The Indian Government has also taken the initiative to export prawn on a large scale to the foreign market, investing a considerable amount of funds, as a result of which many families shifted their economy from agriculture to prawn seed collection and aquaculture. Fishing is the other major activity of people in the region. Fishermen use mechanized as well as non-mechanized crafts. Trawlers, gill-netters, purse seiners, etc. are among mechanized crafts and plank built boats, dugout canoes and catamarans are non-mechanized. A variety of fishing gear are used including trawl nets, purse seines, drift/gill nets, boat seines, fixed bag nets, hooks and lines, shore seines, traps, scoop nets, etc. In India, there are 14 landing centers for capture fisheries in South 24 Parganas district. Of these, Raidighi, Kakdwip Steamer ghat, Kakdwip Akshaynagar, Kakdwip 8 Number lot, Sultanpur fishing harbour, Diamond harbour, Namkhana, Frazerganj fishing harbour, Gangasagar, Beguakhali, Mayagoalini ghat function throughout the year whereas Kalisthan, Frazerganj baliara and Gangasagar west are used for seasonal fishing [57].

As an example of early age transfer of technology, the Portuguese, probably the first Europeans to visit the mangrove forests of the Indian Ocean in the fourteenth century, learned the traditional Indian technique of rice-fish-mangrove farming and transferred this learning to the African countries of Angola and Mozambique [58]. It is well known that the transplantation of rice began early in the Sundarban region and hence, this practice may have been also learnt here and taken across the world.

5.3.1 Carbon Sequestration

Limited attempts have been made to study the productivity and carbon sequestration potential of mangroves owing to methodological shortcomings; however the best estimates suggest that mangrove carbon sequestration is more rapid than other marine and estuarine producers [59]. The productivity of mangrove forests depends various factors [60–62], and it is estimated that approximately half of the carbon dioxide assimilated by mangrove forests is returned back to the atmosphere through respiration [63], while the allocation of carbon in trees varies with many factors including light intensity, species composition, nutrient and water availability, salinity, tide, waves, temperature and climate [61]. A recent analysis on carbon allocation suggests that a higher proportion of carbon is stored in below ground components of trees than above ground [64]. However, another recent study reported that the carbon sink in live biomass is several times greater than that in the sediments but because of a faster turnover rate, the carbon stock in mangrove forests in lower than in the terrestrial tropical forest [65]. Data collected from different Asia- Pacific countries indicate that mangroves have an average 937 t C ha^{-1} which signifies that mangroves are among the most carbon rich forests in the world along with other carbon rich ecosystems such as tropical rain forests [60]. Mangroves account for only 1 % of carbon sequestration by world's forests.

5.4 Management for Biodiversity and Ecology

Purely for the sake of resource exploitation, the Sundarban was declared as a reserve forest by the British in the mid 1800s. Management from the point of view of ecosystem conservation and biodiversity protection began only after India's independence and the partition of India and East Pakistan, when the countries began to take stock of their resources. Approaches to conservation differ considerably between the two countries, as discussed at length by Seidensticker [66]. The Bangladesh Sundarban is managed as a refuge where wildlife is protected in small sanctuaries located within the larger forest tract by providing protection to resource "hot-spots", essential to the maintenance of wildlife populations. This approach assumes that inviolate core areas surrounded by restricted-use buffer zones can ensure the survival of communities and species such as large mammals and birds which depend upon the habitat and resources over larger areas. In India however, the Sundarban conservation strategy involves setting aside areas where the entire life-cycle needs of a community can be met and the ecological needs of wildlife can be linked into the overall management of the system for the Sundarbans. Furthermore, in so doing, the ecological processes upon which wildlife depend become integral values in the management matrix. Active management practices are not yet a part of Protected Area management in either of the countries, and it is unlikely that negative changes in habitat due to various reasons, including invasive species, changing hydrology and salinity levels and degrading water quality cannot be addressed at this point of time.

5.4.1 India

The main, interior islands, forests and water bodies of the Indian Sundarban were rapidly taken under the protective mantle Protected Area designations. Lothian Island (3,800 ha), populated by the Estuarine Crocodile, Olive Ridley Sea Turtle, Spotted Deer, Jungle cat and Rhesus macaque along with other wildlife was designated first, followed by 35,240 ha as the Sajnakhali Wildlife Sanctury in 1960. Tiger hunting was only banned in 1970 following the IUCN listing it as an endangered species. In 1973 the Government of India established a Tiger Reserve in the Sundarban covering 2,585 km^2 under the country wide 'Project Tiger'. Another 241 km^2 was demarcated as a 'Subsidiary Wilderness Area'. The core area of 1,330 km^2 in the Tiger Reserve was later designated as a National Park. In 1976 a Wildlife Sanctuary was established on Haliday Island (595 ha) to protect a unique ecosystem where the spotted deer (*Axis axis*), wild boar (*Sus scrofa*) and rhesus macaque (*Macaca mullata*) are dominant animals in a *Ceriops decandra* dominated forest.

The entire Indian Sundarban area south of the Dampier-Hodges line including 5,366 km^2 of reclaimed lands has also been designated as the Sundarban Biosphere Reserve. Within the Biosphere Reserve several distinct zones are recognised including a Core Zone comprising of the National Park and Tiger Reserve; a Manipulation Zone consisting of 2,400 km^2 of mangrove forests; a Restoration Zone covering 240 km^2 of degraded forest and saline mud flats, and a Development Zone which includes mostly the reclaimed areas.

Of these different zones, only the Core Zone is under strict conservation measures. Income generating activities such as the collection of seeds of black tiger prawn (*Penaeus monodon*), the culturing of oysters and crabs, mushroom cultivation and bee-keeping for honey production are allowed in the Manipulation Zone. Efforts are being made to also rehabilitate certain degraded areas through afforestation. Among faunal species, the estuarine crocodile and the Olive Ridley turtle are receiving some attention by way of captive breeding. The Sundarban Tiger Reserve in India has started a special programme for the conservation of sea turtles (http://www.wildbengal.com/urls/con_ar_sunderban.htm).

In 1987, the Sundarban National Park in India, and in 1997, about 1,400 km^2 of the Sundarban in Bangladesh, were inscribed by UNESCO onto the World Heritage list [67].

5.4.2 Bangladesh

Bangladesh created three wildlife sanctuaries in 1977, on deltaic islands in the Sundarban Forest Division of Khulna District, close to the border with India and just west of the main outflow of the Ganges, Brahmaputra and Meghna rivers [67]. Sundarban East established in 1960 and extended in area in 1996 covers 31,226 ha of natural mangrove system. Sundarban West covering 71,502 ha was established only in 1996, while Sundarban South covering over 36,970 ha was also established in 1996. 23.3 % of the countries protected areas cover mangrove forests [68], while the same forests in the 1980s gave 45 % of the countries extracted timber and wood products [69]. Mukul et al. [70], in a discussion on the effectiveness of Bangladesh's PAs, state that their impact is limited because they are portions of reserved forests which have, in most cases, only been declared after already being degraded heavily by activities such as illegal logging, land clearing, burning, and poaching [70, 71]. Many of country's mammals, birds and reptiles have already been lost, due to the belated awareness of and action for biodiversity conservation and the limited effectiveness of measures implemented to conserve biodiversity. The population size and number of the remaining wild fauna are confined and distributed irregularly in limited forest patches of the country.

Studies have pointed out that various conservation measures may backfire in the long run. Mitra et al. [72] state that official conservation strategies often intervene in the human-mangrove relationship, by imposing restrictions on usage. The paper observes that in the selected study area, restrictions have led to instability of the local economy which may pose a risk of further denudation of forests, making the goal of a holistic management approach unrealizable [72]. Various projects have over time, attempted to make changes in the livelihoods and dependence of people on the Sundarbans, to make the overall system more sustainable and biodiversity friendly. As examples the 'Sundarbans Biodiversity Conservation Project' supported by the Asian Development Bank in Bangladesh has attempted to develop sustainable management practices in the area, while a parallel project in India developed a Sundarban management plan. Since these plans need to be developed and implemented in close coordination with local authorities and government, often the process can show limited outputs over time.

5.5 Drivers of Change

5.5.1 The Hydrology and Salinity Regimes

Throughout the tropics and subtropics, mangroves have developed on deltaic deposits of fine sediments along the estuaries of the rivers which regularly flush the tidal waters back to the sea. The nutrient rich environment with characteristic salinity gradients in the estuarine zone creates the world's most productive habitats. Sundarban owes its existence to the fact that it is fed by two of the world's major rivers which together carry about one billion tons of sediments per year, i.e. 8 % of the total transported by all rivers to the world's oceans. It becomes obvious that any factor causing a change in the nature and amount of sediments or the salinity gradients in the delta also impacts the delta ecosystem adversely. While climate change-induced sea level rise are expected to affect these attributes from the seaward side, human activities on the landward side of the Ganga-Brahmaputra-Meghna basin has regular direct impact on the Sundarban through a complex interaction between hydrology, salinity and sediments.

Salinity in Sundarban estuary varies through the year and three distinct seasons are recognised in the tidal regime [73]. During the season of southwest monsoon, the effect of the flood tides is countered and almost completely nullified by freshwater inflow, and the ebb tides predominate strongly. During the next season lasting from November to February, the strength of the flood tide over the ebb tide becomes minimal. The third season includes the hot and dry months, May and June, just prior to the southwest monsoon, when the effect of the flood tides is much stronger than the ebb tides, and the estuary reaches maximum salinity [37]. In Sundaban, estuary salinity ranges from 11 to 25 PSU, with the highest being in the dry and lowest in the wet season [32].

The salinity regimes of the delta interacting with the characteristic hydrology predominantly govern the ecology and functions of the Sundarban [74]. Freshwater inflows from the rivers themselves exhibit high seasonal variability, and the 2–6 m high tides extend their influence on flooding duration and salinity up to 50 km inland. In the Indian Sundarban about 2,000 km^2 is occupied by tidal river systems. Within this region however only the western section drained by the Hooghly and Muri-Ganga, has a fresh water flow. In the eastern section, the major rivers – Saptamukhi, Thakuran, Matla, Gosaba and Harinbhanga – have lost their upstream connections with the Ganga, and therefore are now mainly tidal in nature [75]. On account of heavy siltation and sewage dumping by Kolkota city, together with the loss of freshwater, the salinity has increased in this sector by 6 psu over the past three decades [75]. The salinity in the western sector, on the hand, has decreased continuously (1.67 psu/decade). In Bangladesh, salinity rises from the east to west, with highest salinity observed in the south-western extremes of the mangroves. Islam and Gnauck [20] monitored 13 main rivers in the Bangladesh Sundarban, to highlight an overall increasing trend in salinity, with the rate of increase being highest in the southern and western rivers, and lowest in the northern section. Overall, these changes in salinity point to a decrease, or relatively stable salinity on the northern, western and to some extent, the eastern extremes of the combined India-Bangladesh Sundarban deltaic region, while the central portion shows a rapid increase in salinity, linked to a continuous decrease in fresh water flows and silting up of rivers and creeks. In estuaries however the salinity usually increases away from a freshwater source such as a river, although evaporation sometimes causes the salinity at the head of an estuary to exceed seawater.

Reduced freshwater flows affect the area in other ways as well. During ebb tides the receding waters cause scouring of the top soil, creating innumerable tiny creeks originating from the centre of the moving islands. The receding waters carry huge volumes of silt, which is deposited along the banks of rivers and creeks during high tides. This increases the height of the river banks, as compared to the interiors of the islands. Over time eroded channels extend further and further into the islands main body, forming muddy blanks which may check penetration of high tides into the interior of the islands. Irregular flooding of these blanks by high tides, coupled with capillary action of the clayey soil and excessive heat during dry periods, results in the deposition of a salt crust on the surface, converting them into saline blanks. Lands on the sea faces are further both continually denuded by tidal waves, and built up by wave action depositing silt back onto the shores [4].

Most of the changes described above are a consequence of the human activities in the basin of the two major rivers. Closest to the delta, most significant has been the construction of the Farakka Barrage (started in 1961, operationalised in 1975) across the river Ganga, in Nawabganj district, about 15 km from the border of Bangladesh. Its object was to divert Ganga waters into the Hooghly for flushing out the heavy silt load in the river that seriously hampered river navigation at the Calcutta port during the 1950s and 1960s. The barrage now diverts over 40,000 cu ft/s of water from the Ganga into the Hooghly during the dry season. This also mitigated the problem of salinity and provided Calcutta and its surrounding areas with irrigation, domestic and municipal water [76]. The barrage created hydrological changes in Bangaldesh during the dry season, impacting hydrology, ecology and livelihoods negatively. It became a reason for conflict until a water sharing agreement was signed between India and Bangladesh in December 1996.

The impacts of Farakka barrage on soils, agriculture, water supply, people, livelihoods and the ecology of Sundarban have been discussed, often passionately, in numerous publications in the two countries. In India, of the greatest concern has been the impact on hilsa fisheries of the Ganga. Hilsa (*Tenualosa ilisha*) is the only major anadromous species migrating in the Indian river systems, and commercially also the most important fishery of the estuary. It ascends to freshwater reaches for breeding during the monsoon season. The volume of monsoon flows and number of flood pulses have direct bearing on the quantum of its migratory population. The fishery fluctuates widely with the rainfall in the catchment areas and the volume of freshwater discharge. The barrage has been an effective barrier to the migration of Hilsa upstream and has resulted in steep decline in the fish catch both upstream and downstream of the barrage. However, freshwater diversion through the feeder canal has altered the hydrology and salinity gradient of the Hooghly estuary, and consequently the extension of the freshwater tidal zone has occurred till further downstream in the Indian Sundarban. With the habitats gradually turning favourable for Hilsa spawners, breeding grounds have shifted to downstream of the Farakka Barrage, showing species adaptation to changing situations. These changes are also accompanied by significant changes in the species composition and yields of estuarine fisheries which are now being slowly but steadily replaced by freshwater species [77]. Similarly, several dams and barrages in the Damodar River basin have resulted in a decreased silt load and lower deposition of detritus in the estuaries downstream.

In contrast, impacts of human activities farther from the Farakka, in all countries of the two river basins, have so far received relatively little attention in the context of the Sundarban and its conservation. However, various land use and water management practices throughout the GBM basin have direct bearing on the mangrove and its productivity. Hydrological alterations that have been already made or are planned in the near future, either for irrigation, domestic supply, flood control or hydropower, will change not only the flow regimes of freshwater into the Sundarban, but also the supply of sediments and to some extent nutrients and pollutants. Large scale storage and diversion of water greatly reduces the water flow during the dry summer season. At the same time, sediments trapped in reservoirs behind barrages accumulate in the river beds causing aggradation. In the absence of sediment transport to the Sundarban, the subsidence processes will not be balanced by accretion and therefore will aggravate impacts of sea level rise. A combination of sustained erosion, subsidence and sea level rise therefore implies that the lower areas and islands will continue to fall below sea level and disappear with time. Severe coastal erosion and areal reduction over the last three decades are resulting in the loss of major areas on the northwest coast. Both numerous households and a significant area of agricultural land and coastal stretches for fish drying have been lost, rendering thousands of people homeless as 'environmental refugess' [78].

During the last 20 years four islands have been submerged entirely, leaving thousands of displaced families. Lohachara Island, located in the Hoogly river in the delta, submerged completely by the 1980s, displacing about 10,000 inhabitants. During the same time period the neighbouring island of Suparibhanga (or Bedford Island) also disappeared. New Moore Island (as it was known in India) also called the South Talpatti (in Bangladesh) emerged from the sea in the early 1970s and disappeared again by 2010. Kabasgadi island partially disappeared and partly attached to the mainland [78, 79]. It is also recorded that most of Ghoramara island has also eroded which has forced the people of this island to migrate to Sagar island, the largest island of the Western Sundarban. Ghoramora is expected to disappear in the near future [80]. In South Sagar there is new land creation known as 'Char' land, whereas the other shores of the island are facing rapid erosion [81–83]. Studies state that about 6,000–7,000 people from Lohachara island have had to move to Sagar Island. Also the New Moore island which was about 50 years old and was about two meters above msl, located about two kilometres from the mouth of Hariabhanga river has completely disappeared from the map [84]. It is estimated that 30,000 people will lose their homes by 2020 as 15 % of the Sundarban's habitable land will disappear.

5.5.2 Land Use Changes Over Time

Among other factors, land use changes within the Sundarban are also important. Over time the mangroves of the Sundarban were inhabited, and slowly converted into agriculture land and aquaculture on a large scale. A few parts of the Sundarban were protected and conserved, initially to extract resources, and later for the protection of biodiversity. In the late nineteenth century, as the rate of agricultural conversion increased, the colonial Forest Department sought to preserve large areas of the remaining Sundarban tidal forest by giving them legal status as Reserved or Protected Forests. These forests were intensively managed to provide a sustainable supply of timber and firewood. Current government policy in both Bangladesh and India supports the use of the Sundarban as forest rather than its transformation to agricultural land. Further expansion of cropland to meet the grain demands of the burgeoning regional population in both countries has largely taken place outside the boundaries of the Sundarban. Overexploitation of these forests for wood products remains a possibility, but large-scale clearing for rice paddies is unlikely under present policies [85]. At the same time wood cutting, fishing, and shrimp farming have caused the deterioration of vast areas of the forest [86–88], and rehabilitation projects in the degraded forests often fail to restore the high ecological values of the original forests [89–91]. Some restoration projects result in monospecific mangrove plantations, predominantly *Rhizophora* species [92].

A clear example of overexploitation of the mangroves is the fast dwindling population of tiger prawn in the Sundarbans. The natural abundance of its seed is reaching a threshold, the primary reason being over fishing at all stages of the life cycle of the shrimp. At its post larval stage in estuaries, it is trapped by fine push and drag nets and fine meshed bag nets (meen jal); juveniles are trapped by bag nets (behundi jal) in estuaries; the juveniles and pre adults are caught in marine waters by large bag nets; the pre adults and adults by trammel nets. Even the spawns is not spared and is caught from open seas using trawl nets [93]. During the collection of prawn seed, fisherfolk remove the *Penaeus monodon* seed and throw away the other 90–95 % of fish and prawn seed which they also catch along with the prawn, ecologically an enormously wasteful practice.

Another sign of degradation of a system is the spread of invasive species [94]. Over the past few decades invasive species have spread significantly across the Sundarban ecosystem. Biswas et al. [95] identified 23 species of aquatic plants and climbers as invasives, of which 19 are native or naturalized to the mangroves. The invasion is still 'controllable' as the abundance, diversity and rate of invasion of the invasive species were highest at the riverbanks and gradually decrease with increased proximity to the forests. A further sign of deterioration of the system is eutrophication along with the presence of toxic Dinoflagellates and Cyanophyceae. Studies on the eutrophication and water quality suggest that this would promote invasive species and cause loss of mangrove character [32].

5.5.3 An Example of Expected Impacts of Climate Change

A World Bank working paper by Dasgupta et al. [96] integrates information on climate change, hydrodynamic models and geographic overlays to assess the vulnerability of coastal areas in Bangladesh to larger storm surges and sea-level rise by 2050. An estimated 27 cm sea level rise and 10 % intensification of wind speed from global warming suggests that the current vulnerable zone increases in size by 69 %, given a +3 m inundation depth and by 14 % given a +1 m inundation depth. At present, Bangladesh has 123 polders, an early warning and evacuation system, and more than 2,400 emergency shelters to protect coastal inhabitants from tidal waves and storm surges. With the changing climate the paper estimates that 59 of the 123 polders would be overtopped during storm surges and another

5,500 cyclone shelters (each with the capacity of 1,600 people) to safeguard the population would be needed. Investments including strengthening polders, foreshore afforestation, additional multi-purpose cyclone shelters, cyclone-resistant private housing, and further strengthening of the early warning and evacuation system would cost more than $2.4 billion with an annual recurrent cost of more than $50 million. However, a conservative damage estimate in the paper suggests that the incremental cost of adapting to these climate change related risks by 2050 is small compared with the potential damage in the absence of adaptation measures.

5.6 Conclusion

Though mangrove ecosystems are receiving increasing attention, we are still far from understanding the energy flow and food web dynamics in the Sundarban, and how the mangroves connect with other ecosystems. There is a need to better understand the effects of environmental change and pollution on local species. Dependent fauna need additional study with respect to larval supply and recruitment.

Although numerous and relevant studies of the Sundarban have been made in recent years, there has been little effort to bring them together and synthesise them for a holistic management plan and implementation action plan. Long term and sustainable conservation of the Sundarban requires that the mangroves be treated as distinct from other forests. Unlike the evergreen or deciduous forests, mangroves experience a highly dynamic environment and are in a state of continued flux. They are governed by their specific flooding regimes which govern the salinity gradients, nutrients and supply of fine sediments. These primary drivers are directly controlled by the biophysical, climatic and anthropogenic processes in their watershed – the basins of the rivers whose delta they occupy. The Sundarban is unique in its size and diversity because of the extent and nature of the GBM basin that also in a way includes the Himalayan ranges which influence the freshwater flows and the sediments. The future of mangroves depends upon the management of land and water resources throughout the Ganga-Brahmaputra basin that in turn requires an integrated river-basin-wide appropriate strategy backed by political will.

References

1. Giri C, Pengra B, Zhu Z, Singh A, Tieszen LL (2007) Monitoring mangrove forest dynamics of the Sundarbans in Bangladesh and India using multi-temporal satellite data from 1973 to 2000. Estuar Coast Shelf Sci 73:91–100
2. Giri C, Ochieng E, Tieszen LL, Zhu Z, Singh A, Loveland A, Masek J, Duke N (2010) Status and distribution of mangrove forests of the world using earth observation satellite data. Global Ecol Biogeogr 20(1):154–159, January 2011 (Article first published online: 17 Aug 2010)

3. Ellison AM, Farnsworth EJ, Merkt RE (1999) Origins of mangrove ecosystems and the mangrove biodiversity anomaly. Glob Ecol Biogeogr 8:95–115
4. Gopal B, Chauhan M (2006) Biodiversity and its conservation in the Sundarban Mangrove Ecosystem. Aquat Sci 68:338–354
5. Chaudhuri AB, Choudhury A (1994) Mangroves of the Sundarbans, vol. 1: India. World Conservation Union, Gland, p 247
6. Guha Bakshi DN, Sanyal P, Naskar KR (eds) (1999) Sundarbans Mangal. Naya Prokash, Calcutta, p 771
7. Hussain Z, Acharya G (1994) Mangroves of the Sundarbans, vol 2: Bangladesh. (ed) World Conservation Union, Gland, p 257
8. Islam MS, Wahab MA (2005) A review of the present status and management of mangrove wetland habitat resources in Bangladesh with emphasis on mangrove fisheries and aquaculture. Hydrobiologia 542:165–190
9. IUCN-BD (2002a) Molluskan Biodiversity of Sundarban – final report. IUCN-Bangladesh, Dhaka
10. IUCN-BD (2002b) Draft Report on the Orchid and Lichen of the Sundarbans. IUCN-Bangladesh, Dhaka
11. IUCN-BD (2002c) Draft interim report on Ornithological Study of Sundarban, IUCN-Bangladesh, Dhaka
12. Naskar KR, Guha Bakshi DN (1987) Mangrove swamps of the Sundarbans: an ecological perspective. Naya Prokash, Calcutta, p 263
13. Seidensticker J, Kurin R, Townsend AK (eds) (1991) The commons in South Asia: societal pressures and environmental integrity in the Sundarbans of Bangladesh. Proceedings from a workshop held 20–21 November 1987 in Washington, DC. The International Center Smithsonian Institution, Washington, DC
14. Anonymous (2009) In-depth recovery needs assessment of cyclone Aila affected areas. Report by Action Aid, Concern World Wide, Dan Church Aid, Muslim Aid, Islamic Relief, Oxfam-GB and Save the Children-UK, p 36. http://www.lcgbangladesh.org/DERweb/Needs%20Assessment/Reports/In-depth%20Recovery%20Needs%20Assessment%20of%20Cyclone%20Aila%20Affected%20Areas_ECHO%20funded.pdf
15. Mukherjee R (2011) Introduction: Bengal and the Northern Bay of Bengal. In: Mukherjee R (ed) Pelagic pathways: the northern Bay of Bengal before colonialism. Primus Books, New Delhi, pp 1–262
16. Eaton RC (1993) The rise of Islam and the Bengal Frontier: 1204–1760. University of California Press, Berkeley, 310
17. O'Malley LSS (1914) The Bengal District Gazeeteers: 24 Parganas. Govt. of West Bengal. Reprinted from the original in 1998 and 2009. Logos Press, New Delhi, p 268
18. Sarkar SC (2010) The Sundarbans: folk deities, Monstors and Mortals. Social Science Press, 69 Jor Bagh, and Orient Blackswan, New Delhi, p 205
19. Ray R, Chakraborty I, Bhattacharyya N (2002) A study among some "Meendharas" of Sundarbans, West Bengal. Anthropologist Special Issue No. 1: 83–89
20. Islam SN, Gnauck A (2008) Mangrove wetland ecosystems in Ganges-Brahmaputra delta in Bangladesh. Front. Earth Sci China 2(4):439–448
21. Hunter WW (1877) A statistical account of Bengal – vol 4: Districts of Bardwan, Bankura and Birbhum. Trubner & Co., London, 468
22. Jha VC, Bairagya HP (2011) Flood plain evaluation in the Ganga-Brahmaputra delta: a Tectonic review. Ethiopian J Environ Stud Manage 4(3):12–24
23. Rahman N Billah MM, Chaudhury MU (1979) Prepoaration of an up to date map of Sundarban forests and estimation of forest areas of the same by using Landsat imageries. Second Bangladesh National Seminar on Remote Sensing (9–15 Dec), Dhaka
24. Sanyal P (1983) Mangrove tiger land: the Sundarbans of India. Tigerpaper 10(3):1–4
25. Alam M (1996) Subsidence of the Ganges–Brahmaputra delta of Bangladesh and associated drainage, sedimentation and salinity problems, Chapter-8, Sea-level rise and coastal subsidence; causes, consequences, and strategies. Kluwer Academic, Dordrecht, pp 169–192

26. Goodbred SL Jr, Kuehl SA (1999) Holocene and modern sediment budgets for the Ganges-Brahmaputra river system: evidence for highstand dispersal to flood-plain, shelf, and deep-sea depocenters. Geology 27(6):559–562
27. Goodbred SL Jr, Kuehl SA (2000) The significance of large sediment supply, active tectonism, and eustasy on margin sequence development: Late Quaternary stratigraphy and evolution of the Ganges–Brahmaputra delta. Sediment Geol 133:227–248
28. Bowman HHM (1917) Ecology and physiology of the red Mangrove. Proc Am Philos Soc 56(7):589–672
29. Davis JH (1940) The ecology and geologic role of Mangroves in Florida. Publications of the Carnegie Institution of Washington, No 517, pp 303–412
30. Ellison AM, Stoddart DR (1991) Mangrove ecosystem collapse during predicted sea level rise: holocene analogues and implications. J Coast Res 7:151–165
31. Biswas H, Dey M, Ganguly D, De TK, Ghosh S, Jana TK (2010) Comparative analysis of phytoplankton composition and abundance over a two-decade period at the land-ocean boundary of a tropical Mangrove ecosystem. Estuaries Coasts 33:384–394
32. Manna S, Chaudhuri K, Bhattacharyya S, Bhattacharyya M (2010) Dynamics of Sundarban estuarine ecosystem: Eutrophication induced threat to mangroves. Saline Syst 6:8. http://www.salinesystems.org/content/6/1/8
33. Mitra A, Banerjee K, Bhattacharaya DP (2004) In: The other face of mangroves. Published by the Department of Environment, Government of West Bengal, India
34. Mukherjee BB (1984) Mangrove forest types and their distribution in Sundarbans, India. In: Proceedings of the Asian symposium on Mangrove environment – research and management, p 7581
35. Sodhi NS, Brook BW, Bradshaw CJA (2007) Tropical conservation biology. Blackwell Publishing, Oxford, p 332
36. Jhingran VG (1977) Fish and fisheries of India. Hindustan Publication, New Delhi, Ch Xv, p 954
37. Chandra G, Sagar RL (2003) Fisheries in the Sundarbans: problems and prospects. zunia.org/post/fishery-in-sundarbans-problems-and-prospects
38. Ray S (2008) Comparative study of virgin and reclaimed islands of Sundarbanmangrove ecosystem through network analysis. Ecol Modell 215(1–3):207–216
39. Gundermann N, Popper DM (1984) Notes on the Indo-Pacific mangal fishes and on mangrove related fisheries. In: Por D, Dor I (eds) Hydrobiology of the Mangal. Dr. W. Junk, The Hague, pp 201–206
40. Lowe-McConnell RH (1987) Ecological studies in tropical fish communities. Cambridge University Press, Cambridge
41. Chaudhuri A, Mukherjee S, Sen S, Chakrabarty S, Homechaudhuri S (2012) A comparison of spatial and temporal pattern of fish diversity of Matla river and adjacent mudflats in Sunderban biosphere reserve, India. Clarion-Int Multidisciplinary J 1:46–55
42. Ram ASP, Shanta N, Chandramohan D (2003) Seasonal shift in net ecosystem production in a tropical estuary. Limnol Oceanogr 48(4):1601–1607
43. Gupta GVM, Thottathil SD, Balachandran KK, Madhu NV, Madeswaran P, Nair S (2009) CO2 supersaturation and net heterotrophy in a tropical estuary (Cochin, India): influence of anthropogenic effect. Ecosystems 12(12):1145–1157, http://www.wildbengal.com/urls/con_ar_sunderban.htm
44. Mukhopadhyay SK, Biswas H, De TK, Jana TK (2006) Fluxes of nutrients from the tropical river Hooghly at the land-ocean boundary of Sundarbans, North-east coast of Bay of Bengal, India. J Mar Syst 62:9–21
45. Gle C, Del Amo Y, Sautour B, Laborde P, Chardy P (2008) Variability of nutrients and phytoplankton primary production in a shallow macrotidal coastal ecosystem (Arcachon Bay, France). Estuar Coast Shelf Sci 76:642–656
46. Jie X, Kedong Y, Lei H, Xiangcheng Y, Ho Alvin YT, Harrison PJ (2008) Phosphorus limitation in the northern South China Sea during late summer: influence of the Pearl River. Deep Sea Res Part I: Oceanogr Res Pap 55:1330–1342. doi:10.1016/j.dsr.2008.05.007

47. Dasgupta S, Huq M, Khan ZH, Ahmed MMZ, Mukherjee N, Khan MF, Pandey KD (2010a) Vulnerability of Bangladesh to cyclones in a changing climate: Potential damages and adaptation cost. World Bank – Development Research Group (DECRG). World Bank Policy Research Working Paper No 5280. World Bank, Washington, DC
48. Khalil GM (1992) Cyclones and storm surges in Bangladesh: some mitigative measures. Nat Hazards 6(4):11–24
49. Murty TS, Flather RA (1994) Impact of storm surges in the Bay of Bengal. J Coastal Res, Special Issue No 12: Coastal Hazards, 149(161)
50. Islam T, Peterson RE (2009) Climatology of landfalling tropical cyclones in Bangladesh 1877–2003. Nat Hazards 48:115–135
51. Dube SK, Chittibabu P, Sinha PC, Rao AD, Murty TS (2004) Numerical modelling of storm surge in the head Bay of Bengal using location specific model. Nat Hazards 31:437–453
52. Mitra A, Halder P, Banerjee K (2011, February) Changes of selected hydrological parameters in Hooghly estuary in esponse to a severe tropical cyclone (Aila). Indian J Geo-Mar Sci 40(1):32–36
53. Mitra A, Mondal K, Banerjee K (2011) Spatial and tidal variations of physico-chemical parameters in the lower Gangetic delta region, West Bengal, India. J Spatial Hydrol 11(1, Spring):52–69
54. Murthy VSN, Rao DP, Sastry JS (1983) The lowering of sea surface temperature in the east central Arabian Sea associated with a cyclone. Mahasagar – Bull Natl Inst Oceonogr 16:67–71
55. Subrahmanyam B, Rao KH, Rao SN, Murty VSN, Ryan SJ (2002) Influence of a tropical cyclone on Chlorophyll-a concentration in the Arabian Sea. Geophys Res Lett 29(22):1–4
56. Shah MAR, Datta DK (2010) A quantitative analysis of Mangrove forest resource utilization by the dependent livelihoods. UNESCO-IHE, Delft
57. Mukherjee M, Roy Choudhury M, Tripathi R (2007) Fishers, fishery and gears in Sundarban wetlands. In: Mukherjee M (ed) Sundarbans wetlands, pp 115–131. Dept. of Fisheries, Aquaculture, Aquatic Resources and Fishing Harbours: Govt. of West Bengal
58. Vannucci M (1997) Supporting appropriate mangrove management. International Newsletter of Coastal Management. Intercoast Netw Spec Edn 1:1–3
59. Duarte CM, Middelburg JJ, Caraco N (2005) Major role of marine vegetation on the oceanic carbon cycle. Biogeoscience 2:1–8
60. Alongi DM (2012) Carbon sequestration in mangrove forests. Carbon Manage 3(3):313–322
61. Clough BF, Ong JE, Gong WK (1997) Estimating leaf area index and photosynthetic production in canopies of the mangrove Rhizophora apiculanta. Mar Ecol Prog Ser 159:285–292
62. Fromard F, Puig C, Mougin E et al (1998) Structure, above-ground biomass and dynamics of mangrove ecosystems: new data from French Guiana. Oecologia 115:39–53
63. Alongi DM (2009) The energetics of mangrove forests. Springer, Amsterdam
64. Berger U, Adams M, Grimm V (2006) Modelling secondary succession of subtropical mangroves: causes and consequences of growth reduction in pioneer species. Pers Plant Ecol Evol Syst 7:243–252
65. Ray R, Ganguly DC, Chowdhury M, Dey S, Das MK, Dutta SK, Mandal N, Majumder TK, Mukhopadhyay SK De, Jana TK (2011) Carbon sequestration and annual increase of carbon stock in a mangrove forest. Atmos Environ 45:5016–5024
66. Seidensticker J (1991) The Bangladesh Sundarbans as wildlife habitat: a look ahead. In: Seidensticker J, Kurin R, Townsend AK (eds) The commons in South Asia: societal pressures and environmental integrity in the Sundarbans of Bangladesh. Smithsonian Institution, Washington, DC, pp 180–194 + tables
67. IUCN (1997) Sundarban Wildslife Sactuaries (Bangladesh): World Heritage Nomination – IUCN Technical Evaluation. http://whc.unesco.org/archive/advisory_body_evaluation/798.pdf, p 12
68. Mukul SA (2007) Biodiversity conservation strategies in Bangladesh: the state of protected areas. Tiger Paper 34(3):28–31
69. Rahman LM (2000) The Sundarbans: a unique wilderness of the world. In: USDA Forest Service Proceedings Rocky Mountain Research Station, Fort Collins, CO, issue 15, vol. 2: 143–148

70. Mukul SA, Uddin MB, Uddin MS, Khan MASA, Marzan B (2008) Protected areas of Bangladesh: current status and efficacy or biodiversity conservation. Proc Pak Acad Sci 45(2):59–68
71. Choudhury JK (2002) Bangladesh country paper on National Forest Policy Review. Country report to the 19th session of the Asia-Pacific Forestry Commission. Ulaanbaatar, Mongolia
72. Mitra R, Bhattaharaya RN, Hazra S, Santra SC (2006) Mangrove conservation efforts and the local economy: a case study. Econ Polit Weekly 41(33):3612–3616
73. Oag TM (1939) Report on the River Hooghly and its headwaters. Commissioner for the port of Calcutta, Calcutta
74. Hughes CE, Binning P, Willgoose GR (1998) Characterisation of the hydrology of an estuarine wetland. J Hydrol 211(1–4):34–49
75. Mitra A, Banerjee K, Sengupta K, Gangopadhyay A (2009) Pulse of climate change in Indian Sundarbans: a myth or reality. Natl Acad Sci Lett 32(1&2):1–7
76. Salman MA, Uprety K (2002) Conflict and cooperation on South Asia's international rivers: a legal perspective. In: The International and National Water Law and Policy Series, Kluwer Law International, p 399
77. Vass KK, Manas KD, Tyagi RK, Katiha PK, Samanta S, Shrivastava NP, Bhattacharjya BK, Suresh VR, Pathak V, Chandra G, Debnath D, Gopal B (2011) Strategies for sustainable fisheries in the Indian part of the Ganga-Brahmaputra river basins. Int J Ecol Environ Sci 37(4):157–218
78. Ghosh T, Bhandari B, Hazra S (2003) Application of a bio-engineering technique to protect Ghoramara island (Bay of Bengal) from severe erosion. J Coast Conserv 9(2):171–178
79. Ali S (2005) Sea level rise and possible solutions to land loss in Bangladesh. Thesis Submitted to the Geology department, Oberlin College, Oberlin, OH, U.S.A.
80. Doyle T, Chaturvedi S (2011) Climate refugees and security: conceptualizations, categories and contestations. In: Dryzek JS, Norgaard RB, Schlosberg D (eds) The Oxford handbook of climate change and society. Oxford University Press, Oxford, pp 278–290
81. Ghosh T, Bhandari B, Hazra S (2001) Assessment of landuse/landcover dynamics and shoreline changes of Sagar Island through remote sensing. Paper presented at the 22nd Asian conference on remote sensing, 5–9 November, 2001, Singapore. http://www.a-a-r-s.org/acrs/proceeding/ACRS2001/PapersF/172HAZRA.pdf
82. Hazra S, Tuhin G, Dasgupta R, Gautam S (2002) Sea level and associated changes in the Sundarbans. Sci Cult 68:309–321, By Indian Science New Association, Calcutta
83. Kumar PKD, Gopinath G, Laluraj CM, Seralathan P, Mitra D (2007) Change detection studies of Sagar Island, India, using Indian Remote Sensing Satellite 1C linear imaging self-scan sensor III data. J Coast Res 23(6):1498–1502
84. Bhattacharjee TT, Chatterjee S, Verya B, Dutta D, Choudhury N (2011) The sinking "beauty" – Sundarbans. Paper presented at the Geo-Spatial World Form under the theme: Directions and DIrmension of the Geo-spatial Industry, pp 18–21. http://www.geospatialworldforum.org/2011/proceeding/pdf/tapaleena.pdf
85. Richards JF, Flint EP (1988) World deforestation in the twentieth century. Duke University Press, Durham
86. Dijksma R, Van Loon AF, Te Brake B, Van Huijgevoort MHJ (2010) Mangrove rehabilitation under climate change: towards a hydrological tool for quantifying the effect of expected sea level rise on mangroves. (Paper to Delta 2011 Conference) http://webdocs.alterra.wur.nl/internet/corporate/nieuws/abstract_Dijksma_delta2011.pdf
87. Dijksma R, Van Loon AF, Van Mensvoort MEF, Van Huijgevoort MHJ, Te Brake B (2010) An extended hydrological classification for mangrove rehabilitation projects; a case study in Vietnam. In: Tropical deltas and coastal zones; food production, communities and environment at the land-water interface. Cabi, ISBN 978-1-84593-618-1, p 475
88. Kairo JG, Dahdouh-Guebas F, Bosire JO, Koedam N (2001) Restoration and management of mangrove systems: a lesson for and from the East African region. South Afr J Bot Pretoria 67(3):383–389
89. De Leon TOD, White AT (1999) Mangrove rehabilitation in the Philippines. In: Streever W (ed) An international perspective on wetland rehabilitation. Kluwer Academic, Dordrecht

90. Lewis RR (2005) Ecological engineering for successful management and restoration of mangrove forests. Ecol Eng 24:403–418, Elsevier Science
91. Sanyal P (1998) Rehabilitation of degraded mangrove forests of the Sundarbans of India. In: Program of the international workshop on rehabilitation of degraded coastal systems, pp 25. Phuket Marine Biological Center, 19–24 January 1998. Phuket
92. Lewis RR (2001) Mangrove restoration – costs and benefits of successful ecological restoration. In review, Proceedings of the Mangrove valuation workshop Universiti Sains Malaysia, Penang, 4–8 April 2001. Beijer International Institute of Ecological Economics, Stockholm, Sweden
93. Mahapatra BK, Chatterjee P, Saha D, Datta NC (1999) Declining trend in the abundance of seeds of tiger shrimp, Panaeus monodon (fabricus) in the Sundarbans with suggestions for its restoration. In: Guhabakshi DN, Sanyal P, Naskar KR (eds), Sundarbans Mangal. Naya Prokash, Calcutta, pp 532–538
94. Biswas SR (2003) Invasive plants of Sundarbans. In: Interim report under SBCP project, IUCN Bangladesh, p 34
95. Biswas SR, Choudhury JK, Nishat A, Rahman MM (2007) Do invasive plants threaten the Sundarbans mangrove forest of Bangladesh. For Ecol Manage 245:1–9
96. Dasgupta S, Huq M, Khan ZH, Masud MS, Ahmed MMZ, Mukherjee N, Pandey K (2010b) Climate proofing infrastructure in Bangladesh: the incremental cost of limiting future inland monsoon flood damage. The World Bank Development Research Group, Environment and Energy Team, WB Policy Research Working Paper 5469, p 32

Wild Mountain goats (IBEX) grazing (Courtesy Geeta Bansal)

Part III
Ganga Water Quality and Flows

View of Ganga at Kaudiyala, near Rishikesh. the stretch from Kaudiyala to Rishikesh is known to be a stretch for adventure sports, like - rafting, beach camping etc. (Courtesy Nitin Kaushal)

Chapter 6
Nature's Cure of the Ganga: The Ganga-Jal

Devendra Swaroop Bhargava

Abstract Ganga had been famous since times immortal for its unique properties of extremely high self-purification and that its water does not purify even after long storages even in air-tight containers apart from many other beliefs like Ganga's water having some medicinal properties, physical and spiritual healing, ensuring salvation after inhalation of a few drops of the Ganga water, etc. The author through his experimental and analysis of data for several years had been able to establish the scientific evidences and logics for the stated special properties of the Ganga. The article also presents a fool-proof permanent strategy for the pollution control of the now severely polluted Ganga at a few places.

Keywords Self purification • Gangajal • Microbes • BOD • Exocellular polymer • Pathogens • Reaeration

6.1 Ganga, the Holy River

The Ganga originates from 'Gaumukh', the cow-head, 3,900 m above mean sea level. At 'Gauri Kund (3,100 m above mean sea level), Ganga makes an only fall of 30 m in its entire length of 2,525 km upto the Bay of Bengal. Because of its supposed descent from Heavens, Hindus regard Ganga as the purest, holiest and most sacred and consider that a dip in it or inhalation of its holy water would purge away their sins and ensure their 'moksha' (salvation) [1–5]. Ganga confluences with Alaknanda at Devprayag, and continues its mountainous course before emerging as a river of

D.S. Bhargava (✉)
IIT Roorkee (retired Environmental Professor), Bhargava Lane, Devpura,
Haridwar 249401, UK, India

Asian Institute of Technology Bangkok (former Visiting Environmental Professor),
Khlong Nueng, Pathum Thani, Thailand
e-mail: dsbhargava@yahoo.co.in

R. Sanghi (ed.), *Our National River Ganga: Lifeline of Millions*,

DOI 10.1007/978-3-319-00530-0_6,

the plains at Rishikesh. It flows through UttaraKhand via Haridwar and UttarPradesh via Narora, Kannauj, and Kanpur, to Prayag (Allahabad) where it is joined by another 'holy' river, the Yamuna at 'Sangam' where trillions of pilgrims bathe on the occasion of 'Kumbh Mela' every twelfth year [1, 6, 7]. From Allahabad, Ganga flows to Varanasi on its way to Kolkata via Patna. The Ganga bifurcates near Kolkata into Bhagirathi (Hooghly) and Padma, both merging into Bay of Bengal at Gangasagar [1].

Haridwar along Ganga was described by the famous seventh-century Chinese traveler Hyun Tsang as a holy place, and the Hindi epic 'Ramcharit Manas' was written by poet Tulsidas on the banks of Ganga at Assighat in Varanasi in the sixteenth century, while his contemporary, the Moghul Emperor Akbar (1542–1605), the greatest Muslim Emperor, regarded Ganga's water as nectar or drink of the Gods, 'Amrit' [1, 8]. Ganga has been used for agriculture, industry, pisciculture, navigation, wastewater disposal, drinking water supply, and recreational uses. People along its entire course use its water for religious bathing, drinking, morning ablutions, and culinary purposes, being seemingly unaware of the quality of the river and without realizing that Ganga has now been badly fouled [1, 9].

6.2 Hindu Convictions and Beliefs

Most Hindu families store Ganga jal in their homes fo religious rites including the offering of a few drops of Ganga jal to a dyeing person for salvation. Ganga jal contained in small tightly closed plastic bottles is now sold. Some religious 'pundits' speak of medicinal properties of Ganga jal apart from the fact that the Ganga jal does not putrefy even after long periods of storage in air-tight containers, that the Ganga jal does not cause any water borne disease even when consumed directly from the Ganga river or even from the polluted Ganga, that the holy Ganga has amazingly high self-purifying abilities, and that that the Ganga water contains some unknown volatile disinfecting material [3, 10]. Such mysterious properties and character of the Ganga and the Ganga jal had been baffling the minds of technologists, scientists, as well as laypersons. Author's experimental studies provide technologically oriented explanations for the stated Ganga mysteries, particularly the extraordinary and amazingly high self-purifying abilities of the Ganga despite its being polluted rather heavily in some of its stretches [3, 4, 11, 12].

6.3 Nature's Care of the Gangajal

The inborn or acquired immunity, immense conviction in the powers of Ganga-jal and the presence of some disinfecting material in the Ganga-jal protect the Indian people from getting infected with some water-borne disease in spite of their drinking (for religious rites) even the sewage-contaminated Ganga water [1, 3].

6.3.1 Literature Survey on Gangajal

Dr. Hanbury Hankin [13] reported that cholera microbes did not survive for long in the Ganga's water. He conducted studies on Ganga at Varanasi, where he collected samples below the cholera-infected dead bodies that had been thrown into the Ganga, and found that the cholera microbes died within a few hours of their contact with the Ganga water. Hankin inoculated cholera microbes in two beakers – one containing Ganga water and the other containing sterile distilled water – and noticed that the microbes died soon (in less than 3 h) in the beaker containing the Ganga water, whereas no great change took place in the living concentration of cholera microbes even after 48 h in the beaker containing the distilled water. Hankin, however, noticed that the Ganga water lost this strong disinfecting property after it had been boiled in an open container. From some of his observations, it is noted that the Ganga water, even if not heated, will, in less than 3 h, kill any cholera microbes that may have been added. However the same water, if heated, lacks this power. Hankin therefore, attributed the disinfecting property of the Ganga to the probable presence in it of some volatile material [13] Around the same time, a French physician, D. Herelle [14], was amazed to find 'that only a few feet below the bodies of men floating in the Ganga water who had died of dysentery and cholera where one would expect to find millions of dysentery and cholera germs, there were no germs at all'. Herelle also noted that, when Ganga water was added to cultures that had been grown from germs taken from patients suffering from the disease, after a period the germs were completely destroyed [1]. Another physician, Dr. C.E. Nelson [14], of what is now called the United Kingdom, observed that Ganga water taken from the Hooghly (one of the dirtiest mouths of the holy river) on board ships leaving Kolkata for England, remained fresh throughout the voyage, whereas ships leaving England for India found that the water which they took on in England would not stay fresh even until they reached Mumbai which was a week in voyage-time closer than Kolkata. None of these workers were, however, able to provide any conclusive scientific explanation for the extraordinarily high and good disinfecting power of the Ganga waters [1, 2, 4].

6.3.2 Research Carried Out by the Author

As far as the present Author is aware, no systematic study has been carried out to determine the bacterial 'die-out' rates and related kinetics in the Ganga, but it is well known that, in an adverse environment in any river, the enteric Bacteria die out naturally in a reasonably short time [1]. The biochemical oxygen demand (BOD) of a river that is subject to contamination by fresh sewage represents the amount of pathogens (of enteric origin) and coliform Bacteria which are present in the river [1].

The above-reported findings showing the strong disinfecting properties of the Ganga's water, and the Author' research showing the Ganga's extraordinarily high self-purifying character [2, 4, 11, 12, 15–19], together with the immunity developed in the Indian people right from a tender age (the Author has witnessed children and people of all ages doing their morning culinary routines in a stretch of the Ganga which was just a few meters downstream of a sewage drain entering the Ganga), backed by a strong religious conviction and faith [11, 20, 21] may explain the probable reasons why such people are not infected with any water-borne disease after drinking sewage-contaminated Ganga water [1, 21]. However, some stray cases of some water-borne diseases do get reported at the religious gatherings along the Ganga.

6.3.2.1 Unmatched Self-Purification of the Ganga

The Ganga river has been endowed with a superfine self-purifying ability [11, 12, 16, 17, 19, 22, 23]. In terms of BOD (the single most significant variable expressing any river's pollution potential), Ganga, according to the Author's recent research, has been vested with the highest-known ability to assimilate it in an extremely short time [1].

The BOD in any stream that is subject to pollution from sewage discharge is always present partly in the dissolved state and partly as colloidal and suspended material. These percentages of BOD would, apart from the ratio of river discharge to sewage discharge, mainly depend on the degree of treatment received by the sewage prior to its disposal into the river. Exocellular polymers, excreted in the endogenous phase of the bacterial life-cycle, act as excellent coagulants [11, 12, 19, 24–26] and provide for coagulation/destabilization of the colloidal material that is discharged with the wastewaters into the Ganga [1]. Pavoni et al. in 1972 found that exocellular polymers excreted by various species of bacteria in their endogenous growth phase are excellent coagulants and are not easily biodegradable. The pollution-deoxygenation-regeneration cycle of a stream would in its last stages involve large populations of microorganisms in an endogenous growth phase due to depleted food resources. Thus, a significant amount of exocellular polymers would be present in the stream to be available for coagulation of colloidal material and it s subsequent removal through settling from the wastes discharged into the streams. To confirm this, equal volumes of distilled water (containing no exocellular polymers) and Ganga water (presumably containing exocellular polymers) were separately mixed with equal volumes of sewage. The mixtures were placed in two different Imhoff cones for observations of sludge settling against time [1, 2, 11, 24]. Three test runs with different ratios of sewage to distill or Ganga water then were made and the results were plotted [3, 11, 12, 23, 24]. It was seen, as expected, that in each test run much quicker and far greater settling occurred in the Imhoff cone containing the Ganga River water. From the plots, it was also observed that as the ratio of Ganga water to sewage is increased, the coagulation and settling increases almost in the same proportion as the ratio of Ganga water to sewage. Some flocculation would also take place due to the stream velocity and bed gradient [11, 12]. These findings, therefore, show that after sewage enters a stream its colloidal material is quickly coagulated and removed as the Ganga is able to provide some mean velocity gradient (G) values which enable the destabilized colloids to flocculate, grow in size, and so be able to settle down at a rate of about 3 m per hour (69 m per day for a 1 mm diameter particle of specific gravity 1.001) and be removed linearly as per the established settling law from the river water [11, 12, 16, 17, 19, 23, 24]. As a result, as much as 60 % of the initial total BOD that becomes settleable after the coagulation and flocculation of the colloidal part of the BOD, gets removed very rapidly in as short a time as from 30 to 60 min [1, 11].

The remaining 40 % of BOD representing the soluble BOD after the discharge of the wastewaters into the Ganga is assimilated exponentially. The BOD rate constant, k, for such an exponential removal, is found to range from 3.5 to 5.5 per day (base e) in the Ganga [2, 11, 12, 19, 22]. When compared with $k = 0.23$ per day (base e), for most commonly reported [27, 28] values for most rivers and streams, the Ganga is seen to assimilate the BOD at a rate which is from 15 to 25 times faster [2, 11, 12, 19, 23] than the values usually reported [1, 2, 11, 12, 18, 19, 23, 27, 28]. Such extraordinary high values of k in the Ganga, brought out for the first time by the Author [11, 12, 16, 17, 19, 24] show the uniqueness of the Ganga River and support the Indian myth about its extraordinarily high self-purifying character. They are attributed to the presence of well adapted microorganisms in the Ganga waters (presence of toxicants inhibits the growth of microorganisms, resulting in a lower value of k, as was the case for Yamuna river at Delhi, where a large number of industrial units discharge toxicants into the drainage basin of the Yamuna) [6, 7, 11], to the very high reaeration abilities of the Ganga to cope with the high rates of oxygen demand for waste stabilization, and to favourable natural environmental conditions in the Ganga against the survival of pathogens and coliform Bacteria, and also to the River's configurations [1, 2, 11, 12, 16, 17, 24, 25].

This dual phenomenon involving the linear decay of the settleable BOD and the exponential decay of the non-settleable BOD has been modeled by the author [11, 12, 16, 17, 19, 23] as $S = S_{o-x}[1 - (v_s/d)t] + S_{o-y}\exp(-kt)$, in which S_{o-x} and S_{o-y} respectively represent the settleable and non-settleable components of S_o, the total initial BOD being in mg/l; S is the total BOD remaining at any time, t (days); v_s is the settling velocity (m/day) of the bioflocculated particles; d is the stream depth in m; and k is the kinetic rate constant (per day) of the non-settleable BOD. The model indicates that the settleable BOD would be completely removed within a striking distance of $T = d/v_s$ (days), after which the BOD exertion would only be due to the non-settleable BOD [1, 11, 12, 16, 17, 23]. The age-old Streeter & Phelps [27] models for BOD prediction and assimilation in streams, which have been in use for the last eight decades, can therefore no longer be considered rational, reliable, or accurate, because they do not take account of the settleable, suspended part of the BOD which is removed linearly after getting bioflocculated in the river system [11, 12, 16, 17, 23]. The Author's BOD predictive models, along with all their details, concepts, and verification, have been presented elsewhere [1, 11, 12, 16, 17, 19, 23].

Correspondingly, the reaeration rate constant, k_r, of the Ganga which works out to about 9 per day (base e), is also higher by an order of magnitude [11, 12, 16, 17, 19, 23], when compared with the usually reported values for other rivers. This ability is attributed mainly to the Ganga's turbulent flow, high width to depth ratios, shallow swiftly flowing waters, etc. As a result, the DO in the Ganga recovers very quickly and prevents undesirable anoxic conditions and consequent fish mortality. Within a distance of from 1 to 3 km from the wastewater outfalls, the DO has been found to recover to reasonably satisfactory levels of about 3 mg/l in the Ganga [8, 9, 11, 12, 16, 17, 23]. No where along its 2,525 km long course, the Ganga ever becomes completely devoid of its DO despite receiving heavy organic pollutions in some of its stretches.

The DO sag in the Ganga has been analysed in great detail. The traditionally used models for evaluating the DO sag-related parameters based on the classic Streeter & Phelps equation, should now be replaced by the Author's new models [11, 12, 16, 17, 19, 23] based on his model for the rational and accurate prediction of a stream's BOD. The derivations and verifications of the predictive models for the DO sag-related parameters, and other models for the polluted Ganga that is subject to extremely fast purification, as well as those for the point and non-point conditions of wastewater discharges into the Ganga, have been presented elsewhere [11, 12, 16, 17, 19, 23].

6.3.2.2 How Ganga Water Is Preserved?

The well tested facts that Ganga jal does not putrefy during its long storage even in sealed air-tight containers and that its drinking by local people directly from the polluted Ganga does not cause any water borne disease in them, involves many aspects such as the presence of pathogens (disease causing bacteria), the foul smelling methane-like gas producing anaerobic bacteria, disinfecting materials originating from the river-bed-soils within the Ganga course and its origin, etc. [3].

The water borne pathogens and coliform group of microorganism originating from the human excretions thrive rather comfortably in the human stomach and intestines in the absence of any oxygen at a temperature of 37 °C. They do not therefore, find themselves comfortable in the waters in an open environment where the various environmental parameters such as the temperature, pH, dissolved oxygen, etc. are not only very different but very hostile to the pathogens and coliforms. The anaerobic bacteria utilize and through decomposition, stabilize the organic material present in water essentially in the absence of dissolved oxygen (DO) to produce methane, the foul smelling gas manifesting staleness or putrefaction. On the other hand, the aerobic type of bacteria consumes and stabilizes the organic material (quantitatively represented through BOD, the biochemical oxygen demand) compulsorily in the presence of DO to produce odourless gases such as carbon dioxide and do not signify any putrefaction. From a consideration of sensitivity to environmental hostility, coliforms and pathogens are thus, closer to the anaerobes rather than to the aerobes [1–3, 10, 11].

The cholera microbes (vibrio, a pathogen) do not survive in Ganga-jal irrespective of the Ganga water being in its natural environment or in an air-tight container but survived up to even a few days in distilled water or in the waters of the river Yamuna [13]. The anaerobes being similar to the pathogens thus, die out (or become inactive) soon in the Gangas jal as do the pathogens. These explain why no putrefactions from anaerobic decomposition of organic material takes place in the Ganga jal even after long storages in closely air-tight containers. Further, due to extraordinary high DO recovery rates in Ganga [2, 3, 11, 12, 16, 17], the DO in Ganga waters is very ably recovered, recaptured and retained. The chances of anaerobic bacteria surviving in the Ganga jal become even more remote and the anaerobes are thus, additionally prevented or eliminated from putrefying the Ganga jal.

6.3.2.3 Religious Bathings In The Ganga At Haridwar

British engineers had built the Ganga canal starting right from the famous religious spot called *Har-ki-pouri* in Haridwar, thanks to thc late Shri MadanMohanMalvia (a British title awardee) for his negotiation efforts despite an opposition from the religious pundits (religiously learned people) who had wanted the canal to start some downstream distance of *Har-ki-pouri* containing the sacred spot, '*brahm-kund*' (the nectar pond where the nectar drop had reportedly fallen). *Har-ki-pouri* is thus not on the real Ganga but is situated on a tributary-canal of the main Ganga canal such that this tributary canal has *nalas* (open drains) and many domestic houses on its right bank discharging all their wastewaters directly and secretly into this tributary canal because the sewerage system and the pumps seldom work regularly or efficiently. On top of this, a *burning-ghat* (group of platforms where the Hindu-deads are cremated) as well as an electric crematorium is also situated on the upstream right bank of this tributary canal and occasionally one can witness ash and small half-burnt bone-pieces at the *brahm-kund* area of which fact, most pilgrims are totally unaware of although many even educated persons ignore this aspect due to their extremely high level of *astha* or *shradha* (belief/trust/reverence) in the holy waters of the Ganga. Haridwar citizens oppose the shifting of this burning *ghat* due to their zero aesthetic sense. Unaesthetic and vulgar-looking bathing by some female pseudo pilgrims at har-ki-pouri and other bathing ghats had been prominently witnessed after our independence. In the streets around har-ki-pouri, it is often seen that the sewage pump house(s) often do not work as a result, the stated streets get flooded with sewage and the cow-looking poor innocent pilgrims mistake this spilled sewage for Ganga-jal and apply it on their fore-heads to ensure reverence to the Ganga. Ganga's high pollution further increases due to the dip by millions of devotees at same time at the same spot and continuously for hours on certain religious days. It is a great environmental engineering challenge to keep the pilgrims epidemic-free during their religious mass-bathing days especially during *kumbh-mela* occasion when many unscrupulous pilgrims even urinate, defecate and spit while bathing and virtually every pilgrim touching the other due to heavy crowds. *Aachman* (direct inhalation of water) is performed by most pilgrims as a religious-rite but it is inadvertently forced upon every pilgrim during a water dip. Thus imagine the dangers (more and more on the downstream side) of epidemics of dysentery, cholera etc. Odour-less ozone, the most potently disinfecting agent can be a possible method to control the said health-hazard through its diffusion in the river-bed. In addition, the pilgrims should be forced to wear water-tight underwear's to prevent leakage of their urinating and defecating material and advised to keep their mouths closely shut while taking a dip.

6.3.2.4 Why Ganga Water Does Not Cause Diseases?

The Author has personally witnessed local people drinking Ganga water directly from the river, apart from doing their morning gargle and culinary routines in the Ganga stretches which are often most polluted from domestic wastewaters. Scientifically and technologically [3], what protects the Indian people from getting infected with some water-borne disease in spite of their drinking (for religious rites) even the sewage-contaminated Ganga-water, is perhaps the immunity that develops in them right from a tender age and the presence of some volatile material responsible to impart the disinfecting action to the Ganga jal apart from a possible strong conviction in the holiness and magical properties of the Ganga-jal held by the Indians, like we have a scientific conviction that a chlorinated water (manifested from a residual chlorine in the tap-water) or the commercially available mineral/bottled available drinking water will not cause any disease even when it is well tested that most of such bottled waters in India are spurious. Likewise, according to an Indian history/mythology, the family of Meera (a royal queen-in-waiting), did not approve of Meera singing and dancing in praise of lord *Krishna* in the streets, so they gave Meera a cup of poison to drink in the name of *prasad* (a God's offering) and Meera, having an immensely strong conviction in her devotion to the lord, drank the poison and nothing happened to her. So that is the power of conviction which perhaps causes some secretions in the human body to quell the ill effects which aspect needs scientific research and investigations. These facts also manifest the Indian belief that Ganga is the most 'holy' amongst rivers and the Hindu belief of regarding the Ganga jal as 'nectar' ('*amrit*') thus certainly 'holds water' [3].

6.3.2.5 The *Gangatva* Comes from the Ganga Course Bed

Today, most often, the real Ganga stream (the '*neel dhara*') no more exists after Haridwar (100–150 km downstream from the Ganga origin point, the 'Gaumukh', in the Himalayan glacier) where all the waters of the upland Ganga is diverted into the famous Ganga canal. The canal system at other downstream locations in the Ganga basin also takes away the most of the Ganga water. As a result, the Ganga after Haridwar most often does not carry the snow-melts from the Himalayan glacier, but carries only the continuously infiltering groundwater that along with tributaries water make up the Ganga river after Haridwar. However, such infiltering groundwater gets recharged from the original snow-melts from the Himalayan glacier, and is added with dissolved chemicals, salts and material present in the ground soil strata. Hankin through his experimental observations had concluded that the Ganga jal lost its disinfecting property (or its ability to kill or rather inactivate the pathogens) when it was boiled. He thus, inferred that some volatile material was responsible to impart the disinfecting action of the Ganga jal. Since, mostly the groundwater constitutes the Ganga of the plains (after Haridwar), the volatile material must be provided by the soils contained in the course of Ganga [2, 3, 9, 29]. This fact clearly manifests that gangatva, meaning abovestated disinfecting and other unique/special properties of the Ganga, does not come from Ganga's origin, the Himalayan glaciers through some medicinal plants as believed by some former Indian scientists, researchers and saints. Some Indian scientists strongly believed that bacteriophages which are considered pathogenic to the bacteria were responsible for the said gangatva which outdated belief can never hold any water because the bacteriophages are present in all waters and rivers but the gangatva and disinfecting power exist only in the Ganga. The soils adjoining the course of other rivers (including that of Yamuna River which is only about 50–100 km away from the Ganga River) do not thus, contain the magic but unknown mysterious volatile material. Bhagirath who according to Indian mythology brought the Ganga River from the 'Heavens' (meaning the Himalayan glaciers in the present context) to flow on the earth (as 'Bhargirathi' River, another name of the upstream/upland Ganga), was indeed a great genius engineer to have selected to canalize the Ganga along such a soil course which contained the magic volatile material responsible for imparting disinfecting properties to the Ganga jal. What magic volatile material, no one knows as yet, however? [3, 20, 30].

6.3.2.6 Water Quality Index and Beneficial Uses of Ganga Water

On the basis of observations taken from a water quality survey of the Ganga River by the Author [6, 8–10, 15, 20, 21, 25, 30], the water quality index (WQI) values for the various beneficial uses, as computed from the modified Bhargava WQI Model [6, 15, 20, 21] are given in the Table 6.1. It is easily seen that the variation in the WQI value is more significant for the uses in which the stated variable has greater relevance, as was desired in the said WQI model concepts [6, 15, 20, 21]. For example, an increase in the chloride content of a water from 50 mg/l to say 80 mg/l would bring the WQI down considerably for agriculture (from 60 to 30), slightly for fish culture and almost negligible for the public water supplies use, because of the unequal relevance of chloride to the different uses. A classification of the Ganga River on the basis of its WQI values for the various beneficial uses in the different stretches of the river is presented in Table 6.2.

Table 6.1 Computed values of the water Quality Index (WQI) along the Ganga [11, 20, 25, 30]

		WQI for beneficial uses Overall											
		Bathing, swimming		Public water supplies		Agriculture		Industry		Fish culture, wild-life		Boating, recreation	
Station WQI		S	W	S	W	S	W	S	W	S	W	S	W
Rishikesh	u/s	51	89	64	100	95	94	51	94	89	100	70	95
	d/s	34	90	46	97	96	94	52	92	85	99	63	94
Haridwar	u/s	30	82	44	90	96	93	54	78	81	99	61	88
	d/s	34	46	47	58	96	93	52	82	80	88	62	73
Narora	u/s	55	43	68	53	94	91	57	55	72	73	69	63
	d/s	52	41	64	51	94	90	61	57	66	72	67	62
Kannauj	u/s	56	75	69	80	89	87	61	57	71	95	69	79
	d/s	45	63	58	67	88	81	51	58	63	81	61	70
Kanpur	u/s	41	58	51	66	89	84	50	57	65	80	59	69
	d/s	11	60	20	70	86	81	49	53	23	81	38	69
Allahabad	u/s	50	49	62	61	82	80	56	52	66	90	63	67
	d/s	35	43	49	54	78	61	57	57	66	86	57	60
Varanasi	u/s	75	55	83	69	71	49	71	62	74	90	75	65
	d/s	33	42	47	50	70	46	72	60	53	71	55	54

S Summer, *W* Winter, *u/s* upstream, *d/s* downstream

Table 6.2 Classification of the Ganga River Water [11, 20, 25, 30]

		Class of the beneficial uses											
		Bathing, swimming		Public water supplies		Agriculture		Industry		Fish culture, wild-life		Boating, recreation	
Station		S	W	S	W	S	W	S	W	S	W	S	W
Rishikesh	u/s	III	II	III	I	I	I	III	I	II	I	II	I
	d/s	IV	I	III	I	I	I	III	1	II	I	III	I
Haridw	u/s	IV	1I	III	I	I	I	III	1I	II	I	III	I
	d/s	IV	III	III	III	I	I	III	II	II	II	III	II
Narora	u/s	III	III	II	III	I	I	III	III	II	II	II	III
	d/s	III	III	III	III	I	I	III	III	II	II	II	III
Kannauj	u/s	III	II	II	II	II	II	III	III	II	I	II	II
	d/s	III	III	III	II	II	II	III	III	III	II	III	II
Kanpur	u/s	III	III	III	II	II	II	III	III	II	II	III	II
	d/s	IV	III	IV	II	II	II	III	III	IV	II	III	II
Allahabad	u/s	III	III	III	III	II	II	III	III	II	I	III	II
	d/s	III	III	III	III	II	III	III	III	II	II	III	III
Varanasi	u/s	II	III	II	II	II	III	II	III	II	I	II	II
	d/s	IV	III	III	III	II	III	II	III	III	II	III	III

S Summer, *W* Winter, *u/s* upstream, *d/s* downstream

6.4 Failure of Ganga Action Plans

While, previous governmental attempts to clean up the Ganga River have miserably failed for political, social, technical, and scientific reasons [1, 2, 5–7, 10], Ganga, India's most revered national river now aspires for a world heritage status to ensure maintenance of its quality status.

The various socio-political and techno-scientific reason for the failure of previous Ganga action plans include:

(i) adoption of unscientific and arbitrary effluent standards instead of working them out scientifically through mass balancing keeping regard of the river's self-purifying abilities (kinetic coefficients). These effluent standards would thus vary for different polluters,

(ii) ignoring Indian conditions like narrow Indian streets disabling sewer laying; many private industrial activities in homes generating toxic wastes; slums in unsewered urban areas; religious mass bathing; open defecation and disposal of dead human and animal bodies; dumping of solid organic wastes including polythenes without practicing a self-supported sustainable strategy (4-S) to keep city streets clean through rag pickers to pick up and sell fruit-skin-vegetable wastes, organic wastes, nails, and newspapers/polythenes, etc. thus also avoiding menace of pigs and cattle on the streets. All this uncollected waste causes 50 % of urban, untapped, untreated wastewater to flow unabatedly into rivers. India's extremely high population, a severe constraint to any developmental action, can effectively be controlled in Indian situations only by limiting the privileges such as voting rights, ration card, free education, reservations, etc. to only one or two children,

(iii) further, the Indian culture and religion demand a very high water quality at Ganga banks where millions of Hindus perform religious rites (including *aachman,* the direct inhaling of waters), thus pollution control strategies successful in western countries cannot be successful or appropriate in India.

(iv) lack of pollution control defensive strategies such as segregation of industrial wastes as in-plant practices, wastewater recycling and reuse, general sanitation, and improved agricultural practices to reduce excessive use of fertilizers and insecticides by the greed of greater yield by the illiterate Indian farmers leading to eutrophication (a bane of green revolution) in the country make river cleaning plans ineffective,

(v) not implementing proactive or offensive strategies such as the regulated release of wastewater to ensure the maintenance of the pre-fixed river standard, artificial re-aeration and in-drain treatment practices using water hyacinths but very carefully that it does not enter the river system.

(vi) comprehensive river pollution control requires professionals with knowledge of several fields including civil engineering, public health or environmental engineering. The lack of experts in these fields (or high dominance of pseudo environmentalists incompetent to identify the relevant literature or real experts) with decision-making responsibilities within the Indian Ministry of Environment has been disastrous for success in previous Ganga clean-up projects. The lobby of pseudo-environmentalists and some opportunistic money-hungry non-engineers in consonance with some other vested interests managed to capture the Indian environmental arena/scene as can easily be manifested from the decision and policy making officers/staff composition in the Indian Ministry of Environment (responsible for country's environmental control and matters) where the real environmental engineers are outcastes and virus which fact is easily manifested from the failure of numerous environmental projects including the several Ganga Action Plans despite all possible efforts at all levels. To evade pseudo-ism in admission into environmental engineering courses at master's level, the author even made an in-person public interest litigation (PIL) to seek stay of admission of non-engineering graduates directly into the post-graduate course in environmental engineering in the hon'ble Supreme Court of India and argued before the ex-honorable Chief Justice of India, Justice Anand (who unfortunately, was himself in a controversy over his birth date records in the UK and India) but the petition was tossed out when the hon'ble apex court was finally asked if an engineering graduate could be admitted directly into a LLM course without having completed a LLB course from any Law Institute of India. This only manifests the severest level of disregard for academics and academic standards in India. Another serious anomaly is in matters of legal cases pertaining to environmental issues, because those who argue as well as those who listen and judge are not fully or well acquainted with or even fully understand the meaning and implications of the many environment related words/terms. Further, even an organization like the union public service commission (UPSC), the apex Indian commission for recruitment, does not differentiate between Master level degree in environmental engineering and environmental science or between real and pseudo environmental engineers and call both of them for interview for the same post together without regard to the nature of duties to be performed. How can the same duty be performed by candidates of both backgrounds ?. On the overall, the Indian authorities do not seem to have any will to remedy the river pollution through the use of the right personnel in the environmental engineering arena.

(vii) political connections or bribery used to secure high positions in government, judiciary, and academia have also stymied efforts to clean up the Ganga. Merit (evaluated through an index representing the integrated effect of all essential requirements), a bygone word in Indian system (infested with severe nepotism, regionalism, casteism, etc.), and right persons not placed in right places together make India a third world developing nation despite its highest qualified hardworking genius manpower in every discipline. As a result, much smaller but merit minded neighboring countries have their leaders/drivers of better merit (manifesting superiority in sports, diplomacy, decision making, politics, academics, etc.) although persons of much higher merit are easily and abundantly available in India who are only forced to make dust bins as their habitat or migrate to even less developed nations to work almost double at half wages compared to their local counter-parts. Thus, in India, right people can not be available to clean the rivers.

(viii) fine collectors for environmental violations demand bribes. A hefty bonus (kind of bribe legalization as well as incentive) to the fine collectors would ensure honest, duty bound and dedicated fine collectors in Indian situations. The corrupt and opaque judicial system makes the legislative approach to river pollution control ineffective through delays and politically motivated judges and lawyers. Administrators deliberately ignore the principles of natural justice to victimize/harass their non-favorite subordinates as they can not be booked till they retire, due to court delays. Indian courts need to be liberal in accepting criticism without any contempt threat, avoid media censuring on any pretext or bitter reporting, classify themselves (subject-wise, like criminal, civil, educational, environmental, service-matters, etc.), appoint mobile magistrates for immediate evidence recording before the evidence is killed or dies, and judges be paid per case as reward for timely disposal of pending 100 million cases in India. Indian Lawyers not understanding pollution and its implications argue such cases only to extort money on pretexts, delay matters for regular income, misuse power of attorney by getting blank papers signed, and hob-knob with opposite parties which add to the continued environmental degradation. The Right to Information Act (RTI), promulgated in 2005, could make government and judiciary decision-making more transparent, however its implementation is excruciatingly slow and tardy and RTI requests have incurred police wrath, harassment, bribe, and assault, apart from being subjected to an easy evasion on any pretext (recently, an information pertaining to the events of the year 1994 being more than 20 years old in the year 2006 has been held right and correct after lengthy court proceedings in Nainital High Court WP No.726/2008). The public is thus severely discouraged to use the RTI Act tool for cleaning of Indian rivers.

(ix) the Indian political system, thriving on muscle-money-caste-region-religion powers, needed for infusing a political will to clean the Indian rivers, can be cured through the adoption of a Presidential form of democracy where the top person commanding the mandate of the entire country will have his own team to work uninterruptedly for the term, and a concept of weighted votes starting from the grass-cutter as one to the highest say 1,000 or 10,000 equivalent votes just as persons of different academic credentials are paid differently. The politics of regionalism (like the anti-Hindi-speaking movement in Maharashtra, Punjab, Assam, etc.) can most effectively be curbed through a 'national integration model' wherein every state will employ/

promote its own natives (the main cause of unrest and fear of native's job being taken by outsiders) and 50 % employees in each state come from all other states to promote national integration, understanding of each others culture, habits etc. In metropolis, entire population is equally (population proportion basis) taken from all states. This will also reflect the real Indian culture to a foreign visitor,

(x) the most rampant corruption in India wastes most of the money allocated to river cleaning works. Former Prime Minister of India Rajeev Gandhi (1984–1989) stated that only 15 % of allocated public funds reach beneficiaries or project execution stage. His son, Rahul Gandhi, 38, the likely future prime minister of India, recently lowered this figure to 5 %. Mammoth river cleaning projects cannot be completed when so many public funds are siphoned off through corruption.

(xi) the Indian public (including saints) has been ineffective in building any pressure to clean the rivers. Increased awareness through seminars, media, non-governmental organizations, and public education could help enhance public support and participation in river clean up campaigns.

For these reasons, the implementation of numerous action plans cannot ever result in a pollution-free Ganga River even with the most modern wastewater treatment facilities

6.5 Permanent Foolproof Prevention of Pollution of Ganga

Instead, a more effective, fool-proof strategy is to prevent wastewater from entering the river by creating a barrier between the river and cities along the Ganga. This barrier could be an embankment ("*bandha*") or dam-like structure, as shown in the Fig. 6.1, constructed so that all urban wastewater is trapped, through a huge sewer or covered canal positioned parallel to the river, with or without any treatment depending on available municipal finances. The wastewater would be carried to a point some two to three kilometers downstream of the city for treatment (in accordance with available funds) and disposal, allowing the Ganga River to naturally purify itself before reaching the next urban center. The distance of this stretch of the river between two cities would depend on the self-purification coefficients of the river, degree of waste treatment in the previous city, river configurations, stream width-to-depth ratio, stream velocity, and other factors. If the distance between two cities is short, then the wastewater treatment at the upstream city should be more intense to produce cleaner effluent.

The stated fool-proof preventive strategy ensuring that not a drop of any wastewater would enter the Yamuna at Delhi would be ideal and only solution for Delhi, the national capital situated along Yamuna, a tributary of Ganga.

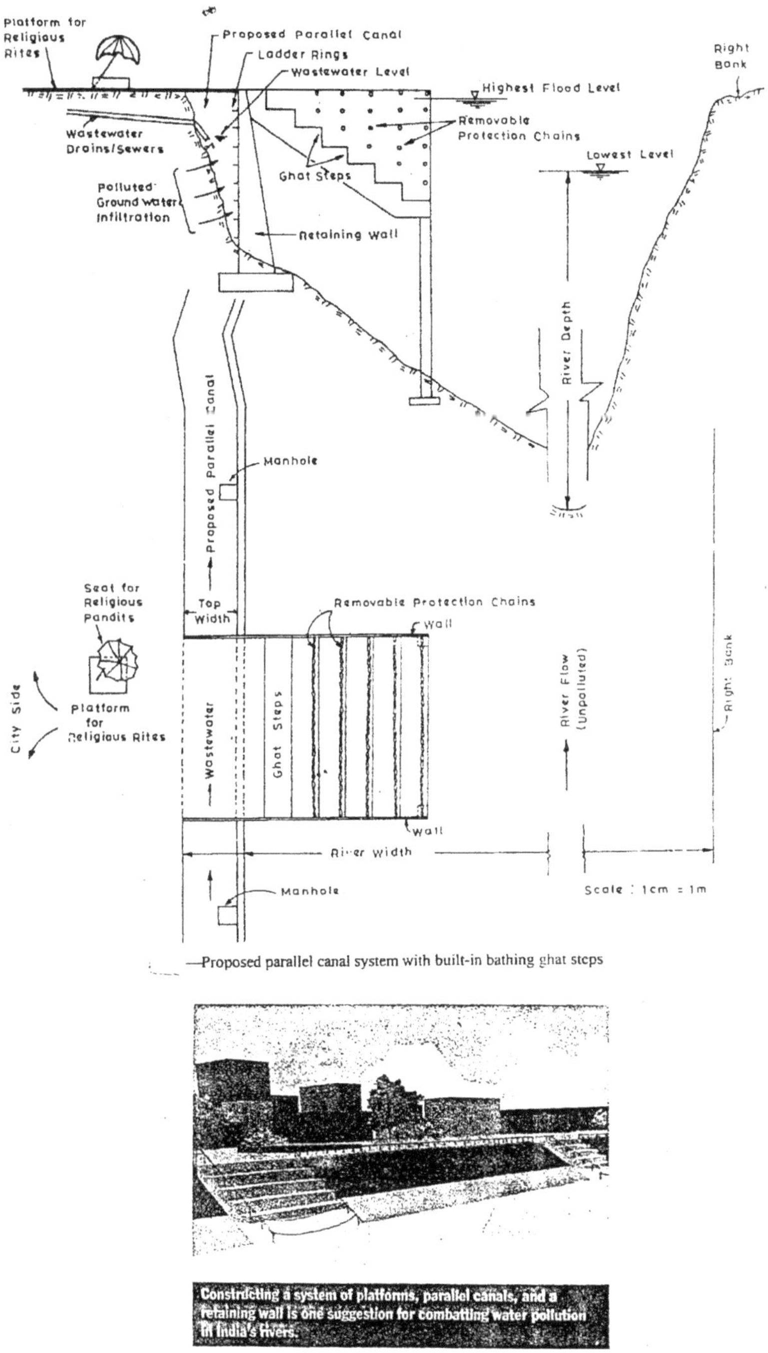

Fig. 6.1 Sketch for the Foolproof Pollution Control of Ganga River

References

1. Bhargava DS (1987) Nature and the Ganga. Environ Conserv Sci J Devoted Glob Surv, Geneva, Switzerland 14(4, Winter):307–318 & 328
2. Bhargava DS (1998) Sacred but unclean. Asian Water Environ J Water Environ Federation USA 1(2):13–16
3. Bhargava DS (2003) Scientific explanations for mysteries associated with holiness of Ganga. Civil Eng Today, Biomonthly Tech J Am Soc Civil Eng India Section (ASCE-IS) XXVI(3, May–June):6–8
4. Bhargava DS (2011a) Ganga most deserves a world heritage status. Aditya Int J Res Technol Manage 1(6):1–16, also printed in the on-line International Journal of Contemporary Practices, as Ganga most Deserves a World Heritage Status – Part I 1(7), pp 1–17
5. Bhargava DS (2011b) Permanent Foolproof prevention of pollution of Ganga (Ganga most deserves a world heritage status – Part II). Aditya Int J Res Technol Manage 1(7):1–19, also printed in the on-line International Journal of Contemporary Practices, as Permanent Foolproof Prevention of Pollution of Ganga (Ganga most Deserves a World Heritage Status – Part II) 1(8), pp 1–19
6. Bhargava DS (1985d) Matching river quality to use: role of monitoring, zoning and classification. Environ Prof, Official J Natl Assoc Environ Prof (NAEP), USA 7(3 (Quarterly)):240–247
7. Bhargava DS (2006) Revival of Mathura's ailing Yamuna river. Environmentalist 26(2):117–130
8. Bhargava DS (1984) Water quality survey of the Ganges. Asian Environ, Philippines, 6(1), First Quarter:4–6 (Also the 14–17 December 1982, Ist international conference on industrial pollution and control, Hotel Equatorial, Republic of Singapore, organized by ECMC, Suite 815, World Trade Centre, 1, Maritime Square, Singapore)
9. Bhargava DS (1985c) Expression for drinking water supply standards. J Environ Eng Am Soc Civil Eng, EED 111(3), Paper no 19773, June:304–316
10. Bhargava DS (1992) Why the Ganga (Ganges) could not be cleaned. Environ Conserv Switzerland 19(2):170–172
11. Bhargava DS (1983a) Most rapid BOD assimilation in Ganga and Yamuna Rivers. J Environ Eng, Am Soc Civil Eng 109(1), Proc paper no 17674, Feb, pp 174–188
12. Bhargava DS (1986a) Models for polluted streams subject to fast purification. Water Res J Int Assoc Water Pollut Res Contr Engl 20(1):1–8
13. Hankin EH (1896) The bacterial action of water of the Ganges and Jumna rivers on cholera microbes. Annales de l'Institut Pasteur 18:511–5517 (Reproduced in Man in India, 18, 1938, pp 135–152)
14. Editorial (1981) Ganga and pollution. Northern India patrika, Lucknow (India), 28 December 1981, p 6
15. Bhargava DS (1985a) Variations in quality in the Ganges. Effluent + Water Treat J Engl 25(2):61–65
16. Bhargava DS (1986b) DO sag model for extremely fast river purification. J Environ Eng Am Soc Civil Eng 112(3):572–585, Paper No. 20696
17. Bhargava DS (1986c) Modelling for compounded DO sags. Trans Inst Eng Aust Civil Eng CE-28(3):222–230
18. Bhargava DS (1994) The Ganga can clean itself up better than any other river, an interview, at Roorkee, by Rakesh Kalshian, of Down to Earth, New Delhi, Ed., Down to Earth (Science and Environmental Fortnightly): Science and Technology – A River with a Difference, 31 January 1994, New Delhi, pp 37–38
19. Bhargava DS (2009) Simplified DO Sag models for rapid BOD removal. Environmentalist 29(3):411–420 (The Netherlands, Online first w.e.f. Thursday, 19 February 2009)
20. Bhargava DS (1983b) Use of a water quality index for river classification and zoning of Ganga river. Environ Pollut Ser B Int J Engl 6(1):51–67

21. Bhargava DS (1985b) Water quality variations and control technology of Yamuna river. Environ Pollut Ser A Int J Engl 37(4):355–376
22. Basin, Sub-basin Inventory of Water Pollution, the Ganga Basin Part II (excluding the Yamuna Sub-basin) – Assessment and Development Study of River Basin Series, ADSORBS/7/1982-83, Central Board for the Prevention and Control of water Pollution, New Delhi, 1982–1983, 204 pp
23. Bhargava DS (2008) Dissolved oxygen sag analysis for a settling fields overlapping type multi-wastewater-outfall. Environmentalist 28(2):128–136 (The Netherlands, Online first w.e.f. Tuesday, 11 September 2007)
24. Bhargava DS (1982) Coagulating capacity of Ganga at Kanpur, at the 9–11 September 1982. In: International symposium on food industries and the environment, held at Budapest, Hungary, Sym Proc, pp 133–141, Publishing House of the Hungarian Academy of Sciences (also in: Hollo J (ed) Developments in Food Sciences, vol 9: Food industries and the environment. Elsevier Science Publishers, P.O. Box 211, 1000 AE, Amsterdam, 1984, Art. pp)
25. Bhargava DS (1983d) A light penetration model for the rivers Ganga and Yamuna. Int J Dev Technol Int Centre Tech Res Engl 1(3):199–205
26. Pavoni JL, Tenney MW, Echelberger WF (1972) Bacterial exocellular polymers and biological flocculation. J Water Pollut Contr Federation USA 44(3):414–431
27. Streeter HW, Phelps EE (1925) A study of the pollution and natural purification of the Ohio rivers. U.S. Public Health Service Bullet 146
28. Velz CJ, Gannon JJ (1962) Biological extraction and accumulation in stream self-purification. In: Ist international conference on water pollution research, London, vol 1, 1962, pp 1–5
29. Bhargava DS (2010) Restoring the pristine waters of the Ganga river. World Water 33(2):28–31, UK (Water Environment Federation, USA)
30. Bhargava DS (1983c) Very low altitude remote sensing of the water quality of rivers. Photogrammetric Eng Remote Sensing J Am Soc Photogrammetry 49(6):805–809

Sunset view at the banks of Ganga in Rishikesh, UTTARAKHAND (Courtesy Nitin Kaushal)

Chapter 7
Water Quality Challenges in Ganga Basin, India

Ramesh Chandra Trivedi

Abstract The river Ganga has a very important economic, environmental and cultural value in India. With fast growth in population and economy in the Ganga Basin the water use has intensified many fold. With highly uneven distribution of rainfall in the basin, it is experiencing severe water scarcity in many parts. Basin's groundwater is over-abstracted leading to lowering of water table at alarming rate which resulted in reducing the flows of most of the tributaries of the Ganga. With fast urbanization and increasing prosperity the wastewater is steeply increasing. Due to paucity of resources the urban local bodies, which are responsible for management of this wastewater, are not able to do so. Thus, a large part of this wastewater is getting into the river Ganga or its tributaries either untreated or partially treated. Reduced level of water flow coupled with increased waste loads pose serious water quality problems in the river. Although, access to drinking water has increased over the past decade due to large number of efforts by the government, still the tremendous adverse impact of unsafe water on health continues. The highest mortality from diarrhoea is said to be among children under the age of five, highlighting an urgent need for focused interventions to prevent and control water quality degradation to avoid diarrhoeal disease.

The water quality monitoring results indicated that pathogenic pollution caused by discharge of untreated or partially treated domestic wastewater is the main cause of water quality degradation in the Ganga river. Most of the tributaries of the Ganga and many groundwater sources are affected by such pollution. Apart from pathogenic pollution organic pollution causing oxygen depletion was observed in many stretches of the Ganga and its tributaries. Nutrient enrichment followed by eutrophication is observed in some stretches of the Ganga and its tributaries. Toxic pollution

R.C. Trivedi (✉)
Expert Member, National Green Tribunal, New Delhi

Former Additional Director, Central Pollution Control Board, New Delhi
e-mail: dr_r_c_trivedi@rediffmail.com

R. Sanghi (ed.), *Our National River Ganga: Lifeline of Millions*,
DOI 10.1007/978-3-319-00530-0_7,

is observed in some pockets of the river or its tributaries, where industries are concentrating. Due to lack of proper transport facilities (sewerage system) for the wastewater in the basin both for domestic and industrial wastewater, a large part of it is percolating in the ground and polluting the groundwater. Such pollution is evident in large urban and industrial areas. The over-exploitation of groundwater in many parts of the basin led to geogenic contamination by underground minerals. Fluorides, iron, arsenic, salinity and hardness is found to be increasing in many parts of the basin.

Keywords Ganga Basin • Water resources • Water quality threats • Pathogenic pollution • Designated best use • Key challenges • Water scarcity • Environmental flows • Sewage treatment • Diffuse pollution

7.1 Introduction

As we know the Ganga river has been respected in India and considered the status of Goddess. People from all over the world come to this river for taking holy dip considering water of the river will rid them of all their sins. However, with passage of time due to increased human activities, the river is gradually getting impure. Geometric increase in population coupled with rapid urbanization, industrialization & agricultural development has resulted in critical shortages of water of suitable quality in several parts of the basin to sustain future growth. The situation warrants immediate redressal through radically improved water resource management strategies. The problem magnified due to conditions of poverty & underdevelopment as also the negative effects of development. The problem of water quality degradation is mainly resulted from discharge of untreated or partially treated domestic wastewater from urban sprawl and city slums resulted from burgeoning population along the Ganga river and its tributaries. The protection of water of this river from pollution present a most fundamental challenge in the scenario of the nation's desire to improve its economy and industrialize faster to be self sufficient in food, and to be capable of fulfilling certain basic needs of the growing population. The situation is further aggravated by highly uneven distribution of rainfall with time and space resulting in water scarcity in many parts of the basin. Thus for restoring the Ganga it is essential to implement more and more efficient water management policies and practices in terms of both quantity and quality. All kinds of human activities have resulted in polluting the river. The pollution is steeply increasing. The major pollution is due to discharge of untreated wastewater from fast growing urban centres located along its bank. Government of India has put several efforts to restore the water quality of this river, but due to steep increase in wastewater generation such efforts have not resulted in improving the water quality of the river. A conscious effort towards reversing the current trends of deterioration of water quality of the Ganga is the need of the hour. However, in the present scenario of fast economic development, industrialization, urbanization and agricultural activities restoration of the Ganga river is a big challenge. The river is not only facing problem of dumping of sewage in it, but also the water in its basin is being over-exploited resulting in reduced flow conditions or even drying of many of its tributaries.

It is increasingly realized that without efficient water management, the future social and economic development and ecological sustenance of the Ganga Basin would be seriously inhibited. Preserving the quality and the availability of the freshwater resources is the most pressing of the many environmental challenges in the Ganga Basin. Perhaps, because water is considered as cheap resources, which are readily available, people fail to realize just how much stress human demands for water are placing on natural ecosystems.

7.2 Water Resources of Ganga Basin

The Ganga Basin is a part of the composite Ganga-Brahmputra-Meghna basin. The basin lies in China, Nepal, India and Bangladesh and drains an area of 10,86,000 km^2. It is bounded on the north by the Himalayas, on the west by the Aravalli as well as the ridge separating it from Indus basin, on the south by the Vindhyas and Chhotanagpur Plateau and on the east by the Brahmaputra ridge. In India the basin covers the whole of Uttarakhand (53,566 km^2), Uttar Pradesh (240,798 km^2), Bihar (143,961 km^2), West Bengal (71,485 km^2) and Delhi (1,484 km^2), and part of Madhya Pradesh (198,962 km^2), Rajasthan (112,490 km^2), Haryana (34,341 km^2) and Himachal Pradesh (4,317 km^2). Basin's climate varies from tropical monsoon in south to temperate in north. The basin is rich with respect of water resources. The annual average rainfall over basin based on data, collected by Indian Meteorological Department is about 1,051 mm [1], thus from rainfall alone the basin is receiving more than 1,000 BCM of water. Of this 3/4 part received only during monsoon. A good part of it is lost through the process of evaporation and plant transpiration, leaving only half of it on the land for use. The rainfall, however, varies from less than 500 mm in Haryana (Hissar) to nearly 1,600 mm in West Bangal (Kolkata, Alipore) and 2,209 mm in upper Himalayan region. During rainy season, availability of water from precipitation is far in excess of natural and man-made holding capacity which results in floods. During the non-rainy period, high evaporation rates coupled with high water demands cause drought conditions requiring import of water. Flood and drought thus constitute the extreme manifestations of hydrologic cycle in the Ganga Basin. The situation is exacerbated due to depleting forest cover in the basin. The Central Water Commission [2] has estimated the basin-wise average annual flow in Indian river systems and for Ganga Basin the estimate is 525 km^3. Due to topographical, hydrological and other constraints, it is assessed that only about 250 km^3 of surface water can be put to beneficial use by conventional methods of development. Analysis of basin wise potential of ground water by the Ministry of Water Resources indicates that the Ganga Basin has the maximum groundwater potential in the country, which is 171 km^3. Other basins are far behind the Ganga that accounts for more than 39 % of total potential in the country. As a consequence of high availability, the groundwater use in the basin is very high as compared to other basins. Net draft though highest in the Ganga Basin, yet ground water development percentage is highest in the Indus basin, where it is 78.9 % and in Ganga Basin it is about 70 %. Since independence, the country has been planning and implementation of retaining major part of rainwater falling in the basin by using engineering innovations such as dams and barrages.

All the important tributaries of the Ganga are not perennial. Many of its tributaries or sub-tributaries like Yamuna, Chambal, Kshipra, Khan, Betwa, Sone, which used to be perennial in the past are now became seasonal in significant part due to over-exploitation of groundwater in the catchment area resulting in lowering the water table and reduced baseflow in non-monsoon period.

7.2.1 Water Use

The water use has been assessed by Central Water Commission, Ministry of Water Resources [2] for fresh water as the quantity of water being abstracted for different uses. It is estimated that total 142.6 BCM of water is being used in the Ganga Basin out of which irrigation alone consumes 134.4 BCM or 94.3 %, domestic use consumes 4.15 BCM or 2.91 % and industrial use consumes 4 BCM or 2.8 %. The consumption of water is bound to increase in future with fast economic growth, urbanization, industrialization and increased standard of living, which will put additional pressure on the water resources of the Ganga Basin.

The over-exploitation of ground water resources is widespread across the basin, and is projected to grow with time. The inappropriate land use practices prevalent in the basin limit the groundwater recharge potential. Accordingly, the ground water levels are receding in some regions of the basin at an alarming rate. On the other hand, due to excessive irrigation and large water storages, some regions have registered rise in ground water levels, at times leading to serious water logging problems.

The irrigation uses far outweigh the domestic and industrial uses. However, the domestic and industrial uses are rising with increase in population coupled with rapid urbanisation and industrialisation. With given situation, future water demands can be met in absolute quantity and quality terms only through pragmatic planning that includes highest priority to water conservation, recycling-reuse of municipal and industrial wastewater, and inter-basin/inter-regional transfers of water.

7.2.1.1 Water Quality Monitoring

The first step towards ensuring water quality is to generate reliable and accurate information about water quality. Several government institutions and departments are involved in water quality monitoring, leading to overlapping of functional areas and duplication of efforts. While the concerned State Pollution Control Board laboratories and the Central Pollution Control Board (CPCB) regularly monitor surface water quality in the Ganga river and its tributaries, the Central Ground Water Board is primarily responsible for monitoring groundwater quality along with the State Drinking Water Mission under the respective public health engineering departments. To regularly monitor the water quality of River Ganga, the CPCB has set up 39 water quality monitoring stations on the main river and 102 stations on its tributaries as in 2009 [3]. Based on the monitoring results, the CPCB identifies polluted river stretches, which needs prioritization in pollution control efforts.

7.2.1.2 Existing Water Quality Status

The CPCB water quality data revealed (cpcb.nic.in) that the Ganga river and its tributaries have retained pristine quality of water in their hilly stretches, as they are least influenced by human interferences. As the rivers enter the plains, they gradually get polluted due to over-abstraction of their water and discharge of treated and untreated wastewater in course of flow. Thus, the middle stretches of the rivers are most affected due to increased water requirement for various consumptive and non-consumptive uses and input of pollution. In the lower stretches, the rivers are generally better due to more water available for dilution and assimilation of pollution.

7.3 Major Water Quality Threats in Ganga Basin

7.3.1 Pathogenic Pollution

One of the major water quality threats in the Ganga river and its tributaries is water borne diseases, which is mainly due to discharge of untreated or partially treated domestic sewage. Due to lack of inadequate transport system for the wastewater, a large part of it is retained in the city forming cess pools leading to promotion of mosquito breeding and groundwater pollution or flows in unlined storm water drains resulting in pollution of groundwater.

7.3.2 Salinity

Salinity is increasing in many parts of the basin particularly in Haryana and Rajasthan part of the Ganga Basin in groundwater and also in surface water. The salinity is increasing mainly due to discharges of industrial wastewaters or agricultural return water. The salinity impaired the fitness of water for drinking or irrigation. It may also affect the ecosystem in surface waters.

7.3.3 *Oxygen Depletion*

As indicated above a large portion of wastewater is discharged into the Ganga or its tributaries. A major portion of it is originated from domestic sources. Such wastewater contains high amount of organic matter. The industries also discharge effluents containing high organic matter e.g. agro-based industries. This organic matter when oxidised in water through microbial activities, consumed dissolved oxygen. Since water has limited availability of oxygen, if consumption exceeds the availability, oxygen depletion is resulted and survival of aquatic life becomes difficult in the river. In stretches of the Ganga and its tributaries particularly Yamuna river in the vicinity of large urban centres massive input of organic matter sets off a progressive series of chemical and biological events in the downstream water. The stretches are characterised by high bacterial population, cloudy appearance high BOD and strong disagreeable odour – all indicating general depletion of oxygen. Masses of gaseous sludge rising from the bottom are often noticed floating near the surface of the water. During monsoon due to flood the sludge deposited in such stretches is flushed and stay in suspension, causing rise in oxygen uptake in the downstream. Due to such sudden oxygen depletion, heavy fish mortality occurs every year during first flushing after onset of monsoon.

7.3.4 *Eutrophication*

The discharge of domestic wastewater, agricultural return water or run-off water and many industrial effluents contribute nutrients like phosphates and nitrates. These nutrients promote excess growth of algae in water bodies leading to a state called eutrophication. In eutrophicated water body during day-time water get supersaturated with oxygen, however during night time due to respiratory consumption the oxygen is depleted to a large extent. This results in large diurnal change in dissolved oxygen. This is not desirable for balanced aquatic ecosystem. The stretches of the Ganga and its tributaries downstream of urban centres often are affected by eutrophication of various level.

7.3.5 *Toxicity*

Due to discharge of toxic effluents from many industries and increased use of chemicals in agriculture and their subsequent contribution to the Ganga river or its tributaries. Presence of toxic substance impairs the water quality by making it unfit for human consumption, aquatic life and irrigation.

7.4 Groundwater Quality

Monitoring groundwater quality remains a prime concern and a major challenge in rural India since it is the predominant source of drinking water [4]. It remains a major monitoring challenge considering the geographical spread of Indian villages and the fact that many of the remote villages are not accessible to regular monitoring by central agencies due to transportation and communication problems. Hence it is the rural population that suffers the most from problems related to fluoride, arsenic as well as microbial contamination [5]. State drinking water mission under the Rajiv Gandhi National Drinking Water Mission [6] and sanitation department through the public health engineering division is mandated to undertake the assessment of all drinking water sources. The National Rural Drinking Water Quality Monitoring and Surveillance Programme was launched in February 2006, which is responsible for monitoring the drinking water quality. It has been provided with water testing kits. However, this program was not very effective in providing the water quality information to the users.

The groundwater quality is highly dependent on nature of aquifer and ambient climate. The groundwater of the basin is calcium bicarbonate type with nearly neutral pH typify the wetter regions of the north and north-east part of the basin, where alluvial sediment constitute the most important aquifer. Salinity is a problem in most of the arid areas of the basin, which is exacerbated by salinity related to irrigation. Soil and groundwater salinization has been documented in Haryana and Rajasthan. Pollutant inputs related to agriculture, domestic wastes and industrial wastes is problem in northern plains of the basin. Coliform organisms are quite common in the water samples collected from large urban areas as per the CPCB's report [7]. Nitrate is emerged as one of the important pollutants in urban groundwater in the basin. Nitrate concentration of above 300 mg/L was reported from the million plus cities in the basin [7].

7.4.1 Major Cause for Water Quality Degradation

The Ganga Basin as rest of the country is urbanizing very fast. Fast urbanizing led to steep increase in sewage generation. Discharge of untreated domestic wastewater is predominant source of pollution of Ganga River and its tributaries. Urban centres contribute most of the sewage in the basin. The smaller towns and rural areas generally do not contribute significant amounts of sewage due to low water supply and predominant practice of open defecation. Whatever, little wastewater generated in these human settlements normally percolates in the soil or evaporates. CPCB carries out regular inventory of water supply, wastewater generation, collection, treatment and disposal in class-I (population >100,000) cities and class-II towns (population between 50,000 and 100,000) of the country. As per the latest inventory [8], there are 179 Class I cities and 148 Class II towns in the

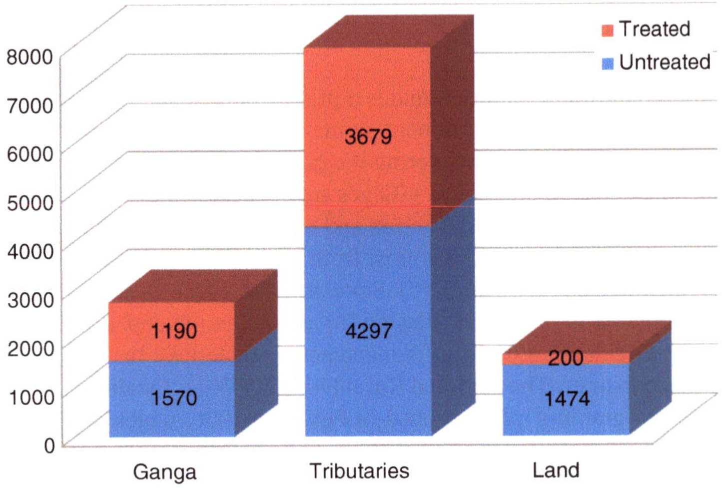

Fig. 7.1 Wastewater generation, treatment and disposal in Ganga Basin, MLD

Fig. 7.2 Wastewater disposal in Ganga Basin, MLD, Total WW 12,410 MLD

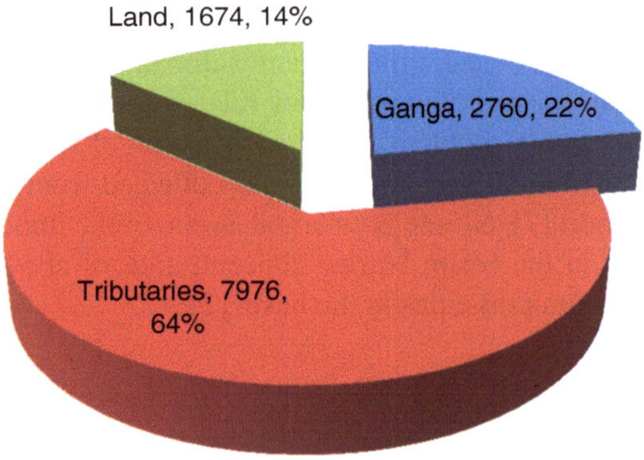

basin, generating about 12,410 million liter per day (MLD) of domestic wastewater, against which treatment capacity exists for only 5,410 MLD, and even this capacity is also not fully utilized (Fig. 7.1). Out of 12,410 MLD of wastewater, 2,760 MLD is discharged directly into the Ganga, 7,976 MLD into its tributaries and 1,674 MLD is discharged on land as shown in Fig. 7.2. As per the CPCB's survey the sewage treatment plants are able to utilize only about 69 % of their capacity and their performance was about 70 % [7]. Thus there is a large gap between generation and treatment. It is observed that although the sewage treatment capacity is continuously being augmented in the basin, the growth of sewage generation is much faster and thus the gap is continuously widening.

It is also observed that there is no proper collection system for sewage and thus, a large part of the sewage is flowing in storm water drains. These drains many times also receive garbage, solid wastes including plastic wastes, which result in their chocking. This leads to retention of large amount of sewage in the city premises, forming cess pools, which are good breeding grounds for mosquitoes and leading to groundwater pollution. Since these storm water drains are ultimately leading to rivers, lakes or sea, the treated, partially treated or untreated wastewater is disposed into natural rivers or lakes or used on land for irrigation/fodder cultivation or to the sea.

7.4.2 *Approach to Water Quality Management*

The government explicitly enacted the Water (Prevention and Control of Pollution) Act, 1974 with the primary objective of prevention and control of water pollution. The Water Act established the Central Pollution Control Board and the State Pollution Control Boards for its implementation. The Water Act empowers the pollution control boards to lay down and maintain water standards, the actual provisions for enforcement such as penalties, imprisonment etc. are largely confined to source-specific standards for individual polluters. The Environment Protection Act, 1986 is an umbrella act providing for the protection and improvement of environment and for matters connected therewith. It authorizes the central government to intervene. The nature of penalties allowed under this act are similar to those authorized under the Water Act. The basic objective of Water Act is to maintain and restore the wholesomeness of national aquatic resources by prevention and control of pollution. The Act does not define the level of wholesomeness to be maintained or restored in different water bodies of the country. The Central Pollution Control Board (CPCB), which was created under the provision of Water Act, 1974 has taken human uses of water as a base to identify the level of water quality to be maintained. It was considered ambitious to maintain or restore all natural water bodies at pristine level. Planning pollution control activities to attain such a goal is bound to be deterrent to developmental activities and cost prohibitive. Since the natural water bodies have got to be used for various competing as well as conflicting demands, the objectives is aimed at restoring and/or maintain natural water bodies or their parts to such a quality as needed for their best uses. Thus, a concept of "designated best use" (DBU) was developed. According to this concept, out of several uses a water body is put to, the use which demands highest quality of water is termed as "designated best use", and accordingly the water body is designated. Primary water quality criteria for different uses have been identified. A summary of the use based classification system is presented in Fig. 7.3 as cmap.

All major rivers in the country were classified according to their designated best uses and a "Water Use Map" was prepared. For identification of the water bodies or their parts where water quality is at variance with water quality criteria, it was felt important to measure water quality of that water body or its part. It would help in preparation of "Water Quality Map" of India. The idea was to superimpose "Water

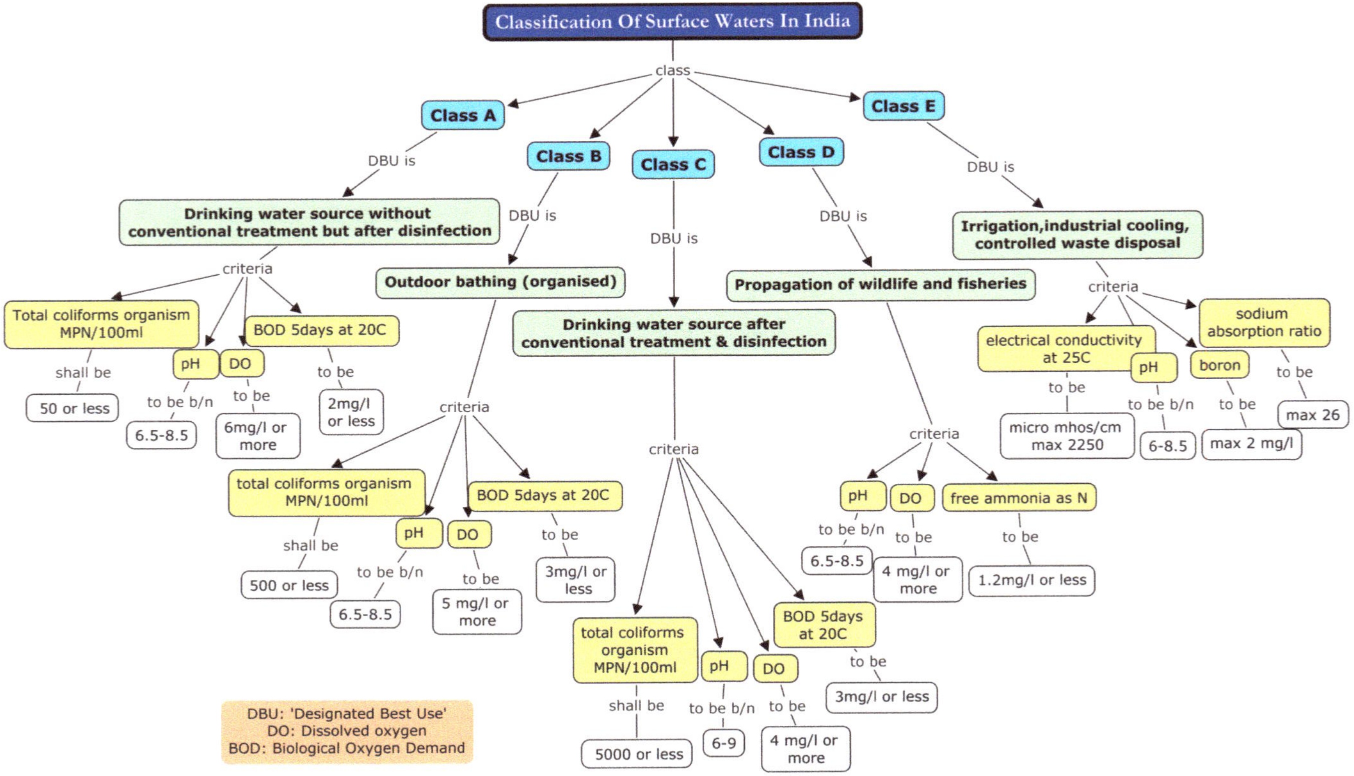

Fig. 7.3 Use based classification of surface waters in India

Quality Map" on "Water Use Map" to identify the water bodies or their parts which are in need of improvement (restoration). Subsequently through a wide network of water quality monitoring, water quality data are acquired. A large number of water bodies were identified as polluted stretches for taking appropriate measures to restore their water quality. Today almost all policies and programmes on water quality management are based on this concept.

7.4.3 Development and Implementation of Effluent Standards

The simplest administrative approach to regulate industrial pollution would be to promulgate permissible limits for various pollutional parameters on a general basis, make them binding on all discharges and prosecute and punish any offenders. Control of pollution at sources is the immediate short-term objective adopted by all the state pollution control boards in India. To control pollution at source the industries must know the extent up to which their effluent or emission must be treated/controlled so that they can discharge the treated effluent to receiving environment without significant effect. The cost of treatment should be such that the industry is able to take the burden (economically feasible). The Central Pollution Control Board has initiated evolving industry specific Minimum National Standards (MINAS) as early as in 1977-1978. At present it has evolved effluent standards and emission standards for almost all major polluting industries. These standards have been notified under Environmental (Protection) Act, 1986, by the Government of India. The MINAS is binding on all the State Boards. No state board is permitted to relax the MINAS, if situation so demand, the state boards may make them more stringent.

7.5 Key Challenges

7.5.1 Water Scarcity

The rainfall, which is the only source of freshwater, is highly un-even in distribution in time and space in India. The rainfall in some area of the Ganga Basin is as low as about less than 500 mm to some areas in Himalayan part of the basin is 2,209 mm. The rainfall is also not evenly distributed throughout the year. It is concentrated to only 3–4 months. The average number of rainy days in a year is about 25–50 in different parts of the basin. Which means the rainfall is concentrated to a very short duration in a year resulting in a major part of it is flowing back to sea. It gets little opportunity to percolate and recharge groundwater. In the remaining dry period all uses of water are being sustained by the water retained by land in form of surface water in lakes, ponds, rivers (dry weather flow) and in form of groundwater. The ever-increasing demand of water for agricultural, industrial and domestic activities, the water resources are over-exploited. This is resulting in shrinking or even drying up of many water bodies for considerable period in a year in the basin. Many rivers in the basin, which used to have considerable flow throughout the year now flow for 3–4 months in a year. The over-drafting of groundwater resulted in lowering the water table below the river bed resulting in low-flow or no-flow conditions in many rivers in the basin. For example River Yamuna has practically very little or no fresh-water release from Tajewala to Wazirabad barrage (about 250 km stretch) after October every year. Similarly, many of its tributaries like Chambal, Parbati, Sindh, Kshipra, which used to flow for whole year just 25 years ago, remain dry for almost 200–250 days in a year. This results in loss of aquatic flora and fauna, leading to loss of livelihoods for river fisher folk, significant impacts on human health from polluted water, increased drudgery for poor, rural women in collecting drinking water from distant water bodies, loss of habitat for many bird species and other terrestrial animals, and loss of inland navigation potential. There are many terrestrial animals dependent on rivers for their drinking water. Once the rivers are dry, they do not get drinking water and left with no choice but to die. Apart from these, many rivers in the basin are inextricably linked with the history and religious beliefs of peoples of India, and the degradation of such river systems accordingly offends their spiritual, aesthetic, and cultural sensibilities.

The direct causes of river degradation are, in turn, linked to several policies and regulatory regimes. While setting Minimal National Standards (MINAS), which are implemented all over the country for regulation of pollution, it was envisaged that at least ten times dilution was available in the receiving water bodies. Unfortunately, in majority of the cases such dilution is not available for considerable time in a year in the recent past leading to water quality degradation. Discharging wastewaters even after treatment in such rivers results in serious water quality problems and maintaining targeted water quality is very difficult. If objectives of Water Act, 1974 or Environment (Protection) Act, 1986 are to be met, minimum level of water needs to be maintained in the water bodies in order to assimilate the residual pollution.

7.5.2 Over-Drafting of Groundwater

India is now the biggest user of groundwater for agriculture in the world [9]. Groundwater irrigation has been expanding at a very rapid pace in India since the 1970s. There is no appropriate policy for groundwater use in agriculture. In order to pursue green revolution, to feed the growing population of the country, and make it self sufficient in food grain, irrigation is promoted all over the country and also in the Ganga Basin. This has led to steep increase in abstraction of groundwater. The number of groundwater structures in the country has steeply increased in last four to five decades. The number was about 3.8 million in 1951 have grown to about 19 million and 7,900-M^3/year water is extracted per structure. This is a phenomenal growth. The farmers do not pay for the amount of groundwater pumped for irrigation. In addition, electricity costs for pumping are very heavily subsidised by many state governments in the basin e.g. Haryana, UP, MP. Accordingly, farmers mostly pump more groundwater than is needed for optimising crop production. In turn, this over-pumping is resulting in a steady decline of groundwater levels in many important aquifers all over the country. As the groundwater levels decline, more energy is needed to pump the same quantity of water, which requires additional subsidising of electricity costs. This has contributed to a vicious cycle of overuse of groundwater, declining aquifer levels, increasing losses to the electricity boards and increasing adverse environmental impacts (like land subsidence), none of which are sustainable on a long-term basis. Thus, major policy changes in the water and energy sectors will be needed in the future to balance water and energy uses, stabilise the levels of declining groundwater tables and reduce electricity subsidies to the farmers. Matthew Rodell et al. [10] used terrestrial water storage-change observations from the NASA Gravity Recovery and Climate Experiment satellites and simulated soil-water variations from a data-integrating hydrological modelling system to show that groundwater is being depleted at a mean rate of 4.0 ± 1.0 cm $year^{-1}$ equivalent height of water (17.7 ± 4.5 km^3 $year^{-1}$) over the Indian states of Rajasthan, Punjab and Haryana (including Delhi), which are part of the Ganga Basin except Punjab. During the study period of August 2002 to October 2008, groundwater depletion was equivalent to a net loss of 109 km^3 of water, which is double the capacity of India's largest surface-water reservoir. The annual average rainfall was close to normal during this period.

7.5.3 Populations Food Grain Requirement

Kalipada Chatterjee [11], an expert from Climate Change Centre estimated the food grain availability at present in India is around 525 g per capita per day. With growing economy and prosperity in the country this may grow. If small raise is made in per capita consumption to 650 g the food grain requirement will be about 390 MT of food grain. Taking the projections of about 1,800 million populations by 2050 AD Kalipada Chatterjee estimated food grain requirement of about 430 MT annually at the present level of consumption, which may need much greater use of available inputs including water and fertilizers. Which may need to increase area under irrigation from present 28 % to roughly 40 % by the year 2050 and adopting drip and sprinkler irrigation in India and Ganga Basin would be the most exploited one.

7.5.4 Drinking Water Quality and Health Impacts

ACCESS to safe drinking water remains an urgent necessity, as 30 % of urban and 90 % of rural households still depend completely on untreated surface or groundwater [12]. More than 300 million episodes of acute diarrhoea occur every year in India in children below 5 years of age (Annual Report, [13]) and major part of these episodes are from Ganga Basin states.

While access to drinking water in India has increased over the past decade, the tremendous adverse impact of unsafe water on health continues [14]. It is estimated that about 21 % of communicable diseases in India is water related [15]. The highest mortality from diarrhoea is said to be among children under the age of 5, highlighting an urgent need for focused interventions to prevent diarrhoeal disease in this age group [16]. Despite investments in water and sanitation infrastructure, many low-income communities in India continue to be bereft of safe drinking water [17]. Regardless of the initial water quality, widespread unhygienic practices during water collection, storage and consumption, overcrowded living conditions and limited access to sanitation facilities perpetuate the transmission of diarrhoea-causing germs through the faecal–oral route [18]. A majority of rivers in the Ganga Basin, which are the sources of drinking water in urban India are also contaminated [3]. While the shift in usage from surface water to groundwater has undoubtedly controlled microbiological problems in rural India [9], the same has however, led to newer problems of fluorosis and arsenicosis [19, 20]. Excess iron is an endemic water quality problem in many parts of eastern Ganga Basin [21]. In 2002, 17 states were affected by severe fluorosis [22] and now the problem exists in 20 states, indicating that endemic fluorosis has emerged as one of the most alarming public health problems of the country [23]. About 62 million people are suffering from various levels of fluorosis, of which six million are children below the age of 14 years; they suffer from dental, skeletal and/or non-skeletal fluorosis [24]. High rates of mortality and morbidity due to water-borne diseases are well known in Ganga Basin. Serious degradation of water quality in urban India has often been attributed to indiscriminate disposal of sewage and industrial effluents into surface water bodies. Although some degree of intervention in terms of chlorination and monitoring of water quality exists in major cities and towns, rural India, which constitutes the bulk (70 %) of the population, is usually deprived of such interventions.

It has been estimated that around 37.7 million Indians are affected by water-borne diseases annually, 1.5 million children are estimated to die of diarrhoea alone and 73 million working days are lost due to water-borne diseases each year [25]. The resulting economic burden is estimated at US$600 million a year [26]. Clearly, the health benefits in terms of reduction in water-borne diseases have not been commensurate with the investments made [27]. Another estimation by Water and Sanitation Program of World Bank along with Asian Development Bank, Australian Government (AusAID) and UKAID the total economic impacts of inadequate sanitation at Rupees 2.44 trillion (US$53.8 billion) a year, which is equivalent of 6.4 % of India's GDP in 2006 or Rs. 2,180 (US$48) per person per year.

7.5.5 Sanitation v/s Water Quality

Although the burden of diarrhoeal disease resulting from inadequate water quality, sanitation practice and hygiene remains high, there is little understanding of the integration of these environmental control strategies [28]. Sanitation and hygiene are often perceived as social and behavioural issues, and water quality as a technical issue requiring technical intervention. In the absence of common verifiable indicators for both water quality, and hygiene and sanitation at the community level, the outcomes of various interventions are seldom quantifiable. Moreover, sanitation is often perceived by both the government agencies as toilet coverage at household level, but seldom emphasis is laid on treating the water emanating from these facilities leading to gross contamination of water sources. Although open defecation remains a major problem in rural India and responsible for contamination of surface water, animal wastes (cattle excreta) also play a significant role in groundwater contamination in rural India in the absence of effective animal waste management [5].

7.5.6 *Inadequate Volumes in Receiving Waters*

Although the MINAS is mainly technology based, some consideration is given to the receiving water body based on the assumption that in the receiving water body at least ten times dilution is available to achieve the ambient water quality objective in the immediate vicinity of the downstream. In the recent past due to over use of surface and ground waters to fulfil the ever increasing demand of agriculture, industrial and domestic uses, the water level in aquatic resources in the Ganga Basin is gradually decreasing. As a result, the assumption of ten times dilution is no more valid in many parts of our country. This poses great difficulties to achieve the ambient water quality objective (wholesomeness) even after implementing MINAS. This is a challenge before the Pollution Control Boards and the polluters. Thus, it has become necessary to achieve progressive reduction or elimination of effluent discharges. It implies adoption of no waste or low waste technologies through new treatment methods which can fulfil the above objective.

7.5.7 *Multiplicity of Organizations*

There are a large number of organizations working on different aspects of water resources. They have their own agenda and their own targets to be achieved. Little cooperation exists between these agencies responsible for the use and management of water resources in different sectors such as environment, public health and agriculture. The result is a planning process which does not take into account the needs of these different interest groups and lacks accountability on the part of any given agency. The major bottleneck in an effective policy formulation and implementation is the current institutional set-up involving various government agencies. Further, there is a separation of responsibilities on the basis of water quality and quantity. As many as 18 agencies are involved in water use and its management. Such a fragmentary approach, both at the central and state levels, results in duplication and ambiguity of functions and discourages unitary analysis of this scarce resource.

7.5.8 Standards and Compliance

Complying with the Minimum National Standards (MINAS) set by the CPCB requires that near-the-maximum effluent reduction technically achievable must be done. It has been found that a significant proportion of small scale industries discharging trade effluents to water stream do not have treatment plants especially those located in residential areas. It has also been observed that a significant proportion of polluting units, which have some treatment mechanism, do not comply with standards. The punitive action by the State Pollution Control Boards (SPCBs) is more or less tied up with litigation and considerable proportion of cases are pending for several years. The SPCBs are, sometimes, not able to exercise the powers to force compliance by stopping electricity supply or water because of interference by powerful pressure groups. Credible action requires formation of sufficiently empowered legal bodies to decide these cases expeditiously.

7.6 Environment Impact Assessment (EIA)

Although EIA processes consider water as component of environment, the impact of a proposed activity is evaluated only on quality. Whereas many developmental activities cause serious impacts on quantity and ultimately on the quality of water, this is not evaluated in integrated and holistic manner, which results in a large number of developmental activities e.g. industries, housing complexes, etc., coming up in an area consuming large quantities of ground and surface water causing their depletion in the area. There is a need for evaluation of impact on quantity and quality in cumulative way of all the activities in an area and its impact zone should be demarcated so that any new proposed developmental activity takes care of already existing impacts.

7.6.1 Ecology and Environmental Flows

The basic objective of Water Act, 1974 and Environment (Protection) Act, 1986 is conservation of rivers by way of restoring their ecosystems to natural structures and functions. Initially it was thought that the pollution of rivers was the main cause for disturbing the ecosystem and that major pollution is coming from the urban centres located along the rivers. The National River Conservation Directorate (NRCD) under Ministry of Environment and Forests, Government of India with the help of concerned State agencies established sewage treatment plants to achieve a reduction in pollution from urban centres. However, at a later stage of implementation it was realized that sewage treatment alone might not be sufficient to restore the river's ecological integrity because it was seriously affected by low-flow or no-flow conditions prevailing over a considerable segments of many rivers.

Many living creatures need perennial flow in the river because they complete their lifecycle in more than a year. They also need a natural flow regime, which means floods and lean flow conditions for different stages of their life cycle. Until this complexity is understood properly, any change in flow regime will cause many vital living creatures to vanish and cause serious ecological damage in the river, thus violating the principal objective of NRCD. Moreover, there are terrestrial animals dependent on river flows for their drinking water. If river is dry, they have no option, but to die as they cannot go far distance to quench their thirst.

The flow is also needed to continuously transport the wastes and to maintain a reasonable capacity in the rivers to assimilate the residual pollution that needs to be discharged even after proper treatment. As no technology, which is presently economically viable in the Indian context, can remove all pollution from wastewater, some residual pollution will be found in the treated effluent, which needs to be assimilated by the river ecosystem. Due to heavy abstraction of surface and groundwater in the catchment area of the rivers, they remain dry for a considerable period in a year. In order to maintain the environmental flows in these rivers, it is important that water resources are first optimally utilised with water conservation as the goal and then augmented water resources in the entire basin in a very large way ensuring that the water table never falls below the river bed in any reach of the rivers. This needs a massive effort in groundwater recharge, revival of the village ponds, watershed development, micro-catchment treatment, optimising cropping patterns and irrigation demands, water conservation integrated in urban planning, recycling of wastewater and overall economy in water use. This would further need significant institutional strengthening to manage and operate all water related activities with full cost recovery.

7.6.2 Pollution Control in Industries

Although, the enforcement framework for industrial pollution control could brought a large change in the behavior of the industries and was successful in reducing industrial pollution in the country in almost all the large and medium industries. However, in case of small scale industries the enforcement is still weak due to various reasons. The most important one is related to their locations, insufficient resources, skilled manpower and space. Moreover, nearly 80 % of the water pollution is caused by discharge of untreated domestic wastewater, which is very often from the Government supported urban local bodies due to lack of technical, managerial and financial capacities leading to such unhealthy practices. Thus even if all industrial effluents are fully treated to the stipulated levels, the water quality objectives cannot be achieved without similar surveillance and strict adherence to water quality protocol

7.6.3 Operation and Maintenance of Sewage Treatment Plants (STPs)

Operation and maintenance of existing plants and sewage pumping stations is a neglected field, as nearly 39 % plants in the country mostly in the Ganga Basin are not conforming to the general standards prescribed under the Environmental (Protection) Rules for discharge into streams as per the CPCB's survey report [7]. STPs are usually run by personals that do not have adequate knowledge of running the STPs and know only operation of pumps and motors. The operational parameters are not regularly analyzed hence the day-to-day variation in performance is not evaluated at most of the STPs. Thus, there is a need that persons having adequate knowledge and trained to operate the STPs be engaged to manage STPs and an expert be engaged to visit the STPs at least once a month and advice for improvement of its performance. In a number of cities, the existing treatment capacity remains underutilized while a lot of sewage is discharged without treatment in the same city. Auxiliary power back-up facility is required at all the intermediate pumping station (IPS) & main pumping stations (MPS) of all the STPs. It is very essential that they be efficiently maintained by the State Governments whose properties and charge they are. Inter-agency feuds and inadequate consideration of which agency would be responsible for what has led to inadequate maintenance of various STPs and other facilities created under various Central or State funded schemes. The maintenance of the sewage system, namely, sewers, rising mains, intermediate pumping stations, etc. should also be entrusted to the nodal agencies identified for the maintenance of the sewage treatment plants and sufficient funds and staff provided to them. Facilities like community toilets, electric crematoria, etc. should be maintained by the local bodies. Also the aspect of resource recovery by way of raising the revenue through sale of treated effluent for irrigation, of sludge as a manure and biogas utilization for power generation wherever provision exist needs to be addressed. Biogas generation, pisciculture from sewage as envisaged in the Ganga Action Plan is still not adequately addressed.

7.6.4 Financing Sewage Treatment

There are about 12,410 MLD of domestic sewage generated in the Ganga Basin out of which nearly 2,760 MLD are directly discharged into the Ganga river, 7,976 MLD is discharged in its tributaries and about 1,674 MLD is disposed on land. The total treatment capacity exists for about 5,069 MLD. Thus there is a gap of about 7,341 MLD. In order to restore the river, there is a need to create additional capacity of treatment for 7,341 MLD ***as of today***, a further addition of sewage in view of fast urbanization needs to be considered. For creating such a capacity a massive investment and a continuing commitment for such investments in the future are needed. Providing such funding from exchequer's money is detrimental to the economy of the country. Thus, it is important that the funding is raised from the beneficiaries following the 'Polluters pay principle'. In all the class-I cities and class-II towns sewage treatment plants needs to be established with adequate resources available for effective O&M of the sewage treatment plants, ensuring continuous flow of fund.

7.6.5 *Diffuse Water-Pollution*

For improving standards of life, running water-supply has been established in most of the towns and even in some villages in the Ganga Basin over the past three decades. This has, in turn, led to flush-latrines and much large use of water in homes for bathing, washing of clothes utensils etc, generating significant amounts of wastewaters. Use of soaps and detergents and amounts of various food materials going to the sink have also grown with improved life standards. Unfortunately, sewerage or improved sanitation was not adequately addressed. Hence sewerage has lagged far behind water supply. A large number of the cities/towns in the basin either do not have any sewerage system or the sewerage system is partially functioning, overloaded or defunct. Thus, the uncollected sewage flows in storm water drains or retained on land to percolate, leach or get washed-off to streams or groundwater. In the recent past, in wake of improved sanitation campaign of the Govt of India, large number of toilets are being constructed based on septic tanks. Such toilets may not be properly designed and maintained leading to their leakage into groundwater and pollute it. This is becoming a serious threat to groundwater quality, which is going to be much more severe in future. Similarly, industrial areas are being developed and large numbers of industries are coming up in the basin. Due to lack of proper conveyance system for the wastewater, a large part of it is retained in the industrial area with same fate as explained above. Even if the wastewater is flowing, it flows in storm water drains often unlined leading to percolation and pollution of groundwater.

7.7 Conclusions

Recent data on the status of water resources management in the Ganga Basin reveal several alarming trends. The basin is urbanizing very fast leading to steep increase in waste generation. Due to paucity of resources the urban local bodies, which are responsible for waste management are not able to cope with the ever increasing problem. In order to achieve higher economic growth, fast industrialization is also leading to put higher demand on water and generating wastes, which are let out into the river or its tributaries causing water quality degradation. The groundwater, which is main source of drinking for nearly 85 % population in the basin is also depleting at alarming rate and its quality is deteriorated due to geogenic and man-made pollution, thus threatened the drinking water security. The situation has dramatically worsened within a short span of last few years. Until a massive efforts are put on water conservation, rainwater harvesting and augmenting effective sewage treatment, the situation would be un-manageable in few years from now in the Ganga Basin.

References

1. Jain SK, Kumar V (2012) Trend analysis of rainfall and temperature data for India. Curr Sci 102(1):37–49
2. CWC (2005) Water sector at a glance. Central Water Commission, Ministry of Water Resources, Govt of India, New Delhi
3. CPCB (2009) Water quality status of India-2009, Monitoring of Indian national aquatic resources series: MINARS/67/2009-10. Central Pollution Control Board, Ministry of Environment and Forests, Govt of India, Delhi
4. World Bank (1999) Groundwater and management in India, p 25
5. Water Aid (2008) Community-based water quality management in India. Technical report, New Delhi, p 40
6. RGNDWM (2006) Guidelines for the national rural drinking water quality monitoring and surveillance programme. Rajiv Gandhi Drinking Water Mission, Ministry of Rural Development, Govt of India, New Delhi
7. CPCB (2006) Status of sewage treatment plants in India, Control of urban pollution series: CUPS/61/2005-06. Central Pollution Control Board, Ministry of Environment and Forests, Govt of India, Delhi
8. CPCB1 (2009) Stattus of water supply, wastewater generation and treatment in class-I cities and class-II towns of India, Control of urban pollution series: CUPS/70/2005-06. Central Pollution Control Board, Ministry of Environment and Forests, Govt of India, Delhi
9. Shah T (2009) Taming the anarchy: groundwater governance in South Asia
10. Rodell M, Velicogna I, Famiglietti J (2009) Satellite-based estimates of groundwater depletion in India. Nature 460:999–1002
11. Chatterjee K (2008) Water resources of India, climate change centre, development alternatives. New Delhi, India. http://climatechangecentre.net/pdf/WaterResources.pdf
12. Kumar R, Singh RD, Sharma KD (2005) Water resources in India. Curr Sci 89:794–811
13. MoHFW (2010) Annual report of ministry of health and family welfare, Govt of India, New Delhi
14. WHO/UNICEF (2004) Meeting the MDG drinking water and sanitation target: a mid-term assessment of progress. WHO, Geneva
15. Brandon C, Homman K (1995) The cost of inaction: valuing the economy-wide cost of environmental degradation in India. Asia Environment Division, World Bank, New Delhi
16. Gadgil A (1998) Drinking water in developing countries. Annu Rev Energy Environ 23:253–286
17. World Bank (2007) Annual report, South Asia, p 2. 6. Montgomery A. M. and Elimelech M., Water and environmental sanitation in developing countries: including health and education, vol 41, pp 17–24
18. Montgomery MA, Elimelech M (2007) Water and sanitation in developing countries: including health in the equation – millions suffer from preventable illnesses and die every year. http://www.yale.edu/env/elimelech/publication-pdf/ES&T_Feature_Montgomery_Elimelech_January_2007.pdf
19. Chakraborty AK, Saha KC (1987) Arsenical dermatoses from tube-well in West Bengal. Indian J Med Res 85:326–334
20. Rao SM, Mamatha P (2004) Water quality in sustainable water management. Curr Sci 87(7):942–947

21. Smedley PL (1991) Groundwater quality problems in coastal Orissa, India. British Geological Survey Technical Report, WD/91/48R, p 33, UK
22. Choubisa SL (2001) Endemic flourosis in Southern Rajasthan, India. Fluoride 34:61–70
23. Teotia SPS, Teotia M (1984) Endemic fluorosis in India: challenging national health problem. J Assoc Physicians India 32:347–352
24. Susheela AK, Majumdar K (1992) Fluorosis control programme in India. In: Water environment and management: 18th WEDC conference, Kathmandu, pp 229–233
25. Divakar H, Nagaraj N (2003) Impact of water pollution on food security and environment, bearing the brunt. Wasteland news August–October 2003
26. Murthy MN, James AJ, Mishra S (1999) Economics of water pollution: the Indian experience. Oxford University Press, Delhi, pp 1–19
27. Hughues G, Lvovsky K, Wheeler D (2001) Environmental health in India: priorities in Andhra Pradesh. South Asian development and environment unit, Washington, DC
28. Eisenberg JNS, Scott J, Porco T (2007) Integrating disease control strategies: balancing water, sanitation and hygeine intervension to reduce diarrheal disease burden. Am J Public Health 97:846–852

Reduced flows due to massive hydropower drive in the upper Ganga region (Courtesy Tripti Singh)

Chapter 8
Issues and Challenges of River Ganga

B.D. Tripathi and Smriti Tripathi

Abstract River Ganga is the basis of life. During past few decades, unplanned urbanization and fast growing industrialization has imposed a serious threat on surface water as well as ground water quality. A general survey of river Ganga was conducted to assess the water quality fluctuation at Varanasi. The pollution load on the river was investigated at different sampling sites. The results revealed that the river Ganga water quality was deteriorated due to presence of sewage, industrial waste, human and animal carcasses etc. It was also observed that the BOD values which are suppose to be less than 3.0 mg L^{-1} raised to 18.5 mg L^{-1} on the bathing ghats and DO content of water which must be more than 5.0 mg L^{-1} is reduced to 2.5 mg L^{-1}. Presence of numerous toxic chemicals such as Iron, Copper, Zinc, Lead, Chromium, Nickel and Cadmium also indicates unhygienic condition of the Ganga water. Microbiological examination of Ganga water has revealed the presence of numerous bacterial species such as Total coliform (25.4×10^3/100 mL), Fecal coliform (20.9×10^3/100 mL), Fecal Streptococci (93/100 mL). Government efforts to clean up the river have failed due to poor planning, technological mismanagement and corruption. Lack of public awareness and poverty fuels the problem. The present paper also suggests some prevention measures to improve the water quality of River Ganga.

B.D. Tripathi (✉)
National Ganga River Basin Authority, Govt. of India,
Varanasi, Uttar Pradesh, India

Centre for Environmental Science & Technology,
Banaras Hindu University, Varanasi 221005, Uttar Pradesh, India
e-mail: tripathibd@gmail.com

S. Tripathi
Centre for Environmental Science & Technology,
Banaras Hindu University, Varanasi 221005, Uttar Pradesh, India

R. Sanghi (ed.), *Our National River Ganga: Lifeline of Millions*,
DOI 10.1007/978-3-319-00530-0_8, © Springer International Publishing Switzerland 2014

Keywords Ganga river • Toxic chemicals • Cremation • Ghats • Sewage • Carcasses • Triple 'P' programs

8.1 Introduction

River Ganga is the basis of life and principle source of economic, sociological, cultural and environmental development of its highly productive and densely populated basin. During past a few decades ecologically unplanned urbanization and fast industrialization has degraded the entire Ganga ecosystem [1–3]. The discharge of city sewage, industrial effluents, cremation of dead bodies, agricultural runoff, disposal of solid and bio-medical wastes, animal bathing, washing of clothes, religious waste materials, extraction of huge quantity of Ganga water through lift canals, construction of dams and deforestation in Himalayan region etc. have imposed a serious threat to the river Ganga ecosystem [4, 5]. Henceforth, holistic ecological approach based on self regulatory mechanism to maintain the homeostasis is required for sustainable development of the basin.

The water of river Ganga is regarded as the most purifier of human body and soul. The river water has its unique and extraordinary ability to retain the oxygen. In Himalayan region flowing from different quality of rocks a number of chemical substances get dissolved into the Ganga water. During runoff from its catchment area water contain lot of alkaloid and other chemical substances as an extract of the medicinal plants growing in this region luxuriantly. It is also reported that *Bdellovibrio* bacteria is found in Ganga water which kills all those bacterial species who are responsible for decomposing the organic matters present into the water. However, during past a few decades water of the river Ganga has started losing its efficacy. It is observed that the BOD values which are suppose to be less than 3.0 mg L^{-1} raised to 18.5 mg L^{-1} on the bathing ghats and DO content of water which must be more than 5.0 mg L^{-1} is reduced to 2.5 mg L^{-1}. Presence of numerous toxic chemicals such as Iron, Copper, Zinc, Lead, Chromium, Nickel and Cadmium also indicates unhygienic condition of the Ganga water [6, 7]. Microbiological examination of Ganga water has revealed the presence of numerous bacterial species such as Total coliform (25.4×10^3/100 mL), Fecal coliform (20.9×10^3/100 mL), Fecal Streptococci (93/100 mL). Discharge of huge quantity of untreated city sewage and toxic industrial influents, cremation of dead bodies on the banks of the river Ganga, agricultural runoff, solid and biomedical waste disposal, animal bathing and washing of clothes, discharge of temple waste material, construction of lift canals and dams as well as deforestation in the Himalayan region are the main reasons for the degradation of Ganga River Ecosystem. The water quality of river Ganga is given in Table 8.1. The author has pioneered researches on Ganga Water Pollution at Varanasi during 1972 and research findings were discussed in Indian Parliament. For the first time on 21st July 1980 the problem of Ganga Water Pollution was accepted in the Parliament and decided to make all efforts to control the problem.

Table 8.1 Water quality of river Ganga at Varanasi (Average of 350 samples, year 2009–2011)

Parameter	Value
pH	7.5–8.8
Ec (μmhos cm^{-1})	185–843
Total alkalinity (mg L^{-1})	260–310
Acidity (mg L^{-1})	55.8–69.3
DO (mg L^{-1})	2.5–9.8
BOD (mg L^{-1})	1.9–87.5
COD (mg L–1)	5.9–170.5
Sulphate (mg L^{-1})	185–230
Chloride (mg L^{-1})	8.2–81.5
Nitrate – N (mg L^{-1})	0.015–0.985
Phosphate (mg L^{-1})	0.005–1.58
Potassium (mg L^{-1})	7.2–18.5
Iron (μg L^{-1})	0.8–1.5
Copper (μg L^{-1})	0.05–0.85
Zinc (μg L^{-1})	2.0–17.5
Lead (μg L^{-1})	0.05–0.07
Cadmium (μg L^{-1})	0.04–0.08
Chromium (μg L^{-1})	0.01–0.06
Total coliform (MPN)	5–375

8.2 Important Degradational Factors of River Ganga

8.2.1 Municipal Wastewater

On the bank of river Ganga there are 29 cities with a population of more than 100,000 and about another 23 cities with populations above 50,000. In addition there are 50 smaller towns with populations above 20,000. From these cities and towns about 3,000 million liter per day (MLD) domestic waste water is generated. There are about 150 identified major industries located near the river bank of river Ganga, of which 85 are considered as grossly polluting. Besides major industries there are thousands of small scale industries located in and around these cities and towns. It is estimated that majority of these major and small scale industries are not having proper treatment plants facilities. Hence, from these industries directly or indirectly untreated effluents along with the domestic wastewaters are released to the river Ganga. Table 8.2 reveals the quality of city sewage generated at Varanasi and Fig. 8.1 shows the important factors responsible for degradation in river water quality.

8.2.2 Dead Body Disposal

In India it is believed that if dead bodies are cremated on the bank of river Ganga the human soul shall directly go to the heaven. Due to this religious faith it has been observed that on only two cremation ground of Varanasi named Harischandra and

Table 8.2 Quality of city sewage at Varanasi

Parameter	Value
pH	8.65
Ec (μmhos cm^{-1})	1,040
Total alkalinity (mg L^{-1})	670
Acidity (mg L^{-1})	76.5
DO (mg L^{-1})	2.35
BOD (mg L^{-1})	310
COD (mg L^{-1})	767
Sulphate (mg L^{-1})	209.3
Chloride (mg L^{-1})	102.42
Nitrate – N (mg L^{-1})	2.64
Phosphate (mg L^{-1})	10.6
Potassium (mg L^{-1})	38.33
Iron (μg L^{-1})	14.37
Copper (μg L^{-1})	12.5
Zinc (μg L^{-1})	11.67
Lead (μg L^{-1})	20.24
Cadmium (μg L^{-1})	18.44
Chromium (μg L^{-1})	23.53
Total coliform (MPN)	14×10^5

Manikarnika ghats about 32,000 dead bodies are burnt every year with the help of 10,000 tons of dry fire wood. Out of this burning it has been observed that 300 tons ash content is released to the river Ganga which contains higher concentration of Nitrogen, Phosphorus and Potassium. During cremation of the dead bodies it has been observed that bodies are not completely burnt. Henceforth, it is estimated that more than 200 tons of half burnt flesh content are also released to river Ganga along with huge quantity of ashes every year. In recent past, due to increase in transportation facilities dead bodies from remote areas are also brought to Ganga river bank for cremation. According to one estimate about 40 % dead bodies are brought from remote areas at Varanasi. Besides burning of dead bodies, about 3,150 human and 6,270 animal dead bodies are thrown to river Ganga without burning, which are being eaten by the carnivorous birds and creates very unhealthy river environment. Table 8.3 shows the enhancement in pollution load due to dead body cremation at different ghats at Varanasi (Fig. 8.2).

8.2.3 Runoff from Agricultural Fields

Ganga river originates from Gomukh and passes through five states i.e. Uttarakhand, Uttar Pradesh, Bihar, Jharkhand and West Bengal. During 2,525 km course of journey Ganga receives water from highly fertile crop fields located in its catchment area. Huge quantity of pesticides, insecticides, herbicides and rodenticides etc. are used for plant protection in these areas are transported to the river through surface

Fig. 8.1 Important factors for degradation in water quality of river Ganga

Table 8.3 Dead body cremation at Harishchandra and Manikarnika Ghat (2009–2011)

Number of dead bodies	32,000/year
Fire wood consumed	10×10^3 tons/year
Wood energy	480×10^8 kcal/year
Ash content released	300 tons/year
Half burnt flesh	200 tons/year
Nitrogen	77,000 kg/year
Phosphorus	48,000 kg/year
Potassium	5,000 kg/year
Floating dead bodies	3,150
Animal carcasses	6,270

Fig. 8.2 Dead body cremation at different Ghats at Varanasi

runoff including chemical fertilizers. Discharge of these chemicals adds another dimension of water pollution.

8.2.4 Disposal of Solid and Bio-medical Waste

In cities and towns located all along the bank of river Ganga generate huge quantity of solid waste materials. Hospital wastes contain a lot of toxic and poisonous materials. These wastes are normally used as landfills near the river banks. It has been observed that huge chemicals through seepage are also discharged to the river. Fractions of solid and bio-medical wastes are directly thrown to river Ganga which pollutes the water.

8.2.5 Animal Bathing and Washing of Clothes

Ganga passing through the cities and towns also received huge quantity of organic and inorganic materials from clothes washing activities. It has been examined that higher concentration of carbonate, phosphate and alkaline materials which are used for washing the clothes are directly discharged to the river. Sometimes pH of water rose to very high i.e. 8.5–9.5 near the washing ghats. Similarly due to scarcity of water majority of bathing ghats are also used for bathing of pet animals such as cows, buffalos, dogs etc. henceforth, these activities also enhance the level of pollution.

8.2.6 Temple Waste Materials

Ganga is regarded as mother hence, worshiped by the Hindus. It has been observed that more than 1,000 tons of garlands and flowers are thrown to river every day. These flowers are not the only source of physical pollution but they decompose in the water and enhance the growth of bacterial population leading reduction in DO content of the water. Lot of plastic bags is also thrown to river that is used for bringing the PUJA materials. This leads physical pollution of the water.

8.2.7 Extraction of Ganga Water Through Lift Canals

There are many dams and lift canals constructed on river Ganga by which huge quantity of river water is extracted and supplied for irrigation, industrial and domestic purposes. In Uttrakhand near Tehri, a dam, has been built on Bhagirathi for hydropower generation At Haridwar, Ganga opens to the gangetic plains, where a barrage diverts a large quantity of its water into the upper Ganga canal. At Bijnore, another barrage diverts water in to the Madhya Ganga canal. At Narora, there is further diversion of water in to the lower Ganga Canal. A number of lift canals are also constructed on Ganga to extract the water and supplies for irrigation of nearby crop fields. Due to heavy extraction of water the flow of water is reduced leading reduction in dilution capacity of the river.

8.2.8 Construction of Dams and Deforestation in Himalayan Region

It is reported that for construction of numerous hydropower projects and dams on Ganga and its other streams in Himalayan region lot of forest plants are removed. Due to heavy blasting in this area not only plants are removed but huge quantity of soil is eroded and through runoff riches to the river. This is the main reason for siltation of the river bed. This leads reduction in the water holding capacity of the river.

8.3 Solutions for Improvement in Water Quality of River Ganga

We are of the opinion that the river Ganga is mainly suffering with the triple 'R' problems i.e.

1. Reduced flow of river water
2. Reduced water carrying capacity of river
3. Reduced quality of river water

River Ganga may be conserved with triple 'P' programs

1. Policy
2. Planning
3. Prevention

8.3.1 Details of Triple 'R'

8.3.1.1 Reduced Flow of River Water

Reduced flow of river water is due to construction of dams, deforestation in Himalayan region, over extraction of river and ground water through pumps and lift canals.

8.3.1.2 Reduced Water Carrying Capacity of River

Reduced water carrying capacity of river is due to high rate of siltation, encroachment of river banks and major constructions in the river catchment area.

8.3.1.3 Reduced Quality of River Water

Reduced quality of river water is due to pollution of water induced by discharge of city sewage, industrial effluents and agricultural runoff, cremation of dead bodies, animal bathing and washing of clothes, solid and biomedical waste disposal.

8.3.2 *Details of Triple 'P'*

8.3.2.1 Policy

Proper policy pertaining to increase in water quantity & maintenance of ecological flow, rain water harvesting, ground water recharging, natural cleaning, extraction of water directly from the river & ground water, land use near banks, disposal of human dead body and animal carcases, water users – drinking, bathing, washing, navigation, aquaculture, plantation, cropping, irrigation, discharge of sewage and industrial effluents, pollution control technologies, land selection for STPs, major constructions in the catchment areas, people's participation in different projects, monitoring of projects, accountability of project leader, adoption of stack holder, biodiversity conservation, utilization of youth power (school, college and university students) and mass awareness.

8.3.2.2 Planning

Short- and long-term planning for water conservation, environmental management based on regeneration ability, socio-economic and cultural development in the Ganga Basin, establishment of specific treatment plants including community treatment facilities, adoption of eco-friendly technologies, launching of time bound projects, people's participation, youth power (school, college and university students) utilization, mass awareness, shifting of flood irrigation system through canal system to modern irrigational practices, development of green belt of resilient plant species on river banks, recycling of waste water, rain water harvesting and ground water recharging technologies, use of solid waste materials as resource for energy production, solar energy harvesting, identification of high dilution zones of the river for discharge of waste water, establishment of Ganga Conservation Societies at ward level in the urban areas and gram sabha level in rural areas ensuring women participation, creation of Ganga Gyan Centres at University level, constitution of local technical expert committees for monitoring and assessment of Projects, encouraging non consumptive activities (fisheries, aquaculture and navigation).

8.3.2.3 Prevention

Prevention is better than cure henceforth, '*GANGA CULTURE*' be developed in the people of Ganga Basin. All preventive steps such as creation of legislation, coordination between Central, State and Local authorities, development of new technologies and mass awareness be adopted. Dos and don'ts materials be prepared and popularise through documentary films, news papers, TV, nukkad nataks, hand bills, slide shows, seminars, symposia and workshops.

8.3.3 *Suggestions for the Sustainable Development of River Ganga*

- Environmental State of the Art for ecologically sound sustainable eco-development.
- Appropriate plantation of mixed plant species near origin of the Ganga river and its basin to check the soil erosion and siltation of the river bed.
- Proper arrangement for clearing the silt deposited on the river bed to increase the water carrying capacity of the river.
- Proper plans for the biodiversity conservation in the river and its basin to save the rare and threatened species.
- Community waste water treatment plants for the treatment of toxic industrial effluents in densely populated areas of the cities.
- Complete ban on discharge of toxic industrial effluents in the city sewers.
- Specific sewage treatment plants for the treatment of sewage mixed with industrial effluents.
- Biological control of waste water for removal of heavy metals from the treated and untreated waste waters.
- Ozonization for the bacterial disinfection of the treated sewage.
- Electric crematoria and incineration plants for the cremation of human dead body, animal carcasses and bio-medical wastes.
- Local technical expert committees for monitoring of work conducted under the Ganga Action Plan.
- Construction of tube-wells in the catchment areas to avoid the over extraction of Ganga water and check the water logging condition and imbalance of ground water recharging.
- Selection of low productive land for construction of Sewage Treatment Plants to avoid the loss of fertile and highly productive lands.
- Mass awareness programmes for people participation.

References

1. Divedi S (1991) Assessment of water quality of river Ganga at Varanasi. PhD thesis, Banaras Hindu University, Varanasi
2. Alley KD (1994) Ganga and Gandagi: interpretations of pollution and waste in Benaras. Ethnology 33(2):127–145
3. Tripathi BD (1991) An overview of the hydrobiological features of the Ganga in the stretch Mirzapur to Ballia. In: Murti CRK, Bilgrami KS, Das TM, Mathur RP (eds) The Ganga. Northern Book, New Delhi, pp 156–160
4. Bhargava DS (1983) Use of water quality index for river classification and zoning of Ganga river. Environ Pollut (Ser B) 6:51–67
5. Shukla P, Tripathi BD (2012) River Ganga at Varanasi: the polluted purifier. Life Sci Leaflet 6:24–28
6. APHA (2005) Standard methods for examination of water and wastewater, 21st edn. American Public Health Association, Washington, DC
7. I.S.I. (1983) Tolerance limits for inland surface water subject to pollution. I.S.O.I.S.10500

A hydropower project under construction on the Ganga(Courtesy Tripti Singh)

Chapter 9
Environmental Flows for River Ganga

Nitin Kaushal

Abstract Globally, human alterations and interventions to the rivers for their usage have led to the creation of science of Environmental Flows (E-Flows); India and specifically Ganga river basin is no exception. This chapter touches upon the concept of E-Flows, its status in Indian context and its assessment through holistic methodologies. The chapter further explains detailed process of assessment of E-Flows with the help of a case study by WWF-India using a holistic methodology, i.e. Building Block Methodology (BBM). The chapter concludes by spelling the challenges and opportunities for E-Flows implementation.

Keywords Environmental Flows (E-Flows) • Assessment of E-Flows • Ganga river basin • WWF-India • GRBEMP

9.1 Introduction

The growing concern about degrading state of rivers due to excessive abstractions and overuse have led to the emergence of sciences of Environmental Flows (E-Flows), i.e. "the flows required for the maintenance of the ecological integrity of rivers, their associated ecosystems and the goods & services provided by them". Globally, the debate around E-Flows started in late 1960s, initially, it used to be considered as a hydrology based process, i.e. the hydrology was the main determinant of E-Flows in a given river. However, since 1990s, the researchers and managers working in the field of water resources started integrating other crucial aspects like ecology, geomorphology, water quality and socio-economics. This integration made

N. Kaushal (✉)
River Basins and Water Policy, World Wide Fund for Nature India,
172-B, Lodi Estate, New Delhi-110003, India
e-mail: nkaushal@wwfindia.net

R. Sanghi (ed.), *Our National River Ganga: Lifeline of Millions*,
DOI 10.1007/978-3-319-00530-0_9,

The Kali river, a crucial tributary of Ganga. Location: Dharchula, District - Pithoragarh, UTTARAKHAND (Courtesy Nitin Kaushal)

the concept of E-Flows a widely acceptable phenomenon across the globe and improved the Environment Flows Assessment (EFA) methodologies from mere hydrology based ones to much more advance and comprehensive set of holistic methodologies.

Many researchers believe that the E-Flows is a crucial step towards the larger aim i.e. Integrated Water Resources Management (IWRM). However, since achieving IWRM is a herculean task, especially in overstressed basins and that too under the *business-as-usual* scenarios, the work towards implementation of E-Flows is something which can be explored. Across the world, countries are taking initial steps towards E-Flows, its assessment and implementation.

Globally, there are examples where before making any other allocation after drinking water, part of the water resources of the countries are considered as 'reserve'. This 'reserve' is allocated for ecological purposes, including E-Flows. For instance – The Government of South Africa, through the South African Water Act, 1998 has introduced the provision of such a 'reserve'. Progress at the field under such reforms is generally gradual and faces lot of teething troubles, which is absolutely normal. The South African Government is facing challenges in implementation of the Act, as the concept of 'reserve' is new within the formal setup and local people. However, it is an indication of being on the right track.

9.2 Concept and Methodology for Assessment of E-Flows

An ideal E-Flows process should includes following steps –

1. E-Flows assessment
2. Trade-offs and negotiations to arrive at a consensus about the recommended E-Flows
3. Implementation of E-Flows

The assessment of E-Flows is an initial but a critical step, as, if this is not done properly, there is a likely possibility that the whole purpose of maintaining E-Flows in its true spirit gets defeated. Like the concept of E-Flows, the science of assessment of E-Flows is also an evolving one and since its evolution, it has been transformed to suit changing dynamics from time to time. It used to be a hydrology based process in the beginning, which is a simple method to arrive at E-Flows by mere considering hydrological features of the river. Ever since the method for E-Flows assessment has undergone various phases of modifications. As a result of that, we now have various set of methodologies for assessment of E-Flows, notable among them include –

1. Hydrology-based and look-up table approaches
2. Hydraulic rating methodologies

3. Habitat simulation methodologies
4. Holistic methodologies

Across the world, all above methodologies are used for E-Flows assessment in some or the other way; however for complex river basins, or the river basins with multiple users or stressed river basins, the holistic set of methodologies is preferred. This is because the holistic set of methodologies offers comprehensive process taking into account all the related aspects while doing the assessment, so that the assessed E-Flows can sustain all the functions and meet the requirements of and from the river. Though the holistic set of methodologies are resource intensive and time taking, but such methodologies provide a participatory approach, where there is ample room for stakeholders' consultation to arrive at describing the present and desired future state of the river while done assessment of E-Flows.

The completion of exercise for assessment of E-Flows leads to next phase which is the most difficult and complex one, i.e. negotiations among various stakeholders and tradeoffs to arrive at a consensus about the recommended E-Flows. This sometimes includes tough and radical decisions.

The last phase is about the implementation of recommended E-Flows, which should ideally be done by the respective regulator or project proponent; however the monitoring should be done in close association with the administration, local communities and other stakeholders. Such a monitoring exercise should not be a one-off affair; it should be an ongoing programme, with a scope for future refinements.

9.2.1 E-Flows in Indian Context

In India, the debate started from the concept of 'minimum flows', i.e. the flows in the river should not be less than the average of 10 days minimum flow of the river in its natural state. This concept was introduced by the Central Water Commission, Ministry of Water Resources during 1992. It was primarily aimed at upcoming river valley projects during that time, but since it did not have any legal or mandatory binding, so this has not been fully adhered to.

However, since 2005, the debate around E-Flows in the country has picked up tremendously. Smakhtin and Anputhas [1] used a hydrology method, based on flow duration curves to calculate E-Flows for a range of Environment Management Classes (EMCs) for several major river basins in India, including the Cauvery, Krishna, Godavari and Mahanadi basins. The EMCs calculated ranged from "*natural*" to "*severely modified*", with the required flow volumes and elements of flow variability of the E-Flows set to progressively reduce, resulting a decreasing level of ecosystem protection. The final environmental water demands (E-Flows) were presented in two forms: as a flow duration curve and as a monthly time series.

Later on, in 2007 the Ministry of Water Resources (MoWR) Government of India, through the Water Quality Assessment Authority (WQAA), constituted a

Working Group which used a modified version of the Tennant method[1] to assess minimum flow requirements in Indian rivers.

The Tennant method takes only a very short time; however, the confidence in the outputs is correspondingly low. The working group classified Indian rivers into two groups, namely "*Himalayan*" and "*Other Rivers*" [2], and made their recommendations.

Since the confidence of the working Group in their findings was low, these recommendations were not accepted, and were never tested.

According to the WQAA,

a. Himalayan Rivers: Minimum flow to be not less than 2.5 % of 75 % dependable Annual Flow expressed in m^3/s; one flushing flow during monsoon with a peak not less than 250 % of 75 % dependable Annual flow expressed in m^3/s.
b. Other Rivers: Minimum flow in any ten daily period to be not less than observed ten daily flow with 99 % exceedance. And where ten daily flow data is not available this may be taken as 0.5 % of 75 % dependable Annual Flow; One flushing flow during monsoon with a peak not less than 600 % of 75 % dependable annual flow expressed in m^3/s.

9.2.2 Challenges the Ganga Faces – The Answer, E-Flows

A river needs water to sustain its own functions, at the same time it serves the society in many ways. An ideal E-Flows assessment would take account of various requirements from the river and for the river. Such requirements, mainly include – geomorphology, biodiversity, socio-economic, cultural and spiritual (if any) and water quality and this is possible by considering a holistic methodology for assessment of E-Flows.

---

[1]Tennant Method – The Tennant method (1975–1976) was developed for 11 streams in Montana, Wyoming and Nebraska in USA in order to find satisfactory discharge levels for fish-passing, since fish were crowded into smaller pool-riffle structures. It has received appeal in regions of British Columbia as a quick and easy-to-understand method of determining habitat and recreational suitability of stream habitat. The method is based on the assumption that flows that are satisfactory for needs of fish and other aquatic biota will also be sufficient for maintaining recreational and aesthetic qualities. Based upon empirical relationships and observations, Tennant suggested that minimum flows at any time of the year must be >10 % of mean annual discharge. Below the 10 % threshold, fish habitat and recreational value will be severely degraded. Above 10 %, habitat and recreational quality increases in a range from fair conditions (10 % in winter and 30 % in summer) to outstanding (40 % in winter and 60 % in summer). An optimum range is considered 60–100 % of mean annual flow at any time of the year. *For details, please visit* http://www.ilmb.gov.bc.ca/risc/o_docs/aquatic/027/final-17.htm and http://upcommons.upc.edu/pfc/bitstream/2099.1/3367/8/36228-8.pdf.

It is reiterated that, E-Flows is not mere for the river or for the biodiversity, but it is also for the human beings, who are dependent on it. So the E-Flows serve two purposes, maintaining the riverine ecosystem and serving the mankind in a sustainable manner, so that they can enjoy the benefits for a long-long time.

The current state of river Ganga is known to everyone, the river faces nested challenges, ranging from reduced flows to domestic and industrial pollution. The key reason for reduced flows are the existing and upcoming hydropower projects in the state of Uttarakhand and the exiting irrigation projects in Uttarakhand (mainly Haridwar) and Uttar Pradesh.

By no means, the idea here is to devaluate the contribution of both irrigation and hydropower to the mankind, however one needs to be sensitive towards requirements of the river to keep it a healthy ecosystem.

The significance of gangetic plains to feed the ever-growing population indeed makes it crucial to have these major canal systems to serve the irrigation demand for today and for the times to come. However, there is a need to work on the exiting irrigation practices and cropping pattern, which leads to over-usage of water resources, diverted from river Ganga or extracted from the Ganga river basin through pumping. The current practises in this regard are unsustainable.

The hydropower projects, whilst leading to significant addition to overall power generation scenario of the country; has also let the rivers either deprived from flows (in case of reservoir based projects) or merely allowed trickle of water in the stretches of rivers (in case of run-of-the-river projects). There has been no consideration of E-Flows assessment and its maintenance under the hydropower projects. The requirement of water for the environment is as essential as the water for hydropower; this need to be understood at all levels, so that there could be water allocations for E-Flows from each of the hydropower scheme. Atleast, there is no reason, why upcoming projects cannot have E-Flows allocations.

Whilst dilution cannot be the solution to pollution, the enhanced flows will improve the assimilative capacity of the river and that is why the water quality plays crucial role in E-Flows.

The key reasons for pollution include inadequate capacity of city utilities to treat municipal sewage, plus industrial waste, which though required being treated to a designated stage, as stipulated by the Central Pollution Control Board (CPCB), remain largely unchecked and continue to pollute the river through drains with harmful chemicals, heavy metals and so on. In case of pollution caused by domestic sewage, the key issue is over-dependence on traditional ways and means of wastewater treatment, i.e. through the Sewage Treatment Plants (STPs). We do need to have STPs, but at the same time we should also explore other alternate technologies, which are equally and in many cases much more effective. Across the world, there has been paradigm shift from usual means of treating wastewater to modern techniques.

In case of industrial pollution, there is a definitive need to enforce the existing regulations.

These challenges are the key motivations for maintaining E-Flows for river Ganga, as by not doing so, we may not be able to save this river for the generations to come.

9.3 E-Flows for River Ganga

The very first question, which needs to be answered to, before undertaking any assessment or implementation of E-Flows is that, '*what the people want their river to do for themselves or how people wish to see their river to look like*'. There are two sets of choices, which the society has to make –

1. Do they want their river to be seen as a resource which can be used for irrigation, cities and industries but not abused for any of these functions? Or do they want to use up the resources to the extent that, it's not available for the generations to come?
2. Do they wish to see their river to be a stream of water with adequate velocity, depth and width, so that it can be called as a mighty river? Or do they wish to see the river to be a trickle of water?

The answer to these questions will set the agenda for the kind of E-Flows one would like to have for its river. Across the world the answer to these questions lies somewhere in the middle, where people wish to use the river in a sustainable manner, so that the river continues to serve them and the upcoming generations.

Any E-Flows assessment and implementation exercise for river Ganga, needs to take care of unique status of this river in cultural and spiritual lives of people of India. So, a holistic methodology, which can take account of this aspect, should ideally be considered for E-Flows assessment for river Ganga, rather than any quick model based approach. An iterative holistic methodology should attempt to take account of the societal values attached to the Ganga River. There should be room for adequate stakeholders' consultations, so that there is scope for refinement.

9.3.1 Assessment of Environmental Flows – An Initiative of WWF-India's Living Ganga Programme

The World Wide Fund for Nature – India (WWF-India), under its Living Ganga Programme (2007–2012) has conducted a detailed and multidisciplinary study to assess the E-Flows in the upper stretch of river Ganga (0–800 km, i.e. from Gangotri to Kanpur). The Living Ganga Programme was part of HSBC Climate Partnership. For this exercise, WWF-India partnered with few reputed international (UNESCO-IHE, the Delft Netherlands and IWMI-Sri Lanka) and national (IIT-Kanpur, Delhi University, INRM-IIT Delhi, PSI-Dehradun, Srinagar Garhwal University and some individual experts) institutions and organization. All the partners played vital role in successful completion of this exercise.

9.3.2 Process of E-Flows Assessment – In Brief

Initially, after deliberating upon various EFA methodologies, the team concluded that selection of methodology should be done amongst holistic set of methodologies, as the Ganga river basin is a complex one with multiple users. The uniqueness of river Ganga in the cultural lives of people prompted the team to select a methodology, which can be tailored to suit the local requirements. As a result of this, the team considered Building Block Methodology (BBM) for assessment of E-Flows. The team improvised the existing version of BBM to integrate unique cultural and spiritual aspects of Ganga into the overall framework of the BBM. Plus BBM is advantageous to work with experts with no past experience of doing E-Flows assessment, *per se*. This E-Flows assessment exercise took about 3 years (2008–2010). Complete process of BBM is illustrated through Fig. 9.1. This figure also briefs about the much more complex task of implementation.

As part of larger process, the team zoned the river from the perspective of homogeneity, through this study four zones were identified and a site under each of the zone was selected, one of the key basic criteria of site selection was its representativity for the respective zone. These zones along with selected site is listed below –

1. Zone I: Gangotri to Rishikesh – Site: Kaudiyala
2. Zone II: Upstream of Garhmukteshwar to Narora – Site: Narora Barrage (*reference zone*)
3. Zone III: Narora to Farrukhabad – Site: Kachla ghat
4. Zone IV: Kannauj to Kanpur: Site: Bithoor

Owing to conduciveness of the conditions at Zone II, it was considered as a *reference zone*. This zone is by and large satisfactory from the perspective of flows and intactness of biodiversity. Though the threats are looming, however current situation is relatively better than in the other zones. The area at upstream of Narora Barrage was considered as the site under this zone, as by virtue of adequate water in that area, it is a reasonably good habitat for gangetic dolphins (the national aquatic animal of India), which is not only a flagship aquatic species, but is also significant from the food chain perspective. It is an indicator of a healthy aquatic ecosystem. Plus WWF has been running a Dolphin Conservation Centre in close vicinity of this location from more than last one and half decade.

Since, the BBM is a holistic methodology and comprised of various aspects associated with a riverine ecosystem, so various thematic groups were constituted to undertake tasks related to respective aspects. Figure 9.2 illustrates various thematic groups and the broad terms of reference for them for the purpose of this study.

After the site selection, various expert groups undertook field surveys/investigations for a complete 1 year, so that all the seasons gets accounted and reported. During this field exercise, the idea was to collect the desired information related to respective perspective, i.e. (i) the biodiversity group was collecting and compiling the information and data pertaining to biodiversity found and impacted by flows in and around the sites plus the habitat requirements; or (ii) the cultural and spiritual aspects group conducted interviews and interactive sessions with pilgrims, tourists, local priests and others to get the information in regard to people's aspirations from Ganga and their thoughts about the river. This group also did thorough research of secondary literature to ascertain the status of river Ganga in historical mythology.

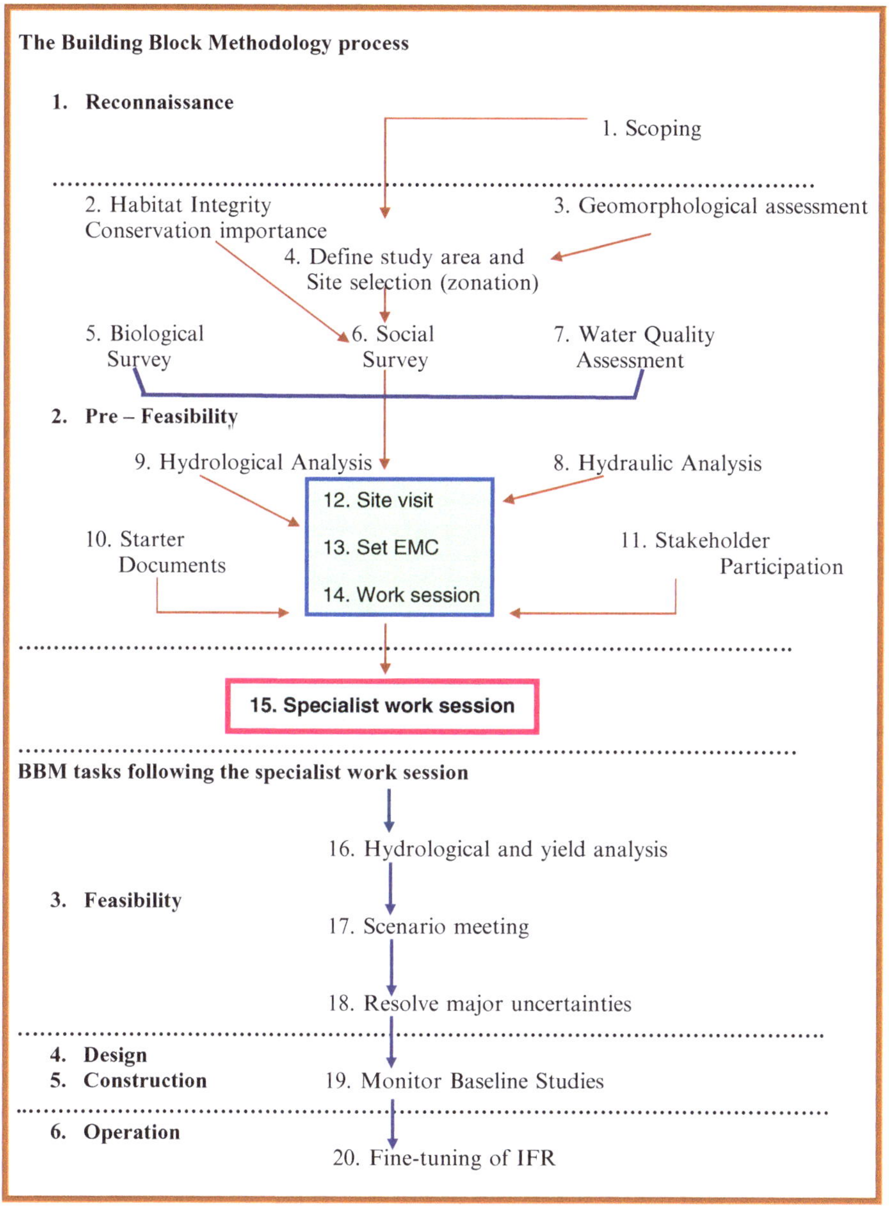

Fig. 9.1 Flow chart illustrating the Building Block Methodology (BBM) (Source: Summary Report by WWF India on Assessment of Environmental Flows for the Upper Ganga Basin by Keeffe et al. [3, 4])

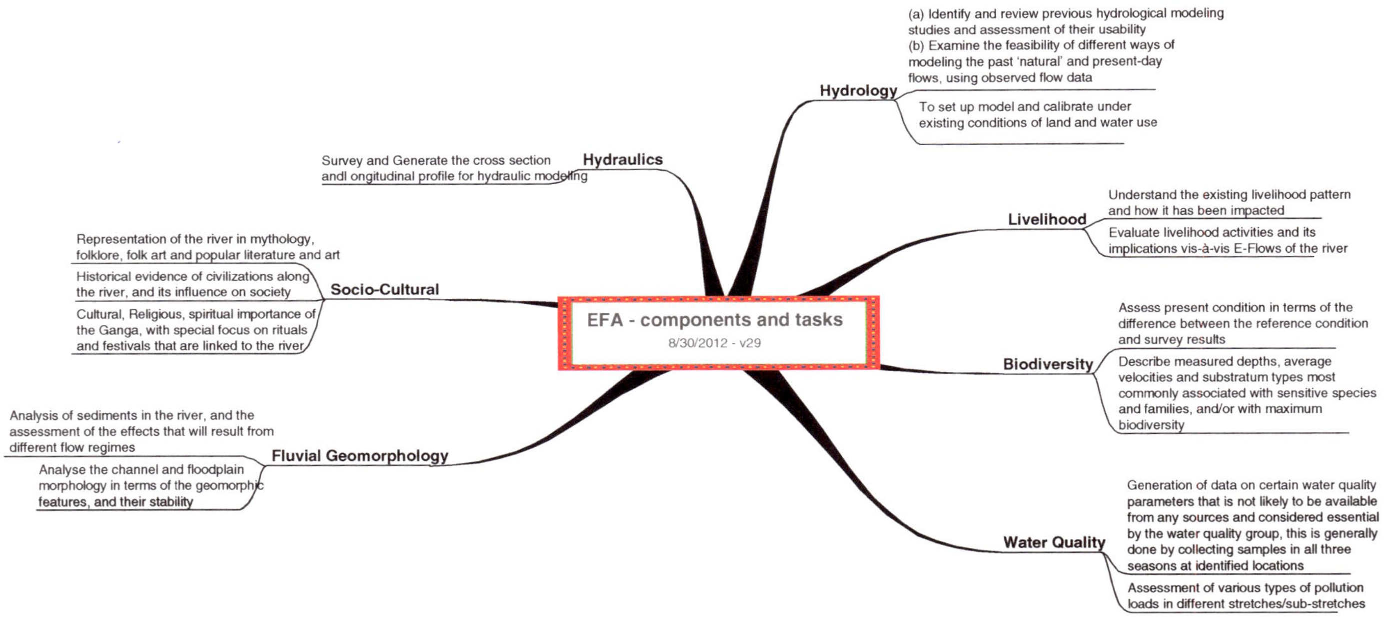

Fig. 9.2 Illustration of aspects studied under E-Flows assessment and key tasks required to be undertaken under each of the aspect
Note: Figure 9.2 Illustration developed from Detailed Report on Assessment of Environmental Flows for the Upper Ganga Basin by Keeffe et al. [3, 4], WWF-India

The outcome of this one year field work was Starter Documents from each of the expert group, explaining the works they did during surveys/investigations. They also came out with set of motivations and repercussions related to required flows. The groups also captured the present and future desired state of the river, based on the Environment Management Class (EMC).

Simultaneously, the hydrology and hydraulics group's contributions are as follows:

1. the hydrology group prepared the report on 'Hydrology of Upper Ganga', which explained various hydrological aspects related to upper Ganga and of specific sites. To do so, the group assessed global datasets and some real-time flow data which was available in public domain. The team could not get access to real-time flow data in time-series manner due to the fact that, Ganga is a trans-boundary river, and flow data of any trans-boundary river is classified by the Government of India.
2. the hydraulics group surveyed the cross sections of the identified sites, they took atleast three cross sections at each of the site, i.e. one at upstream, one at the point and one at the downstream. The entire cross sectional data was documented in the report prepared by the Hydraulics group.

Both the hydrology and hydraulics reports were the backbone of the larger process of assessment of E-Flows, as the remaining thematic groups, i.e. fluvial geomorphology, biodiversity, socio-economic, water quality, and cultural and spiritual aspects, referred to their reports to arrive at E-Flows values for various seasons for the sites, representing respective zones. The E-Flows values suggested by individual groups were debated within the larger group and finally consensus had been arrived at about recommended E-Flows for a site during given season.

After completion of the study, WWF came out with two reports, i.e.

1. a Summary Report, to briefly report the process and results for wider dissemination, including policy makers, relevant state entities and general audience;
2. a Detailed Report, to capture step-by-step details of the process for those who wish to understand the process in detail and for those who wish to use this case study as a guide to do similar studies in their respective river basins.

Both these reports are available on WWF – India website.

9.4 Response to WWF's Study on Assessment of E-Flows

The WWF-India's study on assessment of E-Flows for river Ganga has been a stepping stone in regard to debate around E-Flows in India. Under this study, the team was engaged in number of aspects which are required to be studied in order to assess the E-Flows.

This study in many ways brought lot of awareness of E-Flows; it also raised the existing level of information and database about E-Flows assessment and tried to answer the key question, i.e. what does it means to the society and the stakeholders at large. This initiative was the first of its kind in India, under which something of this scale and nature has been piloted for an iconic river like Ganga.

This study has benefitted the formal process of development of Ganga River Basin Environment Management Plan (GRBEMP), initiated by the Ministry of Environment and Forest. Around the country, the debate around E-Flows is graduating towards the kind of EFA methodology to be followed and how the implementation will take place. This development is indeed satisfying; however the much complex question is about the implementation, which has its own set of challenges.

9.5 Challenges and Opportunities for E-Flows Implementation

An E-Flows implementation in a river basin which does not have competing demands is fairly less complex, in comparison to the river basins which are over-stressed and have competing demands among various set of users. The cost attached to E-Flows implementation in such basins often means reprioritisation of existing water allocations and that is where the real challenge comes into picture. In addition to this, the follow-up question to be answered is from where the water for E-Flows will come; as there were no such considerations in the past hence the set priorities (at that point of time) for allocation of water did not take these requirements into account. This makes the implementation of E-Flows, a daunting task for the water managers and environmentalists. This is a global problem while undertaking implementation of E-Flows.

The biggest motivation to secure E-Flows for river basins is the question discussed earlier, i.e. ‘how we wish to see our rivers’, and from there everything flows. Whilst considering value of water resources for various uses, like – hydropower, irrigation, domestic and industrial; time has come that we also value the same resource for the environment and ecology as well.

There are two definitive avenues, where E-Flows can be made integral part of planning and execution. One is during the hydropower development and another is under the existing irrigation schemes.

1. E-Flows in hydropower projects – as noted earlier, though the existing projects do not take account of E-Flows requirements, however there are few examples, where the projects are taking some steps to ensure mandated flows by the respective state governments. For instance, the project proponents of Karcham Wangtoo hydropower project in Kinnaur district of Himachal Pradesh has made provisions in design that, 15 % of the lean season flow will be maintained throughout the year at downstream of the dam. Though the mandatory requirement at the central government level is 10 % but the Government of Himachal Pradesh in 2006, issued orders for all the upcoming hydropower projects to ensure 15 % flows as described earlier. This 15 % is not the E-Flows *per se*, as there is no sound rational for this figure, which therefore appears to be an arbitrary figure. Nevertheless, the

state has gone one step ahead than others, which itself is a welcome step. Plus, the developer, by ensuring these releases in the design of the project, has exhibited his abidance towards the state orders. However, it is yet to be seen how these flow releases will be monitored, or whether there is going to be any sort of community participation in doing so. From last couple of years or so, some exercises to assess E-Flows in going on in the Ganga Basin and these initiatives are at various stages. Whilst, tangible results out of these exercises are yet to be seen, however it is hoped that, the project proponents. Then the projects proponents will have to abide by the stipulated regulations. Further, options could also be explored to add EFA in addition to usual EIA pre-condition towards the individual projects. Generally, a power purchase agreements (PPA) is made between the hydropower developer and the state and central utilities for power selling, so any reduction in power generation in view of maintaining E-Flows will mean a financial loss to them, therefore maintaining E-Flows in existing projects is difficult. However ways and means can be explored to compensate the hydropower developers to maintain assessed E-Flows. The financial institutions also have to play a significant role in this regard. They can embed EFA guidelines within the funding criteria for the projects.

2. Irrigation projects – the existing projects in the upper Ganga basin including the Upper Ganga Canal, Eastern Ganga Canal, Madhya Ganga Canal, Lower Ganga Canal etc. currently has got no provisions for E-Flows releases for the simple reason that, when these projects were constructed there was no such concept like E-Flows. However, the time is changing and over-abstraction of water resources from rivers is leading to growing concern about river flows. On the other hand the government at both central and state level is striving for improvement in irrigation efficiency. The National Commission on Integrated Water Resources Development Plan (NCIWRDP) in 1999 has advocated for 20 % improvement in current irrigation efficiency level. It assumed that, the current efficiency level of 35–40 % could be improved to about 60 % (Theme Paper – Integrated Water Resources Development and Management [5] by Central Water Commission, Ministry of Water Resources, by Government of India). Part of this saving, if realised, can be allocated to maintaining E-Flows in the river. There is an apprehension that, any saving in irrigation sector will lead to increase in irrigated area or change towards water intensive crops, which in both cases mean that saving will not lead to water for E-Flows, this is potentially the biggest challenge to ensure maintenance of E-Flows, so necessary monitoring and regulation mechanisms are required to be established.

The conservation of riverine ecosystems is critical, especially in wake of our overdependence on them. Maintenance of Environmental Flows is one such mean, through which the larger goal of ecosystem conservation, in particular the riverine one, could be achieved.

Acknowledgements The author is grateful to Mr. Ravi Singh Secretary General & CEO and Dr. Sejal Worah Director-Programmes from WWF-India for their continued support and encouragement. The author is thankful to Prof. Vinod Tare and Dr. Rashmi Sanghi from IIT Kanpur and Mr. Suresh Babu from WWF-India for their valuable comments, which improved the paper immensely.

The author is also thankful to all the partners, as it is because of their support the EFA for Upper Ganga could be successfully accomplished.

References

1. Smakhtin VU, Anputhas M (2006) An assessment of environmental flow requirements of Indian river basins. Research report no 107, International Water Management Institute, Sri Lanka, p 38. This document is available at http://www.iwmi.cgiar.org/Publications/IWMI_Research_Reports/PDF/PUB107/RR107.pdf
2. WQAA (2007) Report of the Working Group to advise Water Quality Assessment Authority (WQAA) on the minimum flows in the rivers, Central Water Commission, Ministry of Water Resources, Government of India, p 105. The report can be reached at http://wqaa.gov.in/reports/Report_on_Min_Flow_in_River.pdf
3. Keeffe J, Kaushal N, Bharati L, Smakhtin V (2012) Summary report by WWF India on assessment of environmental flows for the upper Ganga basin, 24 p. http://awsassets.wwfindia.org/downloads/exec_summary_mail_1_28.pdf
4. Keeffe J, Kaushal N, Bharati L, Smakhtin V (2012) Assessment of environmental flows for the upper Ganga basin. Report by WWF India, 162 p. (http://www.wwfindia.org/)
5. Central Water Commission (2008) Theme paper – Integrated Water Resources Development and Management. Ministry of Water Resources, Government of India, p 23. (www.cwc.nic.in/main/downloads/WWD-2008.doc)

Bhagirathi as captured near Chirbasa whilst on trek to Gaumukh (Courtesy Rashmi Sanghi)

Part IV
Some Vital Challenges Being Faced by Ganga in Changing Times

The unique Kumbh 2013 mood at Sangam, Allahabad. Kumbh 2013 witnessed over 90 million people in the duration of 55 days from 12 January to 10 March 2013 (Courtesy Nitin Kaushal)

Chapter 10
Snow Melt Runoff Status in Part of Ganga Basin

Praveen K. Thakur

Abstract The temporal and permanent snow cover along with glaciers form an important part of Himalayan ecosystem and hydrological cycle. The Ganga River with its origin from glaciers of Gangotri and Alakhnanda valley has significant water contributions from melt waters of these glaciers and snowmelt runoff. The percent contribution of snowmelt and glacier melt water is highest in headwater reaches of Ganga basin. The present chapter has focused on mapping of these snow bound areas using remote sensing and integrating the Snow Cover Area (SCA) with hydrometeorological data in temperature index based snowmelt runoff models. The Bhagirathi river basin upto Uttarkashi and Alakhnanda river basin upto Joshimath are taken for this study. Landsat, Resourcesat-1 and MODIS satellites were used for SCA estimation. WINSRM has been used for snowmelt runoff simulation. The discharge data from central water commission was used to calibrate and validate the simulated snowmelt runoff. The Coefficient of determination, R^2 for Bhagirathi river basin comes in range of 0.76–0.84 for 2002–2007 time periods, similarly for Alakhnanda river it comes about 0.87 and 0.90 for year 2000 and 2008. Overall this study shows that snowmelt runoff can be estimated using temperature index approach in this area and well distributed network of hydrometeorological stations will give better accuracy for snowmelt runoff models.

Keywords Snow melt runoff • Remote sensing • Bhagirathi • Alakhnanda • Hydrometeorological data

---

P.K. Thakur (✉)
Water Resources Department (WRD), Indian Institute of Remote Sensing (IIRS), Indian Space Research Organisation (ISRO), 4-Kalidas Road, Dehradun, India
e-mail: praveen@iirs.gov.in

R. Sanghi (ed.), *Our National River Ganga: Lifeline of Millions*,
DOI 10.1007/978-3-319-00530-0_10,

Bhagirathi river near Uttarkashi, Feb 2011 (Courtesy Praveen Thakur)

10.1 Introduction

Snow is a type of precipitation in the form of crystalline water ice, consisting of a multitude of snowflakes that fall from clouds. The process of this precipitation is called snowfall. Since snow is composed of small ice particles, it is a granular material. It has an open and therefore soft structure, unless packed by external pressure (wikipedia 2012). According to estimates, snow represents about 5 % of the precipitation that reaches the Earth's surface. Snow remains on the ground until it melts or sublimates. In colder climates, this results in snow lying on the ground all winter. When the snow does not melt in the summer and accumulates and becomes a glacier

Snowfall, accumulation and melt are important hydrological processes in watersheds at high altitudes or latitudes. A large portion of snow hydrology groups are pursuing new methods for incorporating snow hydrology into distributed models over complex terrain through theoretical developments, model development and testing with field and remote sensing data sets. Snow hydrology is quite complex and involves both mass and energy balance calculations over a time-varying snow pack which is influenced by spatial location in the watershed, interaction with vegetation and redistribution by winds. Some researchers seek to accurately capture snow dynamics at a point and over a domain as the spatial pattern of snow cover area is readily observable from remote sensing (wikipedia 2012).

Spring snow melt is a major source of water supply to areas in temperate zones near mountains that catch and hold winter snow, especially those with a prolonged dry summer. In such places, water equivalent is of great interest to water managers wishing to predict spring runoff and the water supply of cities downstream. Measurements are made manually at marked locations known as *snow courses*, and remotely using special scales called snow pillows [1]. Many rivers originating in mountainous or high-latitude regions have a significant portion of their flow from snowmelt. This often makes the river's flow highly seasonal resulting in periodic flooding. In contrast, if much of the melt is from glaciated or nearly glaciated areas, the melt continues through the warm season, mitigating that effect.

Snowfall and snowmelt runoff along with melt water of glaciers contributes significantly to Ganga river discharge especially during spring and non-monsoon season. The percent contribution of snowmelt and glacier melt water is highest in headwater reaches of Ganga basin. The contribution of snow and glacier melt reduces as Ganges goes beyond Devpyrag, where it contributes to about 29 % of flow [2].

10.2 Remote Sensing for Snow Mapping and Snowmelt Runoff Modeling

Satellite imagery is successfully used for snow cover mapping and is employed to improve the performance of numerical models that predict snowmelt runoff. Snow cover is one of the most easily recognized features in visible – spectrum satellite image of the Earth's surface and that snow cover can be distinguished from other

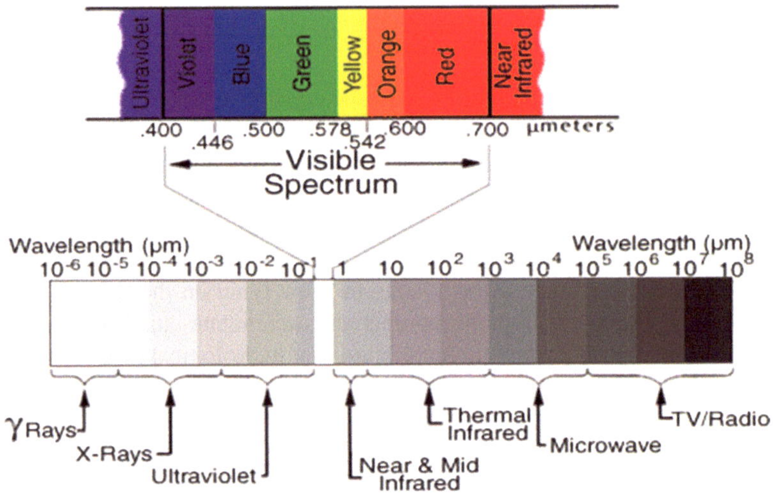

Fig. 10.1 Electromagnetic spectrum (Source: http://www.astro.virginia.edu/class/oconnell/astr121/121supps2-3.html)

objects on the Earth's surface because of its relative albedo. As a result, determination of the extent of the snow cover was one of the early accomplishments of satellite data interpretation. With the help of remote sensing data, various types of snow as well as hydrological parameters can easily be mapped and monitored [3].

10.2.1 Optical Remote Sensing in Snow Cover Area

Remote sensing offers a significant potential for collecting this data in cost effective manner. Because of difficult access, harsh weather conditions and expensive operation of hydrological stations, radar or satellite data are particularly appropriate. However, ground truth data are indispensable in the calibration and verification of remotely sensed data. Aerial and satellite surveys are useful in mapping snow lines. The wealth of observational material obtained by remote sensing can be integrated into models, such as snowmelt runoff models, considerably improving the forecast accuracy. Snow was first observed by satellite in eastern Canada from the TIROS-1 satellite in April 1960. Since then, the potential for operational satellite based snow cover mapping has been improved by the development of higher temporal frequency satellites such as GOES, Landsat, SPOT and IRS series, and NOAA-AVHRR, NIMBUS-SMMR and DMSP SSM/I satellites. Satellite data along with DEM can also be used for snow mapping [4].

Interaction of light and snow in the electromagnetic spectrum should be taken in consideration as important part of using of remote sensing data. Figure 10.1 shows the electromagnetic spectrum and its intervals and Fig. 10.2 shows reflectance spectra of water, soil and vegetation. The behavior of snow in Visible (VI), Near Infrared (NIR), and Thermal Infrared (TIR) is also described below in Fig. 10.3.

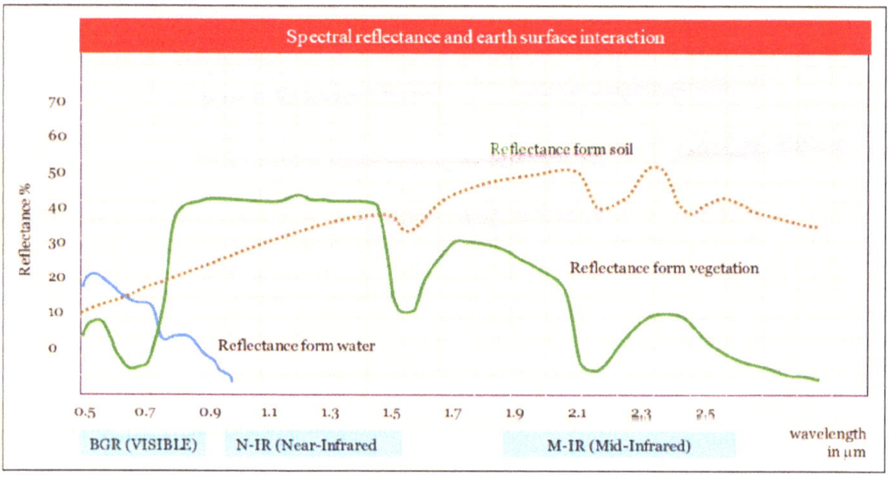

Fig. 10.2 Spectral reflectance and earth surface interaction

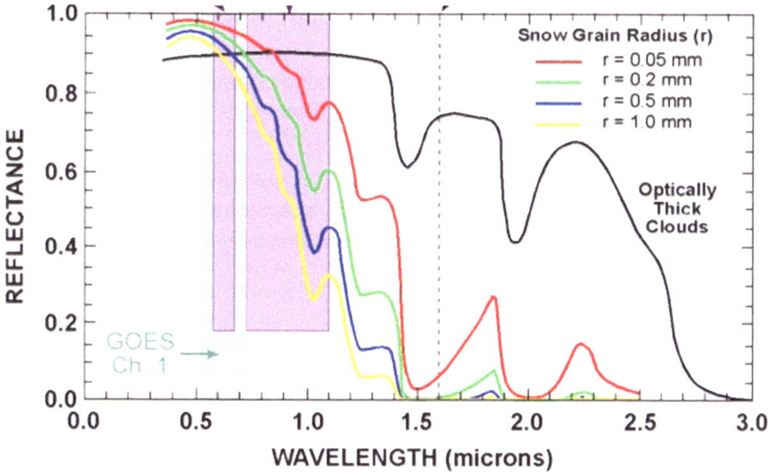

Fig. 10.3 Reflectance Spectra for Snow and Clouds (Source: NSDIC)

10.2.1.1 Visible and Near Infrared

Fresh dry snow appears white to the human eye. That is to say, it is highly reflective, with little variation over the range of wavelengths (approximately 0.4–0.65 µm) to which the eye is sensitive.

The reflectivity of new snow decreases as it ages in both visible and infrared region of spectrum, however the decrease is more pronounced in the curves of snow and ice Infrared. Decreasing reflectivity in the visible region can be attributed to the contaminants such as dust, pollen and aerosols. An advantage of using visible and near infrared data is the easy interpretation of the image. Even though, snow extent can be easily extracted, no information regarding the snow water content can be derived [5].

10.2.1.2 Thermal Infrared

Thermal data have been used less than other remote sensing data for measuring snow characteristics. Since we should recognize snow radiation spectrum to determine temperature of snow. Despite all these limitations, thermal data are of great use to recognize the boundary between snowy zones and no-snowy zones. Like visible and near infrared images, clouds limit the usability of thermal infrared images. If there are clouds, the temperature above them will be measured [6].

One promising method to distinguish between a debris-covered glacier and surrounding bedrock might be analysing surface temperatures, because the underlying ice is supposed to cool the supraglacial debris. This is measurable on the surface with the thermal signature from ASTER and LANDSAT imagery, provided the debris cover does not exceed a certain density and/or thickness. However, the exclusive use of thermal information causes problems for glaciers with massive debris cover. In addition, shaded non-glacier areas have surface temperatures similar to the clean-ice areas, but debris-covered ice with direct radiation turns out to be somewhat warmer. This hampers especially in shaded moraine complexes a clear differentiation of glacier and non-glacier area. An additional draw-back is of course the relatively low resolution of the thermal bands.

The mapping and monitoring of snow using remote sensing data has been done routinely since 1980s [6–8]. The use of remote sensing based Snow Cover Area (SCA) in snowmelt runoff models has also been done regularly since 1970s [9–14].

10.2.2 Snowmelt Runoff Model

Hydrological models uses for simulation or forecasting of stream flow are generally categorized as simple regression models, black-box models, conceptual models and physically based models. In general, snowmelt models can be divided into two types of models, namely energy balance models and index models.

In the energy balance models, the net energy or the heat budget of a snowpack governs the production of melt water. This method involves an accounting for a given period of time of incoming energy, outgoing energy, and the change in energy storage for a snowpack. The net energy is then expressed as the heat equivalent of snowmelt. The presence of cloud cover and vegetation cover significantly affects the energy balance of a snow surface. The seasonal variability in the energy inputs available for melt in general increases towards the poles; the difference between summer and winter is minimized at the equator. The differences in energy receipt on the north and south facing slopes can be critical in influencing the time of snowmelt.

In the temperature index models, the mainly air temperature data is used to infer the snowmelt runoff from a given river basin. Most popularly used method is known as Snowmelt Runoff Model (SRM), which is developed by Martinec in 1975 [15, 16].

This was developed in small European basins originally and was used in most parts of the world in mountain basins of almost any size (so far it has been used from 0.76 to 1,22,000 km^2) and any elevation range [17–20].

10.3 Snowmelt Runoff Studies in Upper Ganga Catchment

Study area for snowmelt runoff study is upper Ganga catchment with main emphasis on Alakhnanda and Bhagirathi Basins. Alakhnanda up to Joshimath and Bhagirathi up to Uttarkashi have been considered in this study (Fig. 10.4a, b and c). The Alakhnanda basin is extended between 30° 0′ N to 31° 0′ N and 78° 45′ E to 80° 0′ E covering an area of about 10,882 km^2, represents the eastern part of Garhwal Himalaya. Out of total area of the basin, 433 km^2 is under glacier Landscape and rest of 288 km^2 is under fluvial landscape. Bhagirathi river basin is situated in the state of Uttarakhand, within geographical co-ordinate 30° 38′ to 31° 24′ North latitude 78° 29′ to 79° 22′ East longitude. The altitude variation is from 1,121 to 7,000 m above the mean sea level and total area upto Uttarkashi is 4,496.21 km^2. Both basins are in Uttarakhand sate of India (Fig. 10.5a, b, c and d).

Uttarakhand lies on the south slope of the Himalaya range, and the climate and vegetation vary greatly with elevation, from glaciers at the highest elevations to tropical forests at the lower elevations. The highest elevations are covered by ice and bare rock. The Western Himalayan Alpine Shrub and Meadows eco region lies between 3,000–3,500 and 5,000 m elevation; tundra and alpine meadows cover the highest elevation. The Western Himalayan sub-alpine conifer forests lie just below the tree line; at 3,000–2,600 m elevation they transition to the Western Himalayan broadleaf forests, which lie in a belt from 2,600 to 1,500 m elevation. There are many small and big rivers in Uttarkashi district. The Yamuna and the Ganges (Bhagirathi) are biggest and holiest and their origin is Yamnotri and Gangotri (Gomukh) respectively. Asli Ganga, Jad Ganga are some of the tributaries of the Ganges. Uttarakhand has a total geographic area of 51,125 km^2, of which 93 % is mountainous and 64 % is covered by forest. Most of the northern parts of the state are part of Greater Himalaya ranges, covered by the high Himalayan peaks and glaciers. The headwaters of the Bhagirathi are formed in the region of the Gangotri glaciers in the Himalaya. From its source, the river flows for about 205 km before meeting the Alakhnanda river at an elevation of 475 m in the town of Devprayag. The provisional capital of Uttarakhand is Dehradun which is also a rail-head and the largest city in the region

The unique Himalayan ecosystem plays host to a large number of animals (snow leopards, leopards and tigers), plants and rare herbs. Two of India's mightiest rivers, the Ganga and the Yamuna take birth in the glaciers of Uttarakhand, and are fed by myriad lakes, glacial melts and streams in the region.

The Alakhnanda basin is characterized predominantly by hilly terrain, deep gorges and river valleys. The region is broadly divided into four major divisions (i) the Great Himalayan Ranges (snow covered regions), (ii) Alpine and pasture land (covered by snow during the 4 months of winter season), Middle Himalaya (characterized by highest population) and (iii) river valleys (characterized by service

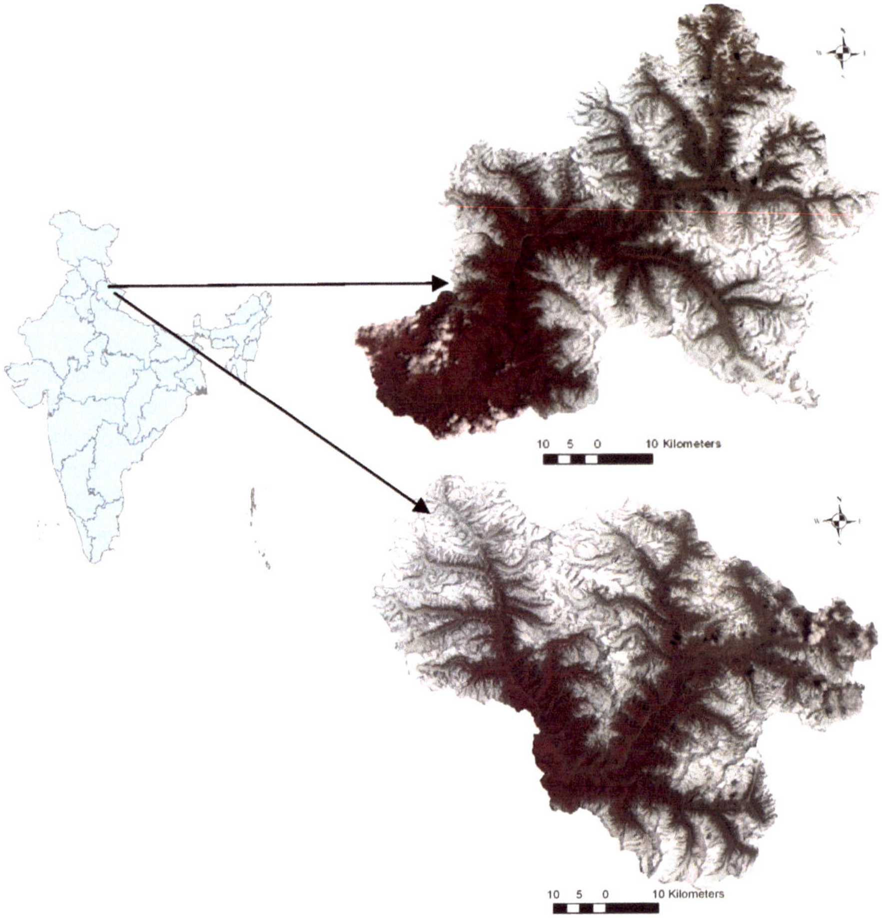

Fig. 10.4 (**a**: *left central*) Location map of India and Uttarakhand with AWIFS image based FCC of (**b**: *upper right*) Bhagirathi river basin up to Uttarkashi (**c**: *lower right*) Alakhnanda river basin up to Joshimath

centers and institutions). Among the major rivers of India, the Alakhnanda river and its tributaries (Dhauli Ganga, Vishnu Ganga, Nandakini, Pindar, Mandakini, and other numerous perennial streams) originate and flow from here. The highest mountain peaks of the Himalayan ranges such as Nandadevi, Kamet, Trisul, and Chaukhamba are located here.

The Alakhnanda river basin is formed by the proglacial Alakhnanda stream and its tributaries, which are mostly fed by snow, ice melt and monsoon precipitation. The sources of the Alakhnanda mainstream are the twin glaciers, Satopanth and Bhagirath Kharak (elevation 3,800 m), 13 km north of the temple town of Badrinath. The higher reaches of the Alakhnanda basin are characterized by active glaciation. The higher reaches of the Alakhnanda basin are characterized by active glaciation. Glaciers, cirque basins, glacial lakes, U-shaped valleys, moraines and avalanche

Fig. 10.5 (**a**) View of Bhagirathi river at Gangotri town, (**b**) Natural view of Tehri reservoir near downstream of Uttarkashi, (**c**) Natural view of Alakhnanda river near Joshimath and (**d**) Snout and confluence of Satopanth-Bhagirath karak glaciers in Alakhnanda river basin

slopes are the common landforms in the region. The river flows for a distance of 240 km from its source (3,800 m a.m.s.l.) to Devprayag (500 m a.m.s.l.). The river in its upper course flows through narrow and deep gorges and then enters its lower course at an elevation of 900 m, 25 km downstream of Joshimath [21].

The word "Bhagirathi" (Sanskrit, literally, "caused by Bhagiratha") refers to a mythological Sagar Dynasty prince who, to gain the release his 60,000 great-uncles from the curse of saint Kapila, brought the goddess Ganga in the form of the river Ganges, from the heavens to the earth (wikipedia 2010). The headwaters of the Bhagirathi are formed at Gaumukh (elevation 3,892 m), at the foot of the Gangotri glacier and Khatling glaciers in the Garhwal Himalaya. It is then joined by its tributaries; these are, in order from the source: Kedar Ganga at Gangotri (elevation 3,049 m), Jadh Ganga at Bhaironghati (elevation 2,650 m), Kakora Gadand Jalandhari Gad near Harsil (elevation 2,745 m), Siyan Gad near Jhala (elevation 2,575 m), Asi Ganganear Uttarkashi (elevation 1,158 m) and Bhilangna River near Old Tehri (elevation 755 m). The Bhilangna itself rises at the foot of the Khatling Glacier (elevation 3,717 m) approximately 50 km south of Gaumukh (http://en.wikipedia.org/wiki/Bhagirathi_River).

The river flows from its source for 205 km (127 mi) before meeting the Alakhnanda River at an elevation of 475 m (1,558 ft) in the town of Devprayag. Downstream of this confluence, the river is known as the Ganga or Ganges River. The Tehri dam lies at the confluence of the Bhāgirathi and the Bhilangna, at 30° 22′ 32″ N 78° 28′ 48″ E, near Tehri. Chaukhamba I is the highest point of the Bhagirathi basin.

10.4 Climate: Temperature and Rainfall

The altitudinal differences coupled with varied physiographic contributes to climatic variations in the Alakhnanda basin. The climate varies from sub-tropical to alpine. Despite diverse physiographic characteristics, sub-regional variations in the average seasonal temperature are not striking. Temperature varies from season to season and from valley regions to highly elevated regions as highest temperature is recorded in Srinagar in the month of June (30 °C) and lowest in Tungnath in the month of January (0.5 °C) (Table 10.1). The Alakhnanda basin receives heavy snowfall about 3–4 months during winter above 2,000 m altitudes [22]. Summers are conducive and favorable for health except a belt extending between Karanprayag to Devprayag comprising (low-lying areas) where average monthly temperature remains about 30 °C. The farming community, during this period migrates to upland for pastoralism. During summer, heavy flow of tourist can be seen in the basin because this basin has two world famous pilgrimages; Badrinath and Kedarnath, five prayags (confluence points of major rivers), and other places of cultural interest. Similarly, there are many natural places of tourist interest. Rainfall mostly occurs from May to September, especially during monsoon season from June to August. It also varies from the valley regions (low) to highlands (high) and north-facing (leeward) to south-facing (windward) slopes.

Rainfall across three stations of the basin (Table 10.2) located in different altitudes was recorded maximum in Karnprayag (144.9 cm) followed by Chamoli (124.7 cm), while lowest rainfall was recorded in Joshimath (122.9 cm). This data reveals that higher the altitude, lower the rainfall and vice-versa. Joshimath is located at leeward direction and recorded lower rainfall.

The basin is characterized by the presence of high moisture throughout the year. The Great Himalayan ranges regulate the climatic conditions; as it occupy 433 km^2 area of the basin. This fact (high percentage of moisture) was noticed in 1987, when

Table 10.1 Mean monthly temperature in the Alakhnanda basin [22]

Name	Altitude	Mean monthly temperature (°C)											
of place	(m)	Jan	Feb	Mar	Apr	May	Jun	Jul	Aug	Sep	Oct	Nov	Dec
Srinagar	550	14	18	20	25	25	30	29	28	25	27	17	15
Mastura	1,800	4	6	12	14	15	20	20	18	17	14	8	4
Joshimath	1,875	2	3	7	11	14	17	18	17	16	10	7	4
Tungath[a]	3,600	0.5	1	3	6	7	12	12	11	5	4	2	1

Sources of data: [a]HAPPRC Srinagar Garhwal (Uttarakhand) & IMD, Pune

Table 10.2 Monthly rainfall in the Alakhnanda basin for 2008 (cm)

	Altitude	Monthly rainfall												
Station	(m)	Jan	Feb	Mar	Apr	May	Jun	Jul	Aug	Sep	Oct	Nov	Dec	Total
Joshimath	1,875	9.5	0.0	1.5	0.0	8.7	32.4	35.6	24.1	6.9	2.8	1.4	0.0	122.9
Chamoli	926	5.6	0.0	0.4	0.0	11.8	35.6	27.9	32.5	5.5	3.9	1.5	0.0	124.7
Karanprayag	883	4.9	0.0	1.0	0.0	8.2	29.2	65.3	29.3	4.1	1.6	1.4	0.0	144.9

Sources of data: IMD, Delhi

the entire India observed drought whereas the impact of drought was negligible in the Alakhnanda Basin [22].

10.5 Data Used

10.5.1 Remote Sensing Data

Estimation of snowmelt requires information of spatial-temporal data on snow cover. This is derived from digital analyses of remote sensing images. The remote sensing images are that of MODIS and IRS-P6 sensors. In addition to these images, the digital elevation model (DEM) of ASTER GDEM (30 m) and SRTM (90 m) has also been used to derive the elevation zones, slope and aspect maps of study basins. The remote sensing data was also used to derive the Land Use Land Cover (LULC) of study area.

For monitoring snow cover from remote sensing images, NDSI (Normalized Difference Snow Index) method was used. NDSI uses the high and low reflectance of snow in visible (green) and shortwave infrared (SWIR), respectively, and it can also delineate and map the snow in mountain shadows. NDSI is defined as:

$$\mathrm{NDSI} = \frac{(\text{Reflectance in Green} - \text{Reflectance in SWIR})}{(\text{Reflectance in Green} - \text{Reflectance in SWIR})}$$

Bands 2 and 4 are mostly used for AWiFS; bands 2 and 5 are usually used for LANDSAT ETM+ [23]. After the NDSI is calculated, the threshold is found by visual inspecting the image and sampling NDSI values. Typical NDSI values for LANDSAT images lie between 0.45 and 0.7 and are different for almost any images due to illumination differences [24–26]. In our study NDSI threshold values for AWiFS 0.4–0.5 and for LANDSAT 0.5–0.8 were taken. Figures 10.6, 10.7, 10.8, and 10.9 show the remote sensing images and calculated SCA for each date of scene of Alakhnanda and Bhagirathi river basins. The Fig. 10.6 shows the Landsat color composite from 2002 to 2003 water year. The landsat data was taken from USGS website (http://landsat.usgs.gov). The snow cover will appear white in the standard False Color Composite (FCC) along with clouds as both have high reflectance in visible an NIR region (Fig. 10.3). The snow reflectance reduces to very low as compared to cloud in SWIR band (Fig. 10.3), that's why SWIR band is used in NDSI for mapping the SCA.

10.5.2 Hydro-Meteorological Data

The air temperature and other hydro-meteorological parameters are very important variable to estimate the importance of the melting of the snow cover when using a global conceptual model such as the snowmelt runoff model (SRM).

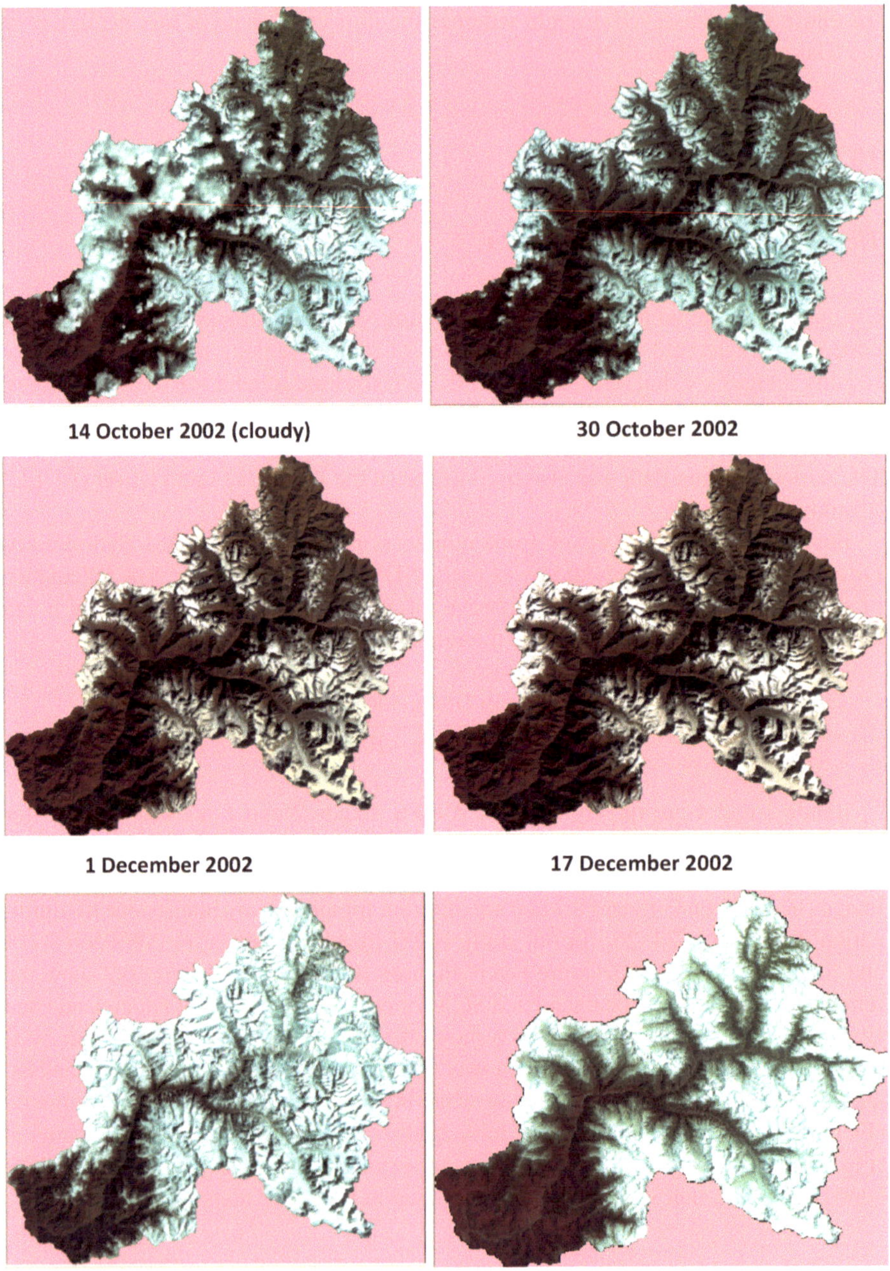

Fig. 10.6 Bhagirathi basin as seen in Landsat images for years 2002–2003

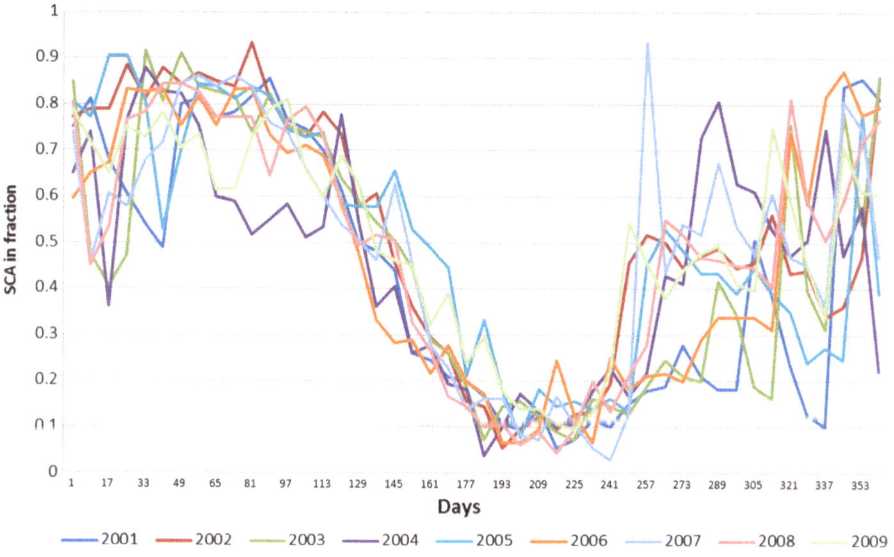

Fig. 10.7 SCA of Bhagirathi basin

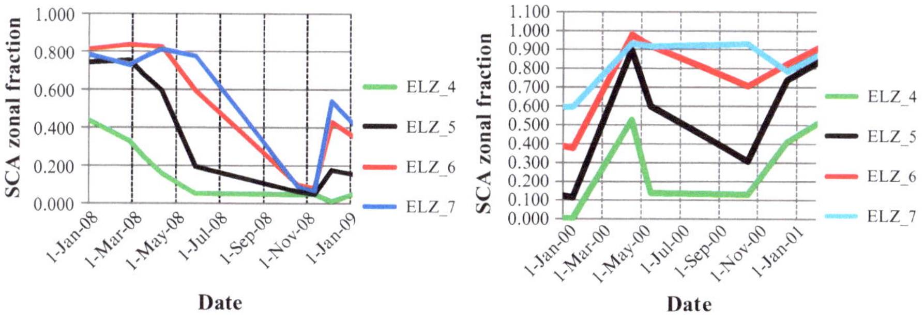

Fig. 10.8 (**a**: *left side*) SCA zonal fraction for year 2008 and (**b**: *right*) for year 2000 of Alakhnanda basin

Hydro-meteorological data requirement for study snow cover area include:

- Maximum, minimum and average temperature; Precipitation; Net radiation;
- Runoff (discharge).

Discharge and meteorological data was obtained from for period 2000–2009 from Central Water Commission (CWC Dehradun office) and India Meteorological Department (IMD) Delhi. Meteorological data was available from Uttarkashi Automatic Weather Station (AWS) and also collected from National Institute of Hydrology (NIH) Roorkee for Bhojbasa field station. The temperature data is one of the most important parameters of SRM model [27].

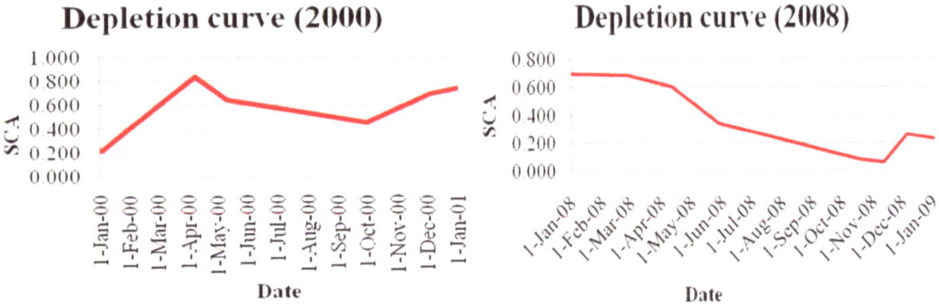

Fig. 10.9 Snow depletion curves of Alakhnanda basin for year 2000 and 2008

10.6 Results and Discussions

10.6.1 Snow Cover Area of Bhagirathi Basin

The Fig. 10.7 shows snow cover area in fraction (0–1) from year 2001 to 2009. After delineating boundary of study area, snow cover in fraction was calculated using daily data by dividing total SCA with basin area i.e., (Number of pixel *5,000*5,000/1,000,000)/Watershed area, and interpolating weekly data to daily snow cover data (Fig. 10.7).

We can see from the graph snow accumulation is maximum from December to march, due to low temperature and snowfall in higher altitude areas, minimum from July to September because of monsoon and high temperature and follow same trend from year 2001 to 2009. During period from October to December snow cover is optimum.

10.6.1.1 Snow Covered Area

The areal extent of seasonal snow cover gradually decreases during the snowmelt season. Depletion curves of the snow coverage for 2008 were interpolated from December 16, 2007 to April 28, 2009 as AWiFS imagery was available and for 2000 – from October 15, 1999 to March10, 2001. For SRM full year 2008 and 2000 were used.

The figures below show the snow cover in the Alakhnanda basin mapped from AWiFS (2008) and LANDSAT ETM+(2000) (Fig. 10.10).

The altitude up to 2,000 m is not under the snow. During 2008 the SCA decreases from January to November and in December due to heavy snowfalls again it increases. The most of snow is concentrated at high altitudes, higher than 5 km. During 2000 the SCA increased from the beginning of the year up to April and at high altitudes did not vary widely during the rest of the year. With the help of temperature and precipitation such depletion curves help to simulate snowmelt runoff.

The snow cover area depletion curves very significantly from year to year and therefore, the snow cover depletion curve has to be made separately for each year under consideration to be used as input to the model (Fig. 10.9). The distribution curve of snow accumulation and discharge achieved from AWiFS 2008 is not normal. It has highest SCA during January and than reduces gradually with time till November.

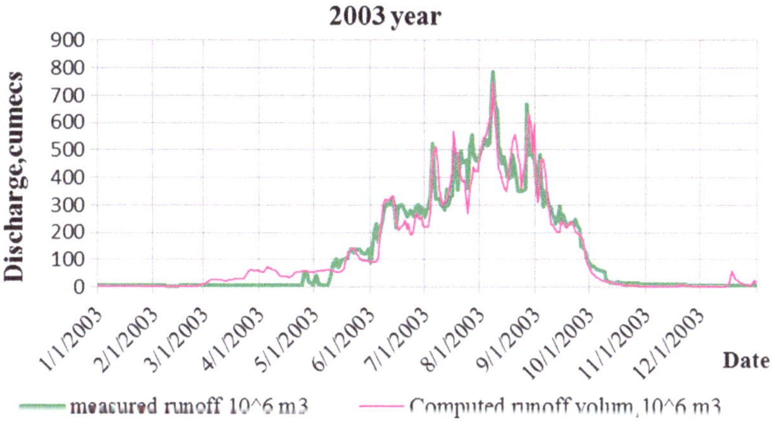

Fig. 10.10 Measured and computed discharge in Bhagirathi basin for year 2003

10.6.2 Result of Snowmelt Runoff

10.6.2.1 Results of Bhagirathi Basin

The snowmelt runoff results of Bhagirathi basin are given in Table 10.3. From the table we can estimate the coefficient of determination for snowmelt season for 2002–2007. The coefficient of determination for snowmelt season for year 2002–0.84, for 2003–0.82, for 2004–0.76 for 2005–0.84, for 2006–0.81 and for 2007–0.74. Measurement runoff for melt season is about 2751.60–4139.83 (10^6 m^3), average runoff 87.01–131.27 (m^3) and computed runoff 2737.50–4119.16 (10^6 m^3), average 86.57–130.62 (m^3/s). Volume difference is varying from 7.54 % for 2006 to 0.48 % for 2002 year.

We can see from Fig. 10.10 that snow melt runoff started from summer season. Maximum runoff observed in August month, and then in winter season from October to February snow melt is decreasing. The SRM parameters are to be calibrated or optimized by historical data. They can be either derived from measurement or estimated by hydrological judgment taking into account the basin characteristics, physical laws and theoretical relations or empirical regression relations. Occasional subsequent adjustments should never exceed range of physically acceptable values.

Runoff simulation has been carried out for the Bhagirathi catchment as illustrated. The simulation was done using SRM. The rainfall and temperature data input the model from station. Reliability and situation of this station was important in using its data for simulation. The simulation is acceptable and it confirms on snowmelt runoff as main water resource in the region. The catchment is free of snow in summer and it has caused the computed runoff to settle lower than measured runoff in this period, also in period of winter especially in February and March there isn't any snowmelt that is because of low temperature and degree-days.

In the monsoon season from mid of June to July rainfall is high because of this computed and measured runoff are high. After July it's coming suddenly down due to low rainfall hence more calibration is required for X and Y coefficient. peak values of

Table 10.3 Snowmelt runoff results of Bhagirathi basin from 2002 to 2007

Period	2002	2003	2004	2005	2006	2007
Measured runoff volume (10^6 m^3)	4,102.93	4,139.83	2,751.60	3,647.82	3,647.33	3,326.63
Average measured runoff (m^3/s)	130.10	131.27	87.01	115.67	115.66	105.49
Computed runoff volume (10^6 m^3)	4,100.97	4,119.16	2,737.50	3,566.63	3,372.16	3,302.83
Average computed runoff (m^3/s)	130.04	130.62	86.57	113.10	106.93	104.73
Volume difference Dv, %	0.05	0.50	0.51	2.23	7.54	0.72
Coefficient of determination, R^2	0.84	0.82	0.76	0.84	0.81	0.74

computed and measured rainfall going up because of runoff coefficient and going down is because of snow coefficient and shifting of peak because of X and Y coefficient.

10.6.2.2 Results of Alakhnanda Basin

For calibration and validation of SRM simulation, 2008 and 2000 were used, respectively. According to model variables (temperature, precipitation and SCA) the principal model parameters the degree-day factor (a) and the runoff coefficients (C_s and C_r) were found. The degree-day factor converts the number of degree-days above critical temperature into snowmelt depth. For entire basin degree-day factor was taken as 0.45. A critical temperature of 0.75 was assumed and kept constant during the year (Table 10.4).

According to the regression between measured and computed runoff for 2000 the correlation coefficient r is 0.84. The catchment in 2008 and 2000 is free of snow in summer but under heavy rainfall that is caused the computed runoff to settle a little lower than measured runoff in this period (Fig. 10.11).

In 2008, period of winter computed runoff slightly higher than measured what is caused by precipitation in January and March. The model simulates the runoff peaks with correct delay related to the causing event. Unfortunately, the simulated peaks in 2008 and 2000 are smaller than measured. This is most probably attributed to the change of runoff coefficient and recession coefficient in time, which could not be accounted for without proper field data.

Numerical accuracy criteria are never perfect. Overall snow melt simulation for both years are good, with, $R^2=0.90$ in 2008 and 0.84 in 2000. In spite of these differences, both simulations can be useful for snowmelt runoff calculation, snowmelt forecasting and changed climate scenario for this area.

10.7 Conclusions for Bhagirathi Basin

- For simulation of snowmelt runoff the Bhagirathi basin was divided on 7 elevation zones range from 1,200 to 7,500 m. More part of area belongs to 5th elevation zone (4,500–5,500 m) and consists 43.67 % (1966.52 km^2), minimum area

Table 10.4 Model parameters for 2000 and 2008

Model parameter	2000	2008
Temperature lapse rate, γ	0.65 °C/100 m	0.65 °C/100 m
Critical temperature, T_{crit}	0.75 °C	
Degree-day factor, a	0.45 cm $°C^{-1}$ day^{-1}	
Lag Time, L	24 h	
Rainfall contributing area, RCA	March to September = 1, else 0	
Coefficient x	1.039	1.039
Recession coefficient, y	0.06	0.01

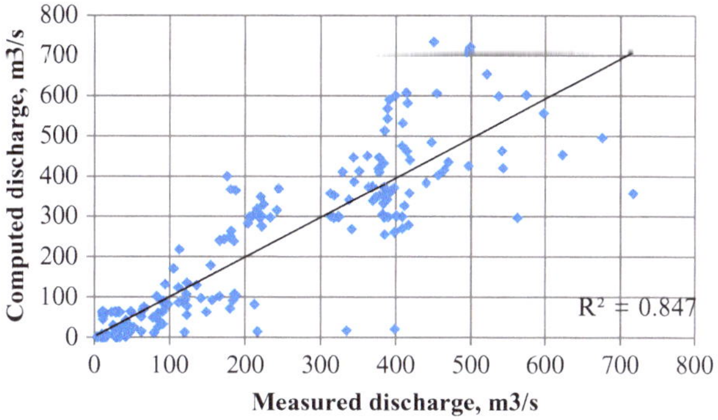

Fig. 10.11 Scatter plot and linear regression fit for measured and simulated discharges of the Alakhnanda river for year 2000

0.39 % (17.58 km^2) is in 7th under range 6,500–7,500 m. Total area of Bhagirathi Basin is 4503.35 km^2.

- For study period 2002–2007 maximum snow accumulation was observed in March 2002–4,200 km^2 and minimum in September 2007. During period from October to December snow cover was about 1,600 to 1,800 km^2.
- Snowmelt runoff for Bhagirathi watershed using SRM modeling was calculated successfully. Maximum discharge in the period of 2002–2007 years in monsoon season (July to August) and then in winter season from October to February snow melt is decreasing.
- The average coefficient of determination for snowmelt season for 2002–2007 is 0.80, measurement runoff is about 3602.69 (10^6 m^3), average runoff 114.20 (m^3) and computed runoff 3,533.21 (10^6 m^3), average 112.00 (m^3/s) and volume difference is 1.93.
- In all 12 months snow is present in higher altitudes. These zones indicate how much snow and rainfall are contributing to runoff. In zone 2nd the only rainfall is contributing on runoff because of low elevation. In zone 4th both snow and rain is contributing to runoff, initial snow gets melted up to April and after that all contribution is because of rainfall. In zone 6th initial snow is assisting more to runoff, followed by fresh snow and very less by rain fall, because of higher altitudes in higher altitudes precipitation is only in the form of snow. The trends for 2002 to 2007 years are similar.

10.8 Conclusions for Alakhnanda Basin

- For Snow Melt Runoff simulation AWiFS 2008 (11 images), LANDSAT ETM + 2000 (7 images) with a spatial resolution of 56 and 30 m respectively and hydro meteorological are most suitable for snow melt runoff studies;
- Snow cover and glacier mapping of satellite data of Alakhnanda river Basin with NDSI method was is most useful technique for SCA mapping and the threshold of 0.4–0.5 for AWiFS and 0.5–0.7 for LANDSAT ETM + imagery was found to be most suitable;
- The Elevation zone approach along with Aspect map of basin improves the quality of SRM simulations. The most sensitive parameters in SRM are Degree-Day Factor, Runoff Coefficients and Recession coefficients. The Snowmelt runoff simulation for Alakhnanda river Basin for 2008 was done successfully with correlation coefficient of (2008) = 0.90; and (2000) = 0.87.

10.9 Recommendations

- Use of daily hydro meteorological data in different elevation zones is preferable;
- Especially temperature, laps rate, precipitation and runoff coefficient are very important and must be selected very carefully.
- For water resources studies in the Bhagirathi and Alakhnanda basin, the snowmelt model of the catchment has to be validated in the next step for a longer period. This validation process may help the further operation use of this model.

Acknowledgements Author acknowledges the valuable support from Dr Y.V.N. Krishnamurthy, Director IIRS, Dr. P.S. Roy, Former Director IIRS, Dr. V.K. Dadhwal, Director NRSC and Dr. S.P. Aggarwal, Head, Water Resources Department, IIRS, for completion of this research. Author also thanks all the post graduate diploma students of IIRS and CSSTEAP for their contributions in this work. Thanks are also due to IMD, NIH and CWC for providing hydro-meteorological data for this project. Funding for this work was provided by National Action Plan on Climate Change (NAPCCC) project "Impact of Climate & LULC change on Hydrological Regime of Ganga River Basin" as part of National Water Mission.

References

1. Singh P, Singh VP (2001) Snow and glacier hydrology. Kluwer Academic, Dordrecht
2. Jain SK (2008) Impact of retreat of Gangotri glacier on the flow of Ganga River. Curr Sci 95(8):1012–1014
3. Engman ET, Gurney RJ (1991) Remote sensing in hydrology. Remote Sensing Applications Series, pp xiv + 225. Chapman & Hall, London
4. Baral DJ, Gupta RP (1997) Integration of satellite sensor data with DEM for the study of snow cover distribution and depletion pattern. Int J Remote Sens 18(18):3889–3894. doi:10.1080/014311697216711
5. Warren SG (1982) Optical properties of snow. Rev Geophys Space Phys 20(1):67–89
6. Roshani MJ, Zouj V, Rezaei Y, Nikfar M (2008) Snow mapping of Alamchal glacier using remote sensing data. In: The international Archives of the Photogrammmetry, Remote Sensing and Information Sciences, vol XXXVII, Part B8, Beijing

7. Andersen T (1982) Operational snow mapping by satellites. In: Proceedings to the Exeter symposium, July 1982, IAHS Publications, 138, pp 149–154
8. Kulkarni A, Rathore BP, Singh SK, Ajai A (2010) Distribution of seasonal snow cover in central and western Himalaya. Ann Glaciol 51(54):121–128
9. Georgievsky MV (2009) Application of the Snowmelt Runoff model in the Kuban River basin using MODIS satellite images. Environ Res Lett 4:1–5
10. Jain SK, Goswami A, Saraf AK (2010) Snowmelt runoff modelling in a Himalayan basin with the aid of satellite data. Int J Remote Sens 31(24):6603–6618
11. McGuire M, Wood AW, Hamlet AF, Lettenmaier DP (2006) Use of satellite data to streamflow and reservoir storage forecasts in the Snake River basin. J Water Resour Plann Manage 132:97–110
12. Mirab SM (1997) Study of snowmelt runoff and water balance in mountainous catchments. Enschede, ITC: 98
13. Rango A, Salomomson VV, Foster JL (1977) Seasonal stream flow estimation in the Himalayan region employing meteorological satellite snow cover observations. Water Resour Res 13:109–112
14. Swamy AN, Brivio AN (1996) Hydrological Modeling of Snowmelt in Italian Alps using visible and infrared remote sensing. Int J Remote Sens 17(16):3169–3188
15. Martinec J, Rango A (1986) Parameter values for snowmelt runoff modeling. J Hydrol 84:197–219
16. Martinec J, Rango A, Major E (1983) The snowmelt-runoff (SRM) user's manual, NASA Reference Publ. 1100. NASA, Washington, DC
17. Najafzadeh R (2004) Application of RS/GIS techniques and SRM model for estimation of Zayand-e-Rood river flow. M.S. thesis, Sharif University of technology, Iran
18. Radchenko U (2009) Snow-melt runoff modeling in Manali sub-basin. Post Graduate Diploma (PGD) report, CSSTEAP, WRD, IIRS Dehradun
19. Sarybekov ER (2004) Snowmelt runoff modeling of Beas basin upto Pandoh dam. Post Graduate Diploma (PGD) report, CSSTEAP, WRD, IIRS Dehradun
20. World Meteorological Organization (WMO, 1986) Intercomparison of models of snowmelt runoff, WMO, No 646; Operational hydrology report (OHR), 23, 440 p. ISBN: 978-92-631-0646-9
21. Singh AK, Hasnain SI (1998) Major ion chemistry and weathering control in a high altitude basin: Alaknanda River, Garhwal Himalaya, India. Hydrol Sci J Sci Hydrol 43(6):825–843
22. Sati VP (2008) Natural resource management and food security in the Alakhnanda basin of Garhwal Himalaya. ENVIS Bull 16(2):4–15
23. Chander G, Markham BL, Helder DL (2009) Summary of current radiometric calibration for Landsat MSS, TM, ETM+and EO-1 ALI sensors. Remote Sens Environ 113:893–903
24. Dozier J (1984) Snow reflectance from Landsat-4 Thematic Mapper. IEEE Trans Geosci Remote Sens 22(3):323–328
25. Dozier J (1989) Spectral signature of alpine snow covers from the Landsat thematic mapper. Remote Sens Environ 28:9–22
26. Dozier J, Marks D (1987) Snow mapping and classification from Landsat Thematic Mapper (TM) data. Ann Glaciol 9:97–103
27. Richard C, Gratton DJ, et al (2001) The importance of the air temperature variable for the snowmelt runoff modelling using the SRM. Hydrological Processes, 18 (Eastern Snow Conference-Canadian Geophysical Union-Hydrological Section), pp 3357–3370

List of Websites

http://landsat.usgs.gov. Accessed in January 2010
http://en.wikipedia.org/wiki/Bhagirathi_River. Accessed in January 2010
http://en.wikipedia.org/wiki/Snow. Accessed in December 2012
http://en.wikipedia.org/wiki/Snow_hydrology. Accessed in December 2012

River flowing from under the glacier (on the way to Gaumukh) (Courtesy Rashmi Sanghi)

Chapter 11
River Bank Erosion Hazard Study of River Ganga, Upstream of Farakka Barrage Using Remote Sensing and GIS

Praveen K. Thakur

Abstract This work has been carried out to analyze and report the river bank erosion hazard due to morphometric change of the Ganga River in the upstream of Farakka Barrage upto Rajmahal. Morphometric parameters, such as, Sinuosity, Braidedness Index and percentage of the island area to the total river reach area were measured for the year of 1955, 1977, 1990, 2001, 2003 and 2005 from LANDSAT and Indian Remote Sensing Satellite (IRS) images. The analysis shows that there is a drastic increase in all of those parameters over the period of time. This study has found that bank failure is because of certain factors like, soil stratification of the river bank, presence of hard rocky area (Rajmahal), high load of sediment and difficulty of dredging and construction of Farakka Barrage as an obstruction to the natural river flow. For the increasing sinuosity, the river has been engulfing the large areas of left bank every year. The victims are mostly Manikchak and Kaliachak-II blocks of Malda district, with a loss of around 1,670 ha agricultural land since 1977. Temporal shift measurements for the river reach between Farakka and Rajmahal has been done with help of 22 cross-sections in this reach. Erosion impact area has also been estimated to emphasize the devastating nature of the hazard.

Keywords Ganges River • River bank erosion • Remote sensing • River bank erosion • River morphometric change • Farakka

P.K. Thakur (✉)
Water Resources Department (WRD), Indian Institute of Remote Sensing (IIRS), Indian Space Research Organisation (ISRO), 4-Kalidas Road, 248001 Dehradun, India
e-mail: praveen@iirs.gov.in

R. Sanghi (ed.), *Our National River Ganga: Lifeline of Millions*, DOI 10.1007/978-3-319-00530-0_11, © Springer International Publishing Switzerland 2014

Winter frozen water and meltwater of Bhagirthi near Gangotri temple, 26 dec 2007 (Courtesy Praveen Thakur)

11.1 Introduction

The river channel migration through time and space is critical to many geo-morphological and river management problems [1–3]. Lateral migration refers to the positional change of a river channel as a response to variations in fluid flow and sediment discharges and is always associated with bank erosion of the stream bed or channel wall under turbulent flow conditions [4]. Lateral migration is therefore a process that can cause catastrophic local or regional changes [5]. Hickin [6] studied the development of meanders in natural river channels and also did the review of river channel changes [1]. In present work, we have studied the river channel changes of Ganga River upstream of Farakka barrage, in India using geospatial methods and recently have been published in detail [7].

River bank erosion and channel shifting is a geo-morphological phenomena which has been studied by various researchers in last few years [8–12]. The river bank erosion study of Mani et al. [10] was done on Majuli, the world's largest river island, which is situated in the middle of river Brahmaputra in Assam. River Brahmaputra flows in highly braided channels most of them are transient in nature, being submerged during high monsoon flows and changing drastically their geometry and location. In this study, they observed the trends of erosion in a small part of Majuli island, the area near Kaniajan village in south Majuli – a stretch of about 11 km, using satellite data of 1991, 1997 and 1998. Erosion and deposition maps of the area were prepared and the erosion of island is measured at various sections at 1 km interval.

Lawler [13] did a review on measurement of the river bank erosion and lateral channel changes. The methods covered by Lawler in this review were, sedimentological evidence, botanical evidence, historical sources, planimetric resurvey, repeated cross-profiling, erosion pins and terrestrial photogrammetry. Muller et al. [14] has provided summary of space based remote sensing data applications to river studies. In this review, all types of satellite remote-sensing data currently available, and their recent uses in studies of river systems has been highlighted.

Milton et al. [2] did the investigations of change in fluvial systems using remotely sensed data. Thoms and Walker [15] did studies on channel changes related to low-level weirs on the River Murry, South Australia. Winterbottom and Gilvear [16] used a GIS-based approach for mapping probabilities of river bank erosion in regulated River Tummel, Scotland. Mani et al. [10] did erosion study of a part of Majuli River-Island using remote sensing data. Thoma et al. [17] quantified river bank erosion with scanning laser altimetry. Kotoky et al. [18] has studied the nature of bank erosion along the Brahmaputra river channel, Assam, India. Matti et al. [19] used remote sensing to detect the riverbank changes along the Mekong River in the Vientiane – Nong Khai area. In the context of the present study, the above cited literature proves that remote sensing along with GIS plays an important role in river morphological and river bank erosion studies.

The Ganga River is one of the major river systems in India. Gangetic plain covers a large portion of India. The River Ganga emerges from the Gangotri glacier, about

4,500 m above mean sea level in the Uttarakhand Himalayas and flows down to the Bay-of-Bengal covering a distance of 2,525 km. Each year the river carries millions of tones of sediment load and deposits in the plain areas. This sedimentation is the source of major water resource related challenge in India each year. The sediment deposition creates many severe problems like the decrease of river depth due to riverbed siltation. But when the discharge increases, the river exerts pressure on its two walls and lateral erosion starts which becomes the root of many flood and river bank erosion hazards in India.

Farrakka Barrage has its special importance for the existence of Calcutta Port. The construction of Farakka barrage was started in 1962 and was completed in 1971. Four more years were needed to build up the feeder canal and the project was dedicated to the nation on the 21st May 1975 [20]. The 2.64 km long Farakka Barrage was designed to divert 40,000 cusecs (or 1,133 cumecs) of Ganga water towards Bhagirathi River with a view to flush the sediment load into the deeper part of estuary and resuscitate the navigational status of the Kolkata port [21]. It was built with the objective of inducing water into the Hugli river The obstruction seems to lead the river to make her own way, mainly in upstream of Farakka barrage. The main sufferers are the two districts of West Bengal, Malda and Murshidabad [20, 22]. This change in Ganga river course and resulting river bank failure is a long-term natural disaster in West Bengal. This has become a chronic problem from early 1960s and it has manifested itself to a formidable magnitude during last four decades as evident from many studies [20, 21, 23–25]. The effect of flood and associated land erosion of socio-economic aspects of this area was studied in details by Iqbal [26]. In the upstream, the major part of erosion victim area comes under Malda district, mainly four blocks viz Manikchak, Kaliachak-II, Kaliachak-III and Ratua-I. A huge area has been eroded by this river bank erosion hazard since 1960s. The rate of river bank erosion is very high and frequent, which causes a huge amount of river bank cutting and population migration from the villages near the river every year [26]. The change of river course was followed by the change in the river morphometry [27], its total width, sinuosity, braiding characteristics etc. It also results in the change in the border shape between Malda and Murshidabad districts, between West Bengal and Jharkhand state, and between India and Bangladesh. It results in a social and political conflict also.

Satellite images give us imprints of past river course. In remote past, the river is said to be passing through Gour, a ruined city in the Malda district of West Bengal, India, on the west bank of the Ganges River, 40 km downstream from Rajmahal (Fig. 11.1a, b). In Fig. 11.1b, the existence of numerous cut-off meanders, abandoned channels with higher moisture content and agricultural suitability proves the early passageway of river Ganga. It can be said that upto Gour, the whole area with the past river imprints is the meander belt of river Ganga. Though historical period, it has been shifted westward and now it tends to come to its earlier position. So, for future, the whole belt upto Gour is the risk zone for river bank erosion. Kalyan Rudra [28] has used historical data along with satellite images to study changes in this area. Rudra observed that since the construction of Farakka barrage in 1971, the Ganga has migrated eastward appreciably and has formed a mighty bend. Trapped sediment load above the Farakka barrage has largely induced the recent change. The continuous oscillation of the Ganga in Maldah district, and also along the Indo-Bangladesh border has posed many problems, including land reallocation, population displacement, and border disputes [28].

Taking into consideration all the earlier studies, this study has been taken up for the Ganga river bank erosion by cross-section and morphological approach. In the 1950s, the river course was near to straight, but afterwards river has increased its sinuosity. The right bank of river is hard rock Rajmahal Hills basalts [29], which comes in eastern Jharkhand state, previously southern Bihar. They consists mainly of quartz-normative tholeiites interbedded with thin bentonites or tuffs [30]. In the Rajmahal Hills the maximum exposed thickness of lava is ~230 m or approximately ten lava flows [31]. In the left bank of the Ganga, there is the Manikchak Ghat, which is a land area having a weak alluvial loose soil structure. Ganges has now started bending its path to its left bank. So, there is need to analyze the whole shifting trend over space-time domain and its severity in terms of impact to the areas

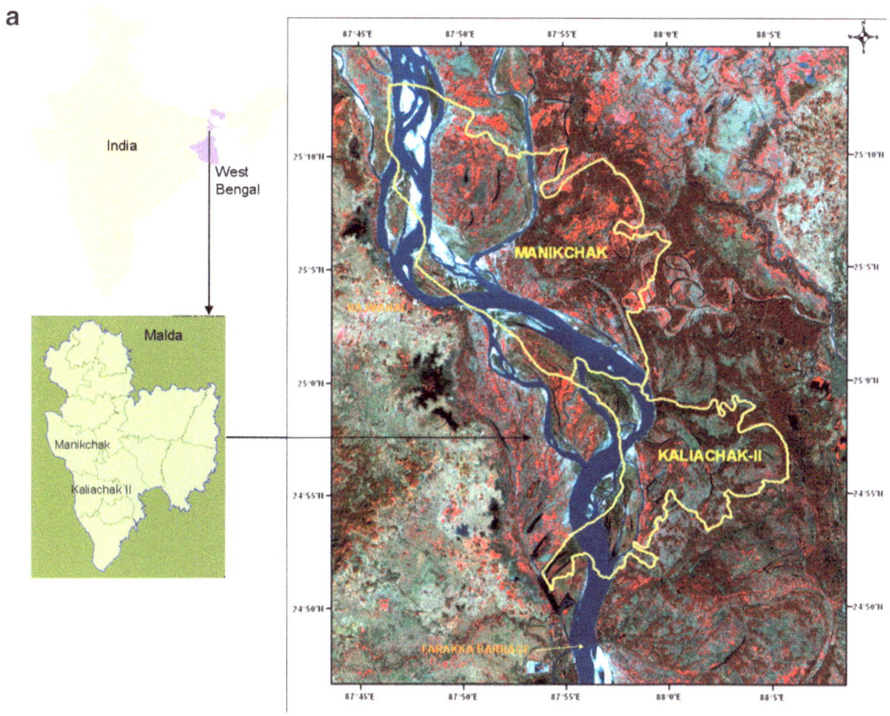

Fig. 11.1 (**a** and **b**) Study area, Ganga River upstreaa of Farakka barrage with block boundaries overlaid over LISS-III image of 2005; (**c**) Left bank of Ganga river showing bank erosion (Source: Thakur et al. [7]); (**d**) Left bank of Ganga river, downstream of Rajmahal (Source: Thakur et al. [7]); (**e**) Temporary huts along road side made by people affected by river bank erosion (Source: Thakur et al. [7])

Fig. 11.1 (continued)

where the concerned authorities can take the preventive steps. To fulfill these targets, this study has been carried out with the following objectives.

To quantify the amount of channel migration or change on spatial and temporal basis, quantification of these change using the morphometric parameters of the Ganga River and overall impact assessment due to river bank erosion.

11.2 Brief Description of Study Area

Malda, a district of West Bengal (Fig. 11.1a), is located almost in the central part of the state, near the Indo-Bangladesh Border. The district total area is about 3,733 km^2 (as per [32]). The district consists of 15 sub-division, commonly known as blocks. One side boundary of the district is drawn by the river Ganges itself. This study is based on the river Ganges left bank erosion on west side of river. So, the study area mainly consists of the two most affected blocks of this natural hazard, namely Manikchak and Kaliachak-II (Fig. 11.1a). The geographical extent of the study area is from 24°51′N to 25°14′N and 87°46′E to 88°6′E, with area of 316.39 and 209.17 km^2 respectively for Manikchak and Kaliachak-II and a total area of about 525.56 km^2 (as on 1977). According to a report, 750 km^2 land area was lost in 30 years in the Manikchak and Kalichak areas [20]. According to 2001 census, the total population of Manikchak was 214,123 where, 110,407 were male and 103,716 were female. In Kaliachak-II, the total population was 211,533 in which 108,955 were male and 102,578 were female. The density of population Manikchak was 677 whereas in Kaliachak-II it was 1,011 in 2001.

11.2.1 Erosion in Study Area

Almost every river course has its own playfield within which it shows its dynamism. The changing is due to some natural factors such as discharge, topography, soil character, geological structure, genesis of the adjoining areas etc. and man-made factors like deforestation, construction of barrages, dams etc. In the study area, bank-failure occurs mainly twice a year. One is pre-flood bank-failure, which is mainly due to high pressure of increasing water on bank walls. The second is post-flood failure, during floods the bank is submerged into water and stagnated water seeps as ground water. Because of a very weak and porous soil profile in the left bank, it weakens and dissects the bank deep-rooted. Thus, after the recession of flood water, left bank collapses in chunks [33] (see Fig. 11.1c, d).

11.2.2 Socio-economic Effect of Erosion Hazard

The river bank erosion hazard has also led to administrative problems. In India village level local administrative units are known as Gram Panchayets (G.P.).

In March 2003, the land area of one such Gram Panchayet named Kakribondha Jhaubona was totally lost by river bank erosion. So the affected persons from this village and there administrative responsibilities were merged with Bangitola Gram Panchayet administration. Earlier Bangitola Gram Panchayet had 11 members and after merging it has increased to 19. Every monsoon season, a huge number of people become landless and lose their livelihood. So, their demand to the government for the reallocation is much more than the available land resources with the government. Therefore, victims have to search land themselves and it creates neo-refugees with many social problems. People of non-affected areas do not want to welcome the victim people. Sometimes because of created poverty, crimes increase. The neo-refugees choose mainly Public Works Department (PWD) roadside for their huts (Fig. 11.1e), because for breaking their huts, local government have to go through a long procedure and these road side huts also results in high rate of road accidents. After reallocation a social groupings as well as conflict among the victims and between victim and non-affected people have started. In many cases, earlier there was mixture of religion, but after reallocation, they choose the same religion and religious grouping starts.

11.3 Methodology and Data Used

The flowchart of the methodology used for this study is given in Fig. 11.2. In this study ortho-rectified image of satellites such as Indian Remote Sensing (IRS) Linear Imaging Self Scanner LISS-III, LANDSAT Multi-spectral Scanner (MSS), Thematic Mapper (TM) and Enhanced TM along with topographical sheets are used for river change analysis (Table 11.1). Image classification was done for making Land Use Land Cover (LULC) maps (Fig. 11.3a) with eight classes. The unsupervised LULC classification has been carried out for 1977 and 1990 image and supervised LULC classification has been carried out for 2005 image with field verification (Fig. 11.2).

The overall accuracy of LULC classification for 1977, 1990 and 2005 are 89.04, 83.54 and 92.11 % respectively. The classified images of 1990 and 2005 were used in change detection model, which gives output in form of the attribute table having 11.2 classes each pixel. One LULC class of 1990 and another class of 2005. Thus, this can be identified which pixels are at same classes, which has become the river pixel and which have changed in other classes (Fig. 11.3b). The second category is of highest concern because that indicated the bank failure zone. Similar approach was used for 1997–1990 time period and impact area due to river bank erosion was calculated for 1977–1990 and 1990–2005 time period (Tables 11.3 and 11.4). The above LULC change maps along with village maps were used in GIS to find the affected villages (Fig. 11.4a, b and Table 11.4).

The Ganga river banks and islands area in study area were digitized and saved as arc coverage GIS files for years 1955, 1977, 1990, 2001, 2003 and 2005 at scale of 1:50,000 for all images and maps. The under-shoot, over-shoot and other

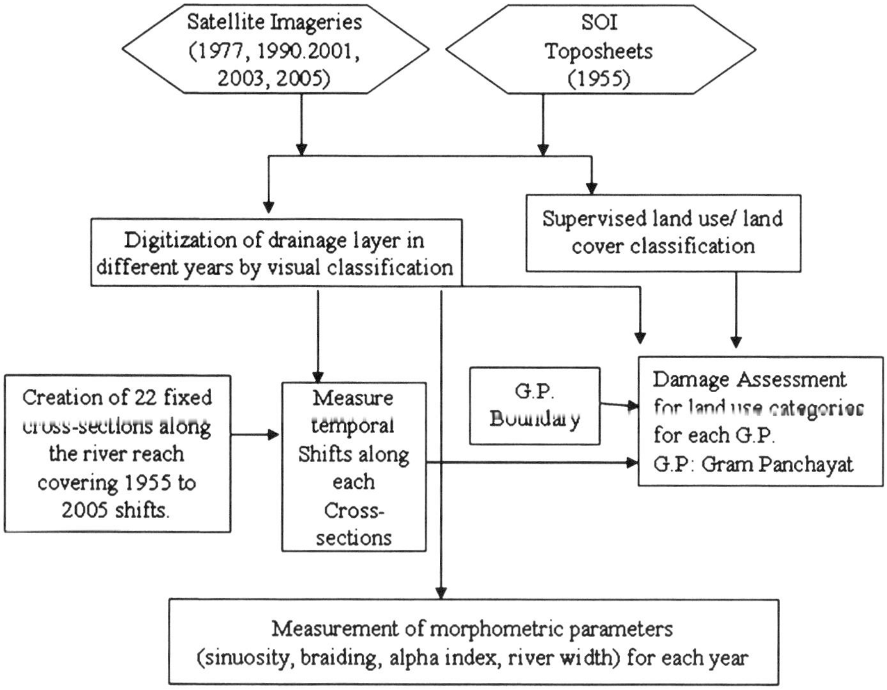

Fig. 11.2 Flowchart of the methodology

Table 11.1 Data used in the study

Data type	Satellite-sensor	Date (dd. mm.year)	Spatial resolution (m) or scale	Source
Remote sensing images	Landsat – MSS	10.02.1977	57	USGS
	Landsat – TM	21.11.1990	30	USGS
	Landsat – ETM+	20.10.2001	30	USGS
	IRS-1D LISSIII	Feb.2003	24	NRSC, India
	IRS- P6 LISSIII	15.11.2005	24	NRSC, India
Topographical maps	72 P/13	1972	1: 50,000	SOI
	NG 45–10, NG 45–11 and NG 45-15	1955	1: 2,50,000	Series U502, U.S. Army Map Service

USGS United States Geological Survey, *NRSC* National Remote Sensing Centre, *SOI* Survey of India

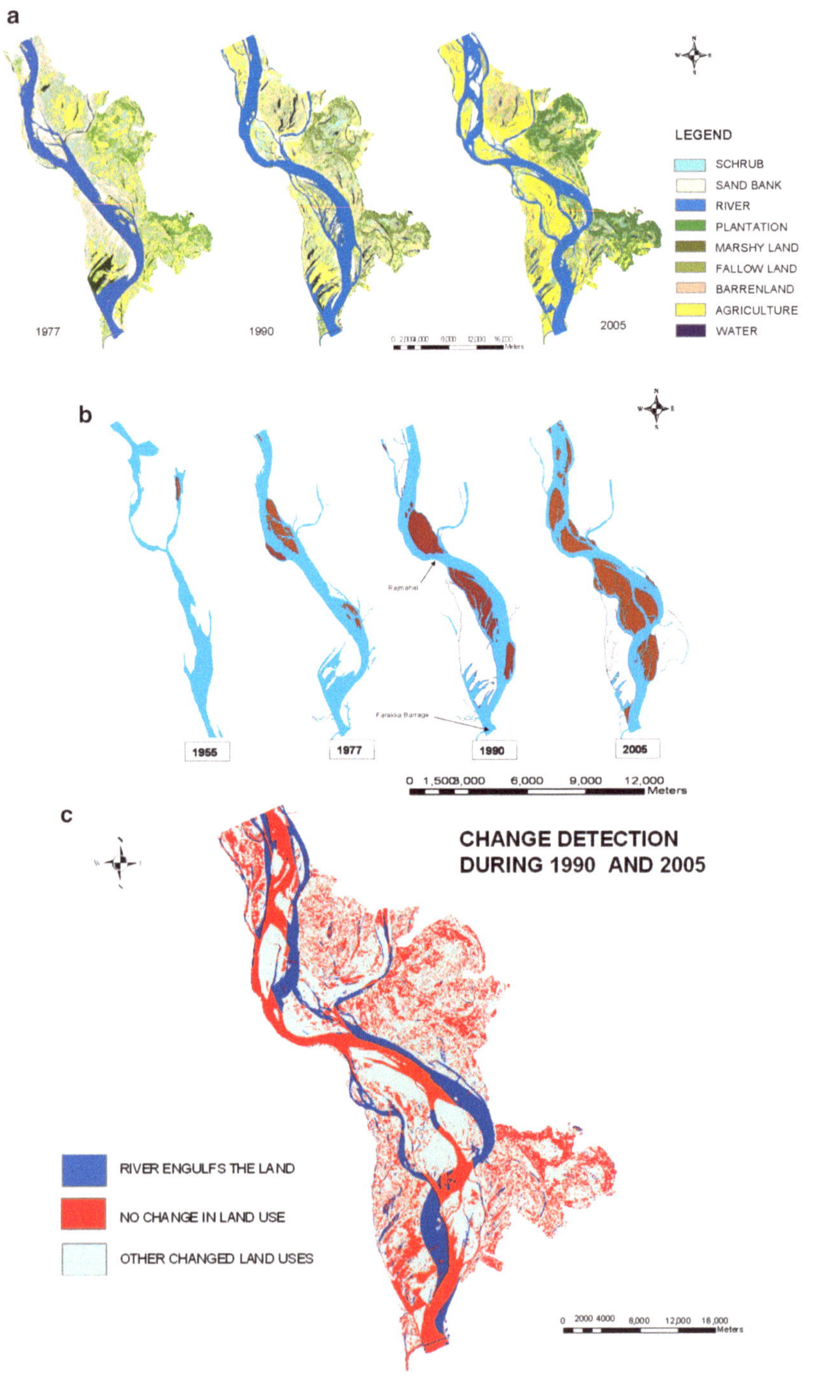

Fig. 11.3 (**a**) Classified land use land cover map of study area in 1977, 1990 and 2005; (**b**) Land Use Land Cover (LULC) change map from 1990 to 2005

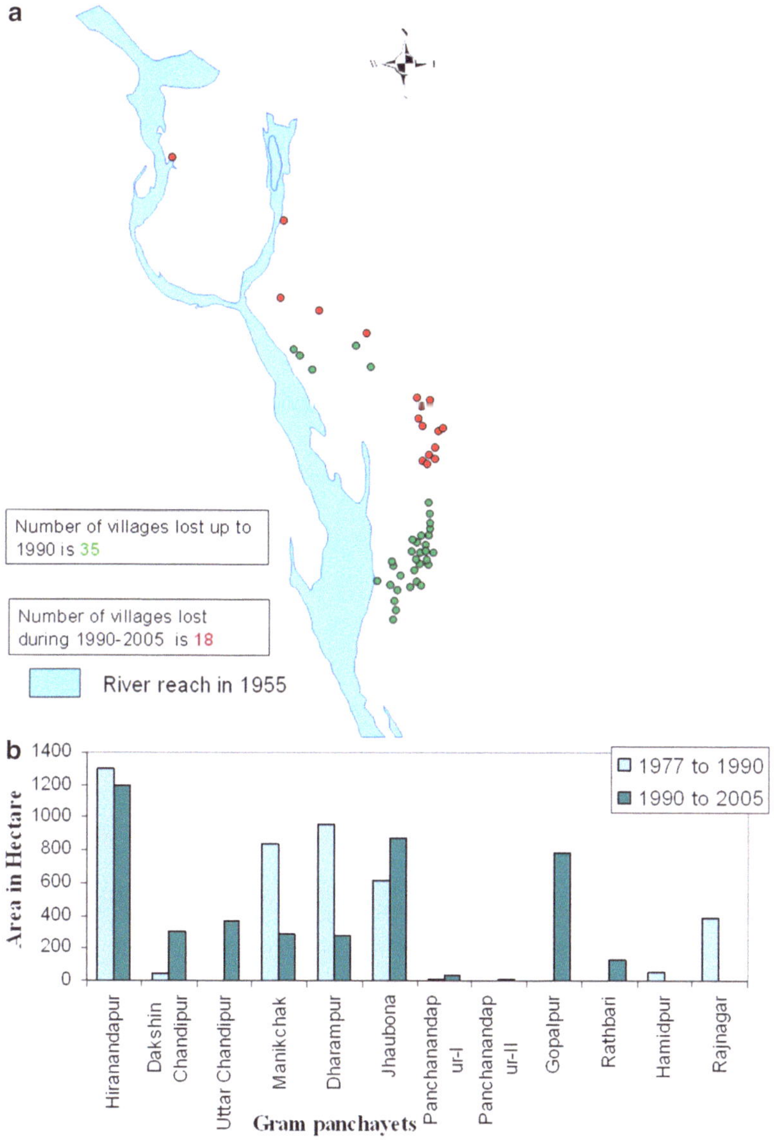

Fig. 11.4 (**a**) Map of impacted villages due to river bank erosion; (**b**) Impact area vs Gram Panchayat (GP)

digitization errors were removed using clean and build operations in vector editor of ERDAS. The 22 fixed cross sections were made with references to 1955 map and 2005 image, considering base and latest position of Ganga River. The cross-sections were created at unequal distances from north to south, giving more consideration to

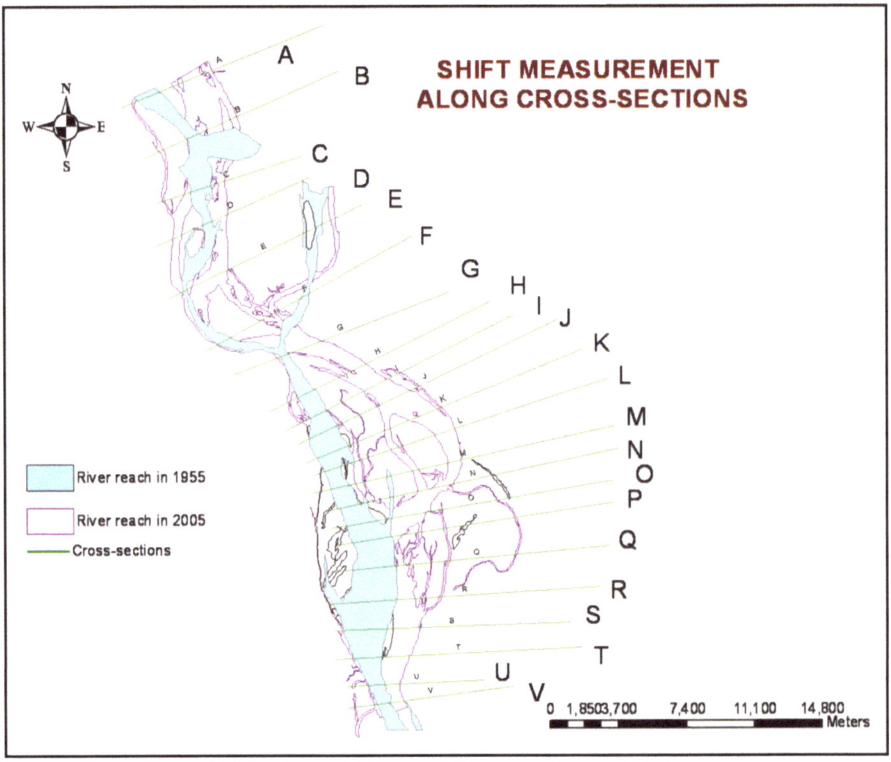

Fig. 11.5 Cross-sections across Ganga for measuring temporal river shift along each cross section

morphometric changes of river (Fig. 11.5). The location of these cross sections is decided on the basis of visual analysis of temporal remote sensing images and major morphometric changes in river, with more dense cross sections at locations where large changes in river bank. The cross-sections were extended more in south-east direction as river is shifting more in this direction. Along those cross-sections, the distance between the left bank of two consecutive years has been calculated and reported in this paper. The block boundaries of Kaliachak-II and Manikchak were taken from Government of West Bengal (GoWB) GIS database. The LULC map and cross sections were intersected with these two block boundaries to compute spatial-temporal changes in river bank and islands. The land and village area impacted or eroded by the river bank erosion was done by sub setting the classified land use/land cover thematic raster map and village map of the two blocks of Malda district, with the shifted area of the left bank during the time interval (Tables 11.4 and 11.5). The eroded area under different land use land cover classes has been measured in GIS. The major river morphometric indices such as Sinuosity and Braidedness index have been used in this study (Table 11.2).

Table 11.2 River morphometric parameters

Year	Braiding index	Sinuosity	Island area (km^2)
1955	0.12	1.09	2.09
1977	0.82	1.12	24.86
1990	1.07	1.10	71.13
2001	1.31	1.14	90.21
2003	2.08	1.21	119.11
2005	2.17	1.25	116.87

11.4 Results and Analysis

11.4.1 Detection of Change in Ganga River Morphometry with Time

11.4.1.1 1955 Ganga River Reach

The river course has been taken from the 1955 SOI toposheet. It was the river course before the establishment of Farakka barrage. The river course was near to straight. The sinuosity index for the entire river reach from Rajmahal to Farakka Barrage was 1.087. There was a very small island during this time in the river reach and the island area is 2.09 km^2, which is 1.72 % of the total river area. So, the braiding nature of the river was also very low. The measured Braiding Index is 0.123. It shows that the sediment load was low and the lateral erosion might be more.

11.4.1.2 1977 Ganga River Reach

The river has been digitized from the orthorectified Land SAT MSS (Table 11.1) data of 10th Feb, 1977. This is the time period during which bending of the river course has become visible, showing the effects on river morphology after the construction of Farakka barrage (Fig. 11.6). The river course got shifted more towards left bank on west side of river, at about 14.25 km far from the barrage. The sinuosity index of the total river reach was 1.12. From 1955 the sinuosity has increased by 0.031. The island area in this year has been also increased. It has increased to 24.86 km^2; which is 12.89 % of the total river area, an increase of 11.17 % island area. The Braiding Index has also increased to 0.821 with the increasing islands.

11.4.1.3 1990 Ganga River Reach

The river course has been digitized from the orthorectified Land SAT TM imagery of 21st November, 1990. In this year, the bend became broader and it was at more distance (about 21.5 km) from the barrage than it was in 1977. The sinuosity became

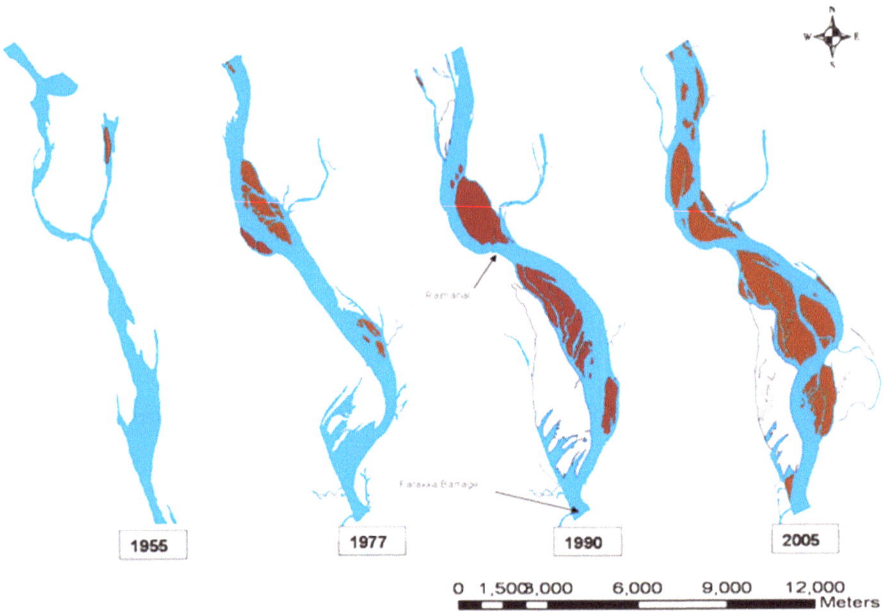

Fig. 11.6 Spatial temporal river shift of Ganga River upstream of Farakka barrage

slightly lower. It was 1.101. The total area of the islands increased to 71.13 km^2, which is 29.4 % of the total river reach area. So, the island area now has increased by 18.23 %. The Braiding Index has also increased to 1.069.

11.4.1.4 2001 Ganga River Reach

Here the river reach has been digitized from the orthorectified Land SAT ETM+ imagery of 20th October, 2001. The sinuosity increased this time to 1.14. That means the river meander has increased. The linear distance between the barrage and the cross section where the highest bend occurred, is about 21.72 km. Island area has also increased more because of the high sedimentation. It had increased to 90.21 m^2, which is 33.78 % of the total river reach area. So, the island area increased by 4.38 % from 1990 to 2001. The braiding index has also increased to 1.315. In year 2001 and upto 2005, a typical character of river morphometry was found in form of creation of two prominent channels, which were created as new branches of the main river. They had diverted mainly from the Manikchak area, near Rajmahal hills in the right bank area. This channel diverted from the opposite of Manikchak Ghat area near Rajmahal hills and joined again with the main channel near a place about 14.70 km upstream of the Farakka Barrage. The another channel called the Central Channel, diverts itself about 7.80 km downstream of the right channel diversion and merges with the main channel about 200 m upstream of the right channel joining point. Among these channels also, there were many interconnected narrow channels (Fig. 11.1b).

Table 11.3 Temporal river changes along each cross section

	Shift in m				
Cross-sections	1955–1977	1977–1990	1990–2001	2001–2003	2003–2005
A	3,053	1,190	310	−85	120
B	680	1,161	1,107	0	164
C	−1,060	1,470	877	0	116
D	−313	1,761	707	283	292
E	4,407	242	3,017	−67	−141
F	4,583	−566	762	−920	490
G	1,480	405	77	206	717
H	1,350	2,588	−166	0	642
I	1,866	2,891	1,988	0	116
J	1,443	3,555	2,216	0	532
K	3,930	2,347	2,285	0	56
L	4,738	1,178	2,178	108	141
M	6,559	94	1,806	235	342
N	5,634	−157	94	46	1,381
O	4,261	−548	−361	0	−180
P	4,671	77	−394	0	0
Q	4,345	62	−62	0	−23
R	2,581	1,069	−54	0	0
S	800	1,885	47	0	0
T	−207	1,917	363	58	0
U	122	402	689	29	0
V	1,165	−98	−49	0	0

11.4.1.5 2005 Ganga River Reach

The river has been identified from the IRS P-6 LISS-III satellite imagery of 15th November, 2005 (Fig. 11.1b). The sinuosity has been calculated as 1.25. The Braiding index has also increased to 2.173. Here the island area has been measured as 116.87 km^2. Though it is less than the island area of 2003, but we can see from the previous analysis that the island area has an increasing trend with time. The decrease in area can be due the reason that, water level in river was higher in November 2005 as compared to that of February 2003,. In this year the island covers about 39.6 % of the total river reach. The Alpha index of the total river reach is 0.14. That means, there is a possibility of development of 86 % islands in future (Refer to Table 11.3 and Fig. 11.7a, b).

So, from the graphical representations (Fig. 11.8a, b) it is clear that there is very strong positive relationship between sinuosity and braidedness. Regression analysis of sinuosity and braidedness with 2nd degree polynomial fit gives R^2 of 0.95. With sinuosity, braidedness of the river reach also increased drastically. Island area has a very positive relation with the braiding tendency. The regression relation R^2 by power fit between these two is 0.97. For the high sediment load and high peak

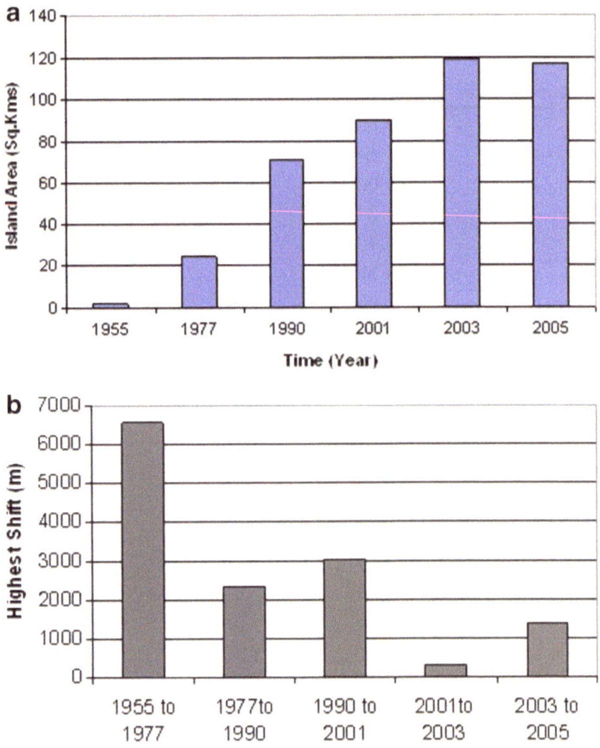

Fig. 11.7 (**a**) River island area (km^2) variation with time (years). (**b**) Highest river erosional shift (m) along cross-sections with time (years)

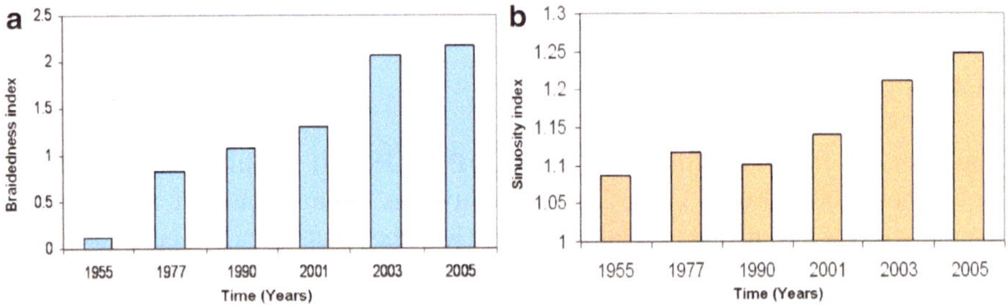

Fig. 11.8 (**a**) Braidedness index variation with time (years); (**b**) Sinuosity index variation with time (years)

discharge over time, and because of an obstruction, sediment deposition is high in this area. This has increased the islands area, with many new narrow channels within and around those, which also increases the braidedness. As the sedimentation increases, the river bed height increases and river goes for the lateral erosion more than the vertical. It increases the river width. As there is Rajmahal hard rock on the right bank, it starts eroding its left bank and that gives birth to the tendency to increasing sinuosity of the river reach.

11.4.2 Temporal Shifts of the Left Bank of the River

The shift in each of the cross-section in 1955–1977 time span, shows that the maximum shift (erosion) was at cross-section M, which was about 6.56 km with respect to left bank site in 1955 (Figs. 11.4 and 11.6). But, the higher range of values of erosion can be seen from K to Q, which mostly comes under Kaliachak-II block. The aerial distance of M in the upper part of Kaliachak-II block from Farakka Barrage is about 20.80 km. And it comes under the Kakribondha Jhaubona G.P. Other most affected gram panchayets are Panchanandapur-I and Hamidpur. Whereas, during 1977–1990, the shift is widespread in entire left bank of Ganga between Rajmahal and Barrage. The main shift area comes under Manikchak block, mainly Hiranandapur and Kakribondha Jhaubona and Rajnagar in Kaliachak-II. The highest shift (K) is about 22.35 km in aerial distance from the Farakka Barrage. During 1990–2001, highly eroded area comes under the cross-sections E, J, K and L. The area under these cross sections comes under some parts of Gopalpur, Hiranandapur and Dakshin Chandipur Gram Panchayets of Manikchak block. The main erosion shift area is at about 36.30 km distance from the barrage. In 2001–2003 time interval, we can see that shift amounts are much less than the previous time intervals. This may be due to the fact that the time interval between two measurements is very less. The highest shift due to erosion was along segment D, which is 283 m. Other cross-sections along which the major erosion took place were G, L and M. The affected gram panchayets were Hiranandapur, Manikchak, Gopalpur of Manikchak block and Kakribondha Jhaubona of Kaliachak-II.

During 2003–2005, a large amount of erosion and a widespread erosional tendency has been found. The highest erosion has taken place along cross-section N, having the value 1,381 m. This shows vast erosion within just 2 year span. The highest erosion took place in Kakribondha Jhaubona and Panchanandapur-I of Kaliachak-II block. By 2005 whole of Kakribondha Jhaubona gram panchayets was eroded. Other erosion affected cross-sections are F, G, H, J and M. So, the other badly erosion affected areas comes under Dakshin Chandipur, Manikchak and Dharampur gram panchayets of Manikchak block. Aerial distance between the highest affected cross-sections from the Barrage is about 20.00 km (Table 11.3 and Fig. 11.5).

11.4.3 Change in River Width

Due to the high river discharge, velocity during heavy flows and floods and river morphometry, the lateral erosion, mainly along left bank of Ganges River has been increasing over time. To account for this shift of river bank, the river width measurement along fixed cross-sections is important to detect the intensity of lateral erosion. The channel width is changing in higher magnitudes over time. One cause may be that due to the sedimentation and high discharge, the lateral erosion has increased. And that higher channel width matches well with the higher island area and higher braidedness, showing the correlation between two parameters.

Table 11.4 Impact area in each Gram Panchayat (G.P: local village administrative block)

Impact area in hectare (1977–1990)		Impact area in hectare (1990–2005)	
Hiranandapur	1,299	Hiranandapur	1,197
Dakshin Chandipur	49	Dakshin Chandipur	307
Dharampur	962	Dharampur	276
Jhaubona	616	Jhaubona	874
Panchanandapur-I	6	Panchanandapur-I	36
Panchanandapur-II	1	Panchanandapur-II	6
Hamidpur	58	Uttar Chandipur[a]	369
Rajnagar	389	Manikchak[a]	286
		Gopalpur[a]	783
		Rathbari[a]	133

[a]New villages has been affected by river bank erosion during 1990–2005

Table 11.5 Impact area for each land use classes due to river bank erosion

Land use category	Impact area (1977–1990) in hectare	Impact area (1990–2005) in hectare
Agriculture	706	970
Marshy Land	61	226
Barren Land	1,345	994
Sand Bank	623	891
River	35	57
Fallow Land	1,268	786
Plantation	41	360
Scrub	286	290

11.4.4 Impact Area Assessment

Impact assessment due left bank erosion has been done from satellite data and village maps as explained in methodology section. Here, in this study, two temporal impact studies have been done, one between 1977 and 1990, and the other between 1990 and 2005. Also the LULC and villages affected by erosion have been identified (Tables 11.4 and 11.5) and shown graphically in Figs. 11.3b and 11.4a, b.

11.5 Discussions and Conclusions

Farakka Barrage was established to maintain the water flow at Calcutta Port. The presence of this barrage followed by other causes like high discharge, weak soil structure, sedimentation, presence of the Rajmahal hilly area in the right bank, the Ganga river left bank erosion has become one of the most severe Geo-hazard in West Bengal. The source of this hazard is the morphometric change of the Ganga River in the study area, i.e., area upstream of Farakka Barrage to upto Rajmahal hills. The previous work in this area [20, 21, 27, 28, 34, 35] had highlighted problem of river bank erosion and its socio-economic effects on community living in this area. The study by Rudra [21] used remote sensing images to map eroded area and changes in river bank, but no river morphological analyses was done and most critically, the study with fixed river cross-sections vis a vis river bank erosion was not studied. But in this work focus has been given on finding major river morphological parameters and using cross-sections approach in GIS to map spatio-temporal changes in Ganga river. The river bank erosion is also very prominent in Ganga-Padma rivers downstream of Farakka Barrage in the Murshidabad district of West Bengal [20, 21, 28]. Further downstream, in Bangladesh, the diversion of Ganga water appears to have reduced the dry season discharge of Ganga and Gorai, the latter being one of the distributaries of the former that supplies water to the southwest region of the country. This reduction is reported to have increased sedimentation and salinity in the south-western part of the country [35]. The present work is focused mainly on river bank erosion in upstream portion of barrage and this approach can be applied to analyze the downstream portion to get the holistic view of the situation.

Most of the images analyzed in this study refer to dry season (October to February) only, during which water level or discharge in Ganga River and also in Farakka pond remains relatively stable. Therefore the errors in comparisons of the island area and river banks during different dates of satellite images are reduced. As the spatial resolution of images is limited upto 24–30 m only, which can cause mixed land-water pixels at land water boundaries, there can be small error in calculation of island area, LULC maps and river bank locations. Within the limitations of these small errors, morphometric change analysis has been done and it was found that the sinuosity has changed drastically over the time period of 50 years (1955–2005). In 1955, it was 1.087 whereas, it has been increased to 1.246 in 2005. Mainly because of the above causes, river shifts towards left bank and this shift erodes a number of villages and land area each year with it. Another increasing morphometric parameter is the Braidedness Index which is highly correlated with the island area. The river with its high discharge, takes lots of sediment load with it from the upper course and when it is suddenly obstructed by the Farakka Barrage, it deposits its sediments and makes islands. This siltation also causes decrease in total cross sectional area of Ganga River and also increase in maximum pond level (21.90–25.40 m) at Farakka [27, 34]. This is significant since the pond level at Farakka was elevated from 15.24 (before barrage) to 21.95 m (after barrage) by impounding 87 million m^3 of water, the hydraulic gradient of the Ganga above the barrage was flattened, leading to increasing tendency of flood and erosion due to change in flow regime [21]. The island area was only 1.72 % of the total river reach area in 1955, with the braidedness index of 0.123. But, in 2005 the braidedness index has been increased to 2.17, when the island area has been measured as 39.6 % of the total river area.

For the completion of this study, 22 fixed cross-sections were made across River Ganga, between Rajmahal hills and Farakka barrage. Because of the lateral erosion, the channel width has increased manifold. Along each of the 22 fixed cross-sections, the temporal shifts of the left bank were measured for each year interval. The "per year highest shift" was found during 2003–2005. Based on this analyses of the trend of river shift for last few decades, we can make an idea about the zone of high risk and that includes the gram-panchayets like Hiranandapur, Dakshin chandipur, Uttar chandipur, Manikchak, Dharampur, Bangalitola, Pawnchanandapur I and II, Hamidpur, Rajnagar, Gopalpur, etc.

11.6 Recommendations

Localized bank protection cannot provide a long-term solution to the problem of execution and subsequent maintenance of protection works as it become very difficult for the erosion and bed stratification of the river. Decreasing water pressure from the left bank by diverting flow from eroding channel, as suggested by a group of experts, may be the only permanent solution under the existing condition. Till the desired diversion of flow is achieved through dredging, localized bank protection works might be necessary as an interim measure. Site-specific conditions limit the possibility of capital dredging by deploying large capacity dredgers. Hydrographic survey is essential for the entire operation- for positioning the dredgers, for evaluating the performance of work of dredging and for assessment of yearly changes. This study has analyzed the temporal changes of the river and that gives us a future synoptic view of the river course. The results and database created during this study, especially the cross-sections, combined with surveyed hydrographic data of river and digital elevation model of flood plain can be used for hydraulic and sedimentation modeling of this river stretch; future flood and river bank erosion risk map can be generated based on this modeling. Thus, this study can become very much helpful for future mitigation and hazard preparedness programs to be taken by the government authority.

Acknowledgements Author acknowledges the valuable support from Dr Y.V.N. Krishnamurthy, Director IIRS, Dr. P.S. Roy, Former Director IIRS, Dr. V.K. Dadhwal, Director NRSC and Dr. S.P. Aggarwal, Head, Water Resources Department, IIRS, for completion of this research. Author also thanks Mrs Chalantika, post graduate diploma student of IIRS for her immense contributions in this work. Thanks are also due to Department of Irrigation and Waterway, Government of West Bengal for support during field work and data collection.

References

1. Hickin EJ (1983) River channel changes: retrospect and prospect. Special Publications of International Association of Sedimentologists, 6, pp 61–83
2. Milton EJ, Gilvear DJ, Hooper ID (1995) Investigating change in fluvial systems using remotely sensed data. In: Gurnell A, Petts G (eds) Changing river channels. Wiley, New York, pp 276–301
3. Petts GE (1995) Changing river channels: the geographical tradition. In: Gurnell A, Petts G (eds) Changing river channels. Wiley, New York, pp 1–23

4. Yang X, Damen CJM, Zuidam AR (1999) Satellite remote sensing and GIS for the analysis of channel migration changes in the active Yellow River Delta, China. Int J Appl Earth Obs Geoinf 1(2):146–157
5. Hickin EJ, Nanson GC (1984) Lateral migration rates of river bends. J Hydraul Eng Am Soc Civ Eng 110(11):1557–1567
6. Hickin EJ (1974) The development of meanders in natural river channels. Am J Sci 274:414–442
7. Thakur PK, Laha C, Aggarwal SP (2012) River bank erosion hazard study of river Ganga, upstream of Farakka barrage using remote sensing and GIS. Nat Hazards 61:967–987. doi:10.1007/s11069-011-9944-z
8. Das JD, Dutta T, Saraf AK (2007) Remote sensing and GIS application in change detection of the Barak river channel, N.E. India. J Indian Soc Remote Sens 35(4):301–312
9. Das JD, Saraf AK (2007) Technical note: remote sensing in the mapping of the Brahmaputra/ Jamuna River channel patterns and its relation to various landforms and tectonic environment. Int J Remote Sens 28(16):3619–3631
10. Mani P, Kumar R, Chatterjee C (2003) Erosion study of a part of Majuli River-Island using remote sensing data. J Indian Soc Remote Sens 31(1):12–18
11. Phillip G, Gupta RP, Bhatatcharya AB (1989) Channel migration studies in the middle Ganga basin, India using remote sensing. Int J Remote Sens 10(6):1141–1149
12. Tangri AK (2000) Application of remote sensing techniques in monitoring the spatial and temporal evolution of fluvio-geomorphic features in Ganga basin with specific reference to their impact on engineering structures. In: Sinha R (ed) Proceedings of the workshop on Fluvial Geomorphology with special reference to floodplains. IIT Kanpur, Kanpur
13. Lawler DM (1993) The measurement of river bank erosion and lateral channel change: a review. Earth Surf Process Landf 18:777–821
14. Muller E, Décamps H, Dobson MK (1993) Contribution of space remote sensing to river studies. Freshw Biol 29:301–312
15. Thoms MC, Walker KF (1992) Channel changes related to low-level weirs on the River Murry, South Australia. In: Carling PA, Petts GE (eds) Lowland floodplain rivers, geomorphological perspectives. Wiley, New York, pp 235–250
16. Winterbottom JS, Gilvear JD (2000) A GIS-based approach to mapping probabilities of river bank erosion: regulated River Tummel, Scotland. Regul Rivers Res Manage 16(2):127–140
17. Thoma PD, Gupta CS, Bauer EM (2001) Quantifying river bank erosion with scanning laser altimetry. Int Arch Photogramm Remote Sens Annap MD 34-3/W4:169–174
18. Kotoky P, Bezbaruah D, Baruah J, Sarma JN (2005) Nature of bank erosion along the Brahmaputra river channel, Assam, India. Curr Sci 88(4):634–640
19. Matti K, Lu XX, Rasphone A, Juha S, Jorma K (2008) Riverbank changes along the Mekong River: remote sensing detection in the Vientiane–Nong Khai area. Quat Int 186(1):100–112
20. Banerjee M (1999) A report on the impact of Farakka Barrage on the human fabric, on behalf of South Asian Network on Dams, Rivers and People (SANDRP). http://www.sandrp.in/dams/ impact_frka_wcd.pdf. Accessed on 12 Feb 2008
21. Rudra K (2004) The encroaching Ganga and social conflicts: the case of West Bengal, India. Independent Broadcasting Associates, Littleton, MA, p 40. http://www.ibaradio.org/India/ ganga/extra/resource/Rudra.pdf. Accessed on 11 Oct 2010
22. Banerjee SN, Chakroborty P (1983) Some observations on recent trends of shifting of the Ganga between Rajmahal and Ahiron. J Geol Soc India 24:318–321
23. Mukhopadhyay S (2003) River bank erosion and land degradation, a study of river Ganga, land degradation and desertification. Rawat Publication, New Delhi, pp 366–373
24. Rudra K (1996) Problems of river bank erosion along the Ganga in Murshidabad district of West Bengal. Indian J Geogr Environ 1:25–32
25. Rudra K (1996) The farakka barrage- an interruption to fluvial regime. Indian J Landsc Syst Ecol Stud 19(2):105–110

26. Showkat I (2010) Flood and erosion induced population displacements: a socio-economic case study in the gangetic riverine tract at Malda District, West Bengal, India. J Hum Ecol 30(3):201–211, © Kamla-Raj 2010
27. Parua PK (2002) Fluvial geomorphology of the river Ganga around Farakka. J Inst Eng 82:193–196
28. Rudra K (2010) Dynamics of the Ganga in West Bengal, India (1764–2007): implications for science–policy interaction. Quat Int 22:161–169. doi:10.1016/j.quaint.2009.10.043
29. Kent WR, Malcom SP, Muller DR, Saunders DA, Ghose CN (2002) $^{40}Ar/^{39}Ar$ Geochronology of the Rajmahal Basalts, India, and their relationship to the Kerguelen Plateau. J Petrol 43(7):1141–1153
30. Sengupta S (1998) Upper Gondwana stratigraphy and paleobotany of the Rajmahal Hills, Bihar India, Geological survey of India monograph (Paleaontologica Indica) 98. Geological Survey of India, Calcutta, p 180
31. Kent WR, Saunders AD, Kempton PD, Ghose CN (1997) Rajmahal basalts, eastern India: mantle sources and melt distribution at a volcanic rifted margin. In: Mahoney JJ, Coffin MF (eds) Large igneous provinces: continental, oceanic and planetary flood volcanism, Geophysical monograph, 100. American Geophysical Union, Washington, DC, pp 145–182
32. Census of India (2001) Provisional population totals, West Bengal, Table – 11.4". Maldah District (06). Government of West Bengal. http://web.cmc.net.in/wbcensus/DataTables/02/Table4_6.htm. Retrieved 21 July 2011
33. Keshkar G et al (1996) Report of experts committee for bank erosion problem of river Ganga-Padma in the districts of Malda and Murshidabad, West Bengal, Planning commission Govt. of India, pp 1–71
34. Parua PK (1999) Erosion problems of the river Ganga in the districts of Malda and Murshidabad in West Bengal. Civil Engineering Today XIII(2):3–20, ASCE, Calcutta
35. Sen HS (2010) The drying up of River Ganga: an issue of common concern to both India and Bangladesh. Curr Sci 99(6):725–727

List of Websites

http://glcf.umiacs.umd.edu/index.shtml
http://wrmin.nic.in/index3.asp?subsublinkid=714&langid=1&sslid=296
http://wrmin.nic.in/writereaddata/linkimages/anu141200246187.pdf
http://www.lib.utexas.edu/maps/ams/india/
http:/en.wikipedia.org/wiki/Gaur,_West_Bengal
http://malda.nic.in/

Winter snowmelt at near gangotri temple, 26 dec 2007 (Courtesy Praveen Thakur)

Chapter 12
The Developments, Policies and Assessments of Hydropower in the Ganga River Basin

Kelly D. Alley

Abstract This article reviews the practices, policies and assessments of hydropower facilities in the upper Ganga basin and explores the main developments in hydropower infrastructure and accompanying land use changes over the past several decades. It introduces the policy framework and also includes a review of the social and political movements for and against hydropower development. It concludes with suggestions for alternative ways to measure energy and externalities that give importance to other water priorities and uses in the basin.

Keywords Hydropower • Hydroelectric • Dams • Run of river • Impact assessment • Vishnugad Pipalkoti

12.1 Introduction

The Ganga river basin is a place of majesty where glaciers hug the world's tallest mountains, snow melt and precipitation combine to form the water of a vibrant river system, and millennia of cultural diversity guide human life ways. The towering mountains of the Himalaya house the great water storages of Asia in their vast glacial formations. Over the last century these natural storages and river systems have provided water to sustain millions of people across South Asia. Today the flows of these rivers have doubled in value for human populations, as they provide not only water but the power to fuel the desires and needs of twenty-first century civilizations.

K.D. Alley (✉)
Anthropology Program, 7030 Haley Center, Auburn University,
Auburn, AL 36849, USA
e-mail: alleykd@auburn.edu

R. Sanghi (ed.), *Our National River Ganga: Lifeline of Millions*,
DOI 10.1007/978-3-319-00530-0_12,

Courtesy Kelly alley

Given the Ganga's water wealth, all religions of the region have granted the tributaries and main stem of this river a revered position in cultural narratives and practices. The river Ganga and her religious significance as a Mother Goddess and eternal purifier are unparalleled in human history. The water wealth of this river has been worshipped by humans as part of the overall engagement that is necessary for human life. From these great attachments have emerged understandings of the river that revere and thank her. But as water enters a new phase of global commodification, even more is at stake for this sacred river, its tributaries and the human populations that depend upon the river system.

In addition to the Ganga's deep and vast ecological history, the concentrations of water in the snowfields and glaciers of the Himalayas add to its value and complexity. The glaciers are a valuable storage and frontier resource, especially at a time when nation-states are vying for more water to meet growing demands and populations [1–3]. But what will happen to these storages if the planet warms? How fast will glaciers melt and how will this accelerated melting affect the flows of the river and its tributaries? These are questions propelling a new wave of policy and assessment on river management and engineering in the region. The availability of water storage in the glaciers and the assumption that these glaciers might be melting faster are motivating a push for hydropower across the Ganga and in the other eight river basins that receive their water run-off from the Himalayas.

Many decisions related to water governance occur at points of intersection, where interests in some water uses may drive out other water needs that are critical to human survival. In particular, the rapid development of hydropower is posing new challenges for human adaptation to a changing climate and to long term water availability. The public-private race into the hydropower market entails covert forms of influence and decision-making but produces overt ecological consequences that may impact a broad sector of water users. Additionally, hydropower facilities must be legitimized and made acceptable by river basin policies and assessment reports to justify the ecological and human effects that may ensue from a rapid proliferation of facilities and supporting infrastructure.

This paper looks at the practices, policies and assessments of hydropower facilities in the upper Ganga basin and explores how they work together to create the complex human uses of water. To do this, this paper reviews the main developments in hydropower infrastructure and sketches the accompanying land use changes over the past several decades. This brief history will include a review of the main social and political movements for and against hydropower developments. In addition, the paper will introduce the policy framework that has emerged from four key reports and their peers. These reports issued by FAO [4], USAID [5], and Duke University [6] are described in terms of their overlapping foci. In addition to these policy reports, the paper will describe a major assessment report produced by the Alternate Energy Centre of IIT, Roorkee [7]. These policies and assessments are important to examine alongside hydropower developments because they are written to set out and legitimize allowable resource uses within the context of competing strategies and plans.

12.2 The Socio-economic Complexity of the Ganga Basin

Worldwide, glaciers provide the concentrated mass to supply melt water, stream flow and sediment to river valleys. In the Himalayas, the glacial system provides water and sediment to the intensively tilled valleys of the Indus, Amu Darya, Ganga (Ganges), Brahmaputra, Yellow, Yangtze, Sutlej, Mekong and Nu/Salween, and these river systems nourish food production and sustain the lives of millions. The Indian Himalayan ranges sit within two mega basins, the Ganga-Brahmaputra-Meghna basin and the Indus; both have raised great river valley civilizations through human advances in hydraulic engineering. If we take one sub basin in this paper, the Ganga-basin, we can focus on the major water and energy interests at work to represent what is occurring across the river systems of the Himalayas.

The Ganga's main stem and tributaries drain more than one million square kilometers of China, Nepal, India and Bangladesh. The Ganga basin in India, which includes the Yamuna sub-basin, covers one fourth of India's geographical area. From the confluences of the Bhagirathi and the Alaknanda tributaries in the Himalayas, the river Ganga gains additional flow from Nepal's tributaries, glacial snowmelt and monsoon rainfall. Now the basin's sediment loads, which are integral to the river system, are driven by the deforestation of the Gangetic plains and the Himalayan foothills.

12.2.1 The Sacrality of the River and River Banks

For at least two and a half millennia, the river Ganga has nourished human civilizations and great dynasties, and the Hindu and Buddhist pilgrimage traditions have grown up along the riverbanks. By the fourth century BCE, Pataliputra (now near Patna, the capital of the state of Bihar) was one of ten ancient capital cities of India. At the headwaters of the Ganga in the Himalayas, sacred shrines at Gangotri, Kedarnath and Badrinath have marked the sources of the river's sacred power in the Hindu traditions. The temples of Kedarnath and Badrinath also celebrate their position at the snouts of Himalayan glaciers. Farther downstream in the sacred towns of Uttarkashi and Rishikesh and along the plains at Haridwar, Allahabad (Prayag), Banaras, Vindhyachal, Nadia and Kalighat people worship Ganga's waters through rituals of purification.

The Ganga has been worshipped as a river goddess by Hindus across India and the world. According to the Hindu view, sacred spaces are not detached from ecology and the built environment but are embedded in them; Hindu texts and rituals explain this conjunction of divine power and the physical world. In this integrated view, Ganga is a goddess who absolves worldly impurities and rejuvenates the cosmos with her purificatory power. She is also a mother who cleans up human sin and mess with loving forgiveness. Hindus show their respect to her in oil lamp rituals (*arati*) performed on the riverbank and in temple worship (*puja*). Most importantly, devotees seek spiritual purification by doing ritual ablutions (*snan*) in the river.

12.2.2 Water Quality Problems

Today the Ganga basin holds over 800 million people. From the Himalayas to the Bay of Bengal, the Ganga passes by more than 30 major cities of more than 300,000 residents and the river borders many other smaller towns. The Ganga has provided municipal and industrial water for these cities. India's Central Pollution Control Board reports that three-fourths of the pollution of the river comes from the discharge of untreated municipal sewage draining from these urban centers (Central Pollution Control Board at http://cpcb.nic.in/water.php [8]). The Upper Ganga plain in the state of Uttar Pradesh is home to sugar factories, leather tanneries, textile industries of cotton, wool, jute and silk, food processing industries related with rice, dal and edible oils, paper and pulp industries, heavy chemical factories, and fertilizer and rubber manufacturing units. Industrial wastewater is discharged by all these industries and contains hazardous chemicals and pathogens. Four major thermal power plants depend upon water from the Ganga [9].

These industrial and urban effluents have led to a very serious deterioration of river water quality across the plains. Moreover, groundwater levels are declining across India and especially in northwestern India from over-pumping for agriculture and domestic activities [10, 11]. As surface water quality declines, residents turn to groundwater for a good portion of domestic, municipal, agricultural and industrial needs. The groundwater supply will need recharge from adequate river flows to continue to meet such high demands. River flows that are altered by hydroelectric dams and canals and that divert water to needy urban centers are affecting this recharge rate. In the warming climate, faster glacial melt may bring more water into the river system at some times of the year but can lead to flash floods especially in riverbeds that have become disembedded from ecological and hydrological systems by dams and diversions [12]. Increased rainfall and glacial melt may help to recharge groundwater and dilute pollution in the river's flow but both can lead to dangerous and deadly flooding.

12.2.3 Changes to Flow Regimes from Hydroelectric Dams

People living across the Ganga basin face extreme fluctuations in water availability and river basin conditions according to an annual weather cycle. The weather alternates between high water availability – through extreme rainfall and flooding during the monsoon – and extended low flow during the 9 month dry season. With the use of hydropower technology, the water source and availability is also modified in time and space through storage ponds and reservoirs, to meet year round demand. In addition, hydropower is attractive for contemporary societies because it serves as an add-on to coal and nuclear power through its capability to meet needs for "peaking power." While large storage dams such as the Tehri dam can hold a massive amount of water behind a barrage and facilitate water redistribution and reallocation to agriculture and urban centers, run of the river dams halt the river flow for a short period, hold water in a small storage pond and then release it through a head race tunnel to generate power on demand. Especially with run of the river projects, the downstream flow regime alternates between diminished flow at some hours of the day and rushes of water at others. Residents living downstream face seasonal flooding from glacial melt and monsoon rains and in addition see changes in stream flow from the hydropower projects, which may also create flood effects. This means that residents living downstream from one or many dams and diversions will be witnessing and adapting to all these changes in the river's rate and direction of flow, which create cumulative requirements for human adaptation [3, 5, 13–15].

More generally over the last century, the people of South Asia have engaged with the land and its resources intensively, to meet growing demands for food, bioenergy and urban development. Human populations have converted forest, grassland, and shrubland to cropland at a rapid rate, making it the dominant landscape in most regions of South Asia today. More than 70 % of total land area is now under cultivation. Irrigation, use of fertilizers and double cropping have also increased since the 1950s. This agricultural expansion and intensification have triggered carbon and greenhouse gas emissions, land degradation, soil erosion, and loss of biodiversity and freshwater storage [16, 17]. Carbon loss through deforestation and phytomass degradation has dominated the terrestrial carbon balance in the twentieth century. In a region governed by a monsoon climate, the shrinkage of natural vegetation weakens the sustainability of systems and makes the region more vulnerable to extreme climate events, such as flooding.

12.3 Hydropower Developments in the Ganga Basin

Hydropower is an important energy strategy that reshapes the ecological functions and services of a river system. Large dams were built just after Indian Independence as part of national development and significant resistances to these large dams developed in the following three decades [18–22]. The current wave of dam investment is motivated by the twenty-first century interest in industrial growth and urban expansion, and by expanded water consumption. In 2002, the Government of India announced a *50,000 MW initiative* to narrow the gap between supply and the growing demand for power. This hydropower push has focused on the Indian Himalayas where the steep drops of tributaries to the Indus, Ganga and Brahmaputra rivers have the potential to generate larger outputs of power. The sites of current development are located across the northern region of India, in the states of Jammu and Kashmir, Himachal Pradesh, Uttarakhand, Uttar Pradesh, Sikkim, Arunachal Pradesh and Assam.

12.3.1 Hydropower Across the Himalayas

Along the northwestern tributaries of the Ganga river in the State of Uttarakhand, the Tehri dam and several run of the river dams were completed in the first decade of the twenty-first century to provide energy and water supply to the northwestern

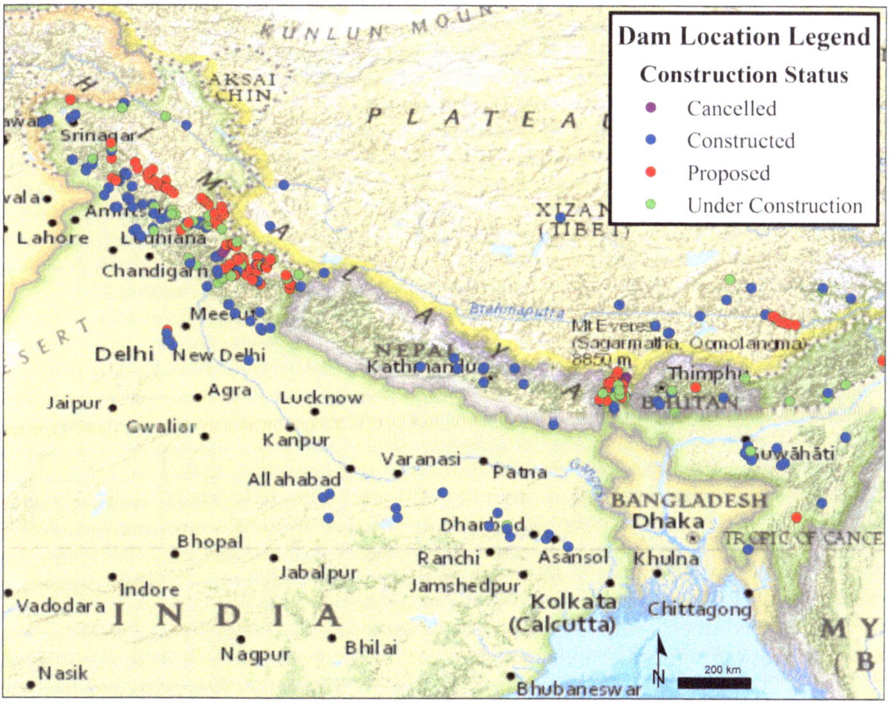

Fig. 12.1 Dam sites (constructed, under construction, and proposed) within the Ganges and Brahmaputra basins (Source: National Geographic basemap in ArcGIS. Map created by: Ryan Hile (permission granted) [file name: South Asia dams])

states of Uttar Pradesh, Delhi and Rajasthan. The development has been fierce and controversial with energy and industrial interests in water pushing out allocations and uses for farmers and residents, and citizens have mounted various campaigns and movements against specific projects [22–27]. The practices in the Ganga basin are similar to those playing out along all the other river systems sharing the Himalayas. For example, along the Beas and Sutlej rivers that flow into the Indus river system, several hydropower projects have been constructed and many are underway. There are also local and regional protests over these projects. In Sikkim a cascade of dams is proposed along the Teesta river to augment the existing two. In the northeastern state of Arunachal Pradesh, the government has sketched up a blitz of projects along the main tributaries of the Brahmaputra, along the Siang, Subansiri, Lohit, and Dibang rivers [28] (see Fig. 12.1).

The current push for hydropower across the Himalayas is supported by assessments that only a small portion of the power potential has been tapped in the region [29–31]. Investors have been lured by new incentives for open access and the freedom to sell power on a merchant basis, the possibility of transferring hydrological risks to the public, and recent trading in Clean Development Mechanism (CDM) carbon credits [28, 32, 33]. But contrary to expectations, hydropower does not always result in an increase in energy for people living in these river basin cities and towns; generally local citizens get the end of the trickle down effects of an increase in power supply. The bulk of energy generated is sold to high end users such as industries and urban facilities [22, 34]. In addition to their energy usage, the high end users also withdraw significant amounts of water for industrial and urban processes and return large amounts of wastewater to the river system.

12.3.2 The Establishment of an Eco-Fragile Zone

In the Ganga basin, the short term effects of run of the river dams have come to light after pressure from a number of groups forced the final cancellation of three projects in 2010. Shortly after the government's announcement of the 50,000 MW initiative, the Ministry of Power charted out an over-ambitious plan to dam all the tributaries of the river Ganga at more than 60 places in Uttarakhand state. Maps of these plans began circulating through civil society networks and people in and outside the state grew worried about the cumulative effects of these dams on water availability downstream, and general water quality in low flow situations (see South Asian Network for Dams, Rivers and People at sandrp.in). Local resistance movements formed and then pressure was exerted on government through exercises of resistance fasting. Civil society or environmental activists use the Gandhian ritual of fasting to push government officials toward a final decision on an issue or project, and to draw the attention needed to register the official decision in the public media. These public records can be used by citizens to enforce accountability when a government agency attempts to backpedal or reverse a decision later in time. To oppose the Loharinag Pala dam, a retired professor from a top engineering school and former chairman of the Central Pollution Control Board began a fast unto death. He started and stopped the fast several times before his final stretch in the summer of 2010. When he was critically ill, the Environment Minister announced that the dam would be scrapped and an eco-sensitive zone would be established in the area. This fasting by an important figure was the tipping point for the decision to cancel a problematic project [24, 35]. A year later another sadhu-sant fasted for the cause of the river Ganga but there was little attention to his protest. He perished without any response from the authorities.

In 2010, the Indian magazine, *Frontline*, ran an announcement on the report titled "Performance Audit of Hydropower Development through Private Sector Participation" [36]. In the report, the Comptroller and Auditor General (CAG) argued that the government of Uttarakhand had pushed the state toward a major environmental catastrophe by following a highly ambitious hydropower policy. After the cancellation of run of the river dams at Loharinag Pala and two in the advanced planning stage (Pala Maneri and BhaironGhati), the Ministry of Environment and Forests issued the *Notification for an Eco-Fragile Zone on the Upper Bhagirathi* to protect the upper Bhagirathi and ban additional

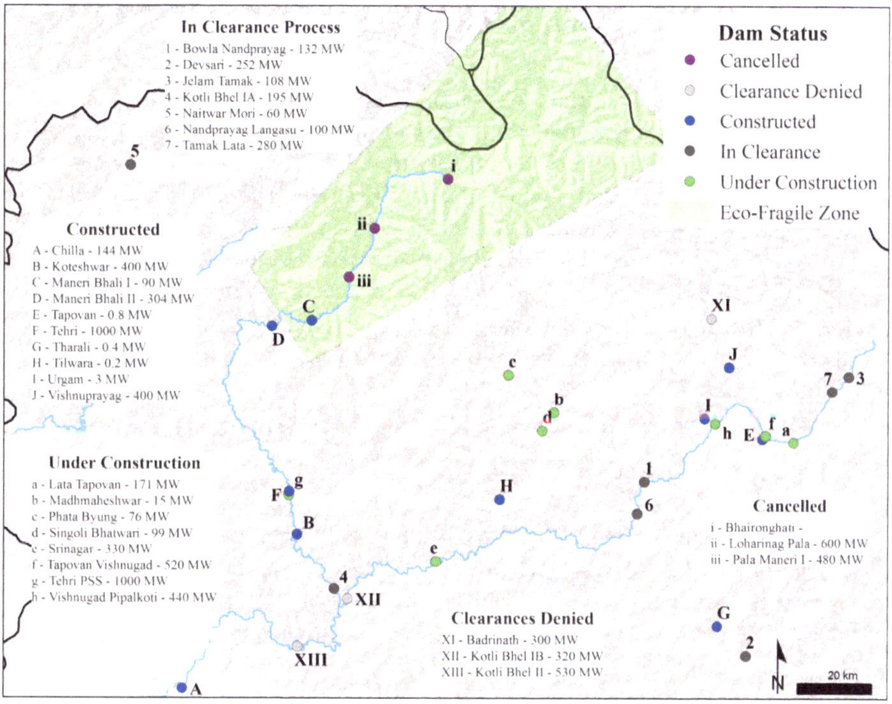

Fig. 12.2 Dam sites within the upper Ganges basin above Devprayag (Source: National Geographic basemap in ArcGIS and data from South Asian Network for Dams, Rivers and People, www.sandrp.in). Map created by: Ryan Hile (permission granted) [file name: SpringerUttarakhandmap]

hydropower projects, indicating that some policy makers may have realized a threshold for altering the river stream and flow regimes in the upper reaches of the Bhagirathi. A new study on recent flooding in Pakistan states that the aggradation of river channels caused by water withdrawals and dam construction may be reducing the width of downstream channels, making river beds less elastic to extreme flows in the rainy season [12]. The loss of river beds and the carriage of sediment outside the channel may worsen flood peaks. Yet despite this understanding of the risks associated with emerging patterns of climate change, hydropower projects remain on the execution list for the 12th year plan. The government has theoretically closed the upper Bhagirathi to additional dam construction, but continues to grant permits to projects on the Mandakini, Dhauli Ganga and Pinder rivers which flow into the Alaknanda river and eventually into the Ganga. On the Alaknanda tributary, a cascade of four dams is under construction with a mix of private and public sector financing and management. All the dams currently under construction – Vishnuprayag, LataVishnugad, Vishnugad Pipalkoti and Srinagar – are located in the fragile upper reaches of the Ganga basin. In total 13 dams have already been constructed in Uttarakhand and 57 more are approved or in various stages of construction [7, 37] (see Fig. 12.2).

12.3.3 *Hydropower and the Policy Framework*

Climate change is a conundrum as a cultural-geophysical phenomenon; it is also a discursive net that captures a range of discussions and interests, some that contradict the core values driving concern for climate in the first place. The language of climate change is evolving and climate planning is an experiment with language and ultimately knowledge. The climate policy discussion creates new meanings for the terms mitigation, adaptation, and resilience and new measurements and incentives for human activity in terms of carbon footprints, clean development mechanisms, and climate funds. Climate change planning and funding look like old North to South capital and technology aid and control regimes, and are intensively reflective of shifting international coalitions.

12.3.3.1 Four Key Reports

Hydropower is a key mitigation strategy in climate change policies across the world. There are four reports that are most relevant to the river Ganga and the Himalayas on the subject of climate change and adaptation. Together they provide a policy framework and justification for hydropower development in the region. These reports are:

1. the FAO report titled, Climate Change, Water and Food Security [4],
2. the USAID report titled, Changing Glaciers and Hydrology in Asia [5],
3. the Duke University report titled, Water, Climate Change and Adaptation: Focus on the Ganges River Basin [6], and
4. the Indian Institute of Technology report titled, Study on the Cumulative Impacts of Hydropower in the Alaknanda and Bhagirathi basins up to Devprayag [7].

In these reports, climate change in the Himalayas is tied to discussions of water security and to concerns about the status and future of water, its availability, its quality and its distribution. The FAO report, *Climate Change, Water, and Food Security*, by Turral et al. focuses on water security in the Himalayas. Other reports examining water security have been issued by the China Dialogue and Humanitarian Futures Programme [3], Malone [5], and the Government of India [29, 38, 39] and together they form a larger circle of opinion. The FAO report elucidates the way research and aid organizations are categorizing climate patterns around the world [4]. The Himalayan region is considered a "snow melt system" where river flows are expected to increase over the next 20 years. After that, reduced flow is predicted. The short term (20 years) adaptation plan for the Himalayas is to build water storage infrastructure and capture the increased flow ([4]: xix). The FAO report notes, "storages will have to cope with more variable and extreme flows, and are likely to be set in more environmentally sensitive landscapes" ([4]: xxii). This report also mentions that acquiring good water accounting practices (hydrological analysis of water resource availability and actual use) and developing robust and flexible water allocation systems "will be a first priority" ([4]: xxii). The Duke

University report by Hosterman et al. [6] considers storage dams an adaptation strategy and notes that "the full spectrum of natural and artificial storage options" must be included. Additionally, adaptation strategies "must consider approaches for protecting and restoring natural storage capacity in the basin and updating existing infrastructure."

12.3.3.2 The Effects of Storage Dams v. Run of the River Dams

The call for storage dams extends the process of disembedding water from ecological and hydrological systems and imposes ideologies of control and redistribution that favor commercial and private interests [40]. Specifically in this region, the call for water storage is justified by the broader characterization of Himalayan rivers as "melt water basins"; in light of climate change and melting glaciers, storage dams must catch additional melt water and store it for use in the dry season when water availability is low. However, run of the river dams do not use large ponds for storing water, so they are not catering to needs for storing and distributing water throughout the year. Instead they are created for the single purpose of generating electricity and are an issue of energy security. Large storage dams, on the other hand, hold a greater volume of water and can distribute this water over a longer period of time and especially during the dry season. However, storage dams can also alter river flows and shape the distribution of water in ways that may create shortage or scarcity for some water users.

In national policy and planning, this spurt in hydropower is justified by two general interests, a primary interest in additional energy generation and a secondary concern with reducing CO_2 emissions by moving away from coal. The state of Uttarakhand along with Himachal Pradesh and Jammu and Kashmir are the fastest growing hydropower markets in the country. The steep drops of the high Himalayas are prime sites for hydropower where the greatest power of the river's flow resides in the drops of the feeder streams and tributaries of the upper reaches. Hydropower interests travel uphill toward glacial snouts and concentrate in the most fragile ecological zones. This is where high altitude flora and fauna temper glacial melt and monsoon rains, as if they are Shiva's braided hair. The expectation is that the greatest amount of power can be generated during the period of spring and summer, when melt water and monsoon rains greatly increase river flows. Power generation would then taper off during the lean season from November through February. During the spring and summer in particular, hydropower can meet needs for "peaking power," that is demanded at high load times of the day.

River flows are affected by large storage dams and run of the river projects. Since the majority of dams now under construction and planned for Uttarakhand are run of the river projects, it is anticipated that the flow regimes for all these rivers will be greatly impacted. A 'run of the river' dam differs from a storage dam in that river flow is halted for a short period behind a small barrage, held in a small storage pond and then released into an earthen tunnel. The tunnel then guides the river flow toward a steep drop and the sheer force of the engineered waterfall turns the turbines that generate power. The downstream flow of the river then alternates between reduced flow while water is halted and collected in the storage pond and a flashy flow at other times, when water is released into the tunnel and down the shaft. This alternation in flow creates somewhat unpredictable rivers for downstream residents and can increase erosion along the riverbanks. Hydropower engineers argue that this flashy flow can be leveled out by creating a series of run of the river dams in a cascade, so that each storage pond would stop the rush of flow from the tunnel above it. Engineers assert that alteration to the total flow will remain minimal. A new assessment report notes, "The downstream fluctuation will get moderated by valley storage effect and lateral flow joining the river" ([7]: 65).

Therefore residents living downstream from these cascading hydropower projects will be "adapting" to an altered flow regime. This is not adaptation to ecological changes brought about through a complex process of climate change alone, but adaptation required by a specific set of engineering projects that will compound the experience of climate instability. River valley residents will undergo serial adaptation. In some climate change policies, this is called "maladaptation." These multiple and staged effects can occur across a range of offset projects created to reduce carbon emissions. These facilities and land use changes will create another set of impacts for communities and especially for subsistence resource users. Together, climate change and the programs established to mitigate climate processes make communities more vulnerable to changes in their water systems over time [5, 15].

12.3.3.3 Minimum Flow Requirement

Recently, India's Ministry of Environment and Forests announced that rivers in the Ganges river basin should maintain a minimum flow in the river stream of 10–15 % of the total seasonal flow. The draft National Water Policy under consideration advocates for minimum environmental flows. This means that at any reach 85–90 % of the flow can be diverted away from the river bed. The requirement to keep 15 % in the river has become a nebulous estimate and assertion and NGOs argue it is not rooted in any specific legislation or regulation [41]. Recently, the World Bank set a greater minimum flow for the Alaknanda river in its project appraisal for the Vishnugad Pipalkoti dam [42]. The World Bank report states, "The impact of the Project on downstream water quality is expected to be negligible. A minimum flow of 15.65 cumecs of water will be left in the river at all times to sustain the aquatic health of the river. This is equivalent to approximately 45 % of the average lean season flow of the river."

12.3.4 Hydropower and Impact Assessment

The central paradox in evolving concerns about climate change is the "business as usual" nature of its activities on the ground. Environmental impact assessments are needed to create justifications for specific resource uses. Environmental assessments that must weigh a project's ecological costs and benefits may be used to grant project clearances and permits. A cumulative impact assessment is now required for projects that have many components spread across a region or river basin. A cumulative impact assessment may therefore become a clearing mechanism for many projects all at once, avoiding the need of individual assessments for each component of a large development plan. This makes the idea of a cumulative assessment suspect, and current institutions of governance at national or international levels cannot answer to the principle of intergenerational equity. So measures now used to calculate cumulative impacts and future scenarios are inadequate.

12.3.4.1 The Cumulative Impact Assessment as Living Policy

In 2009, the Uttarakhand High Court responded to a citizen petition demanding a cumulative impact assessment for all the hydropower projects planned and under construction in the upper Ganga river basin. The Court ordered a scientific study to analyze land use changes and basin-wide ecological problems, and to predict the effects of a rapid and prolific development of hydropower facilities. The final report brought new science and data into the public domain but had the effect of endorsing all the planned projects without finding a single one dangerous to ecosystems and services. In its conclusion, the report also argued to reopen three projects the Ministry of Environment and Forests cancelled in 2010. This science report also contradicted the Notification for an Eco-Fragile Zone on the Upper Bhagirathi, a legal document issued in 2010 to protect the upper tributary from additional hydropower and urban development [39]. In late 2011, the Government of India announced a committee to perform final financial closure for the three canceled dam projects, suggesting that it would not reconsider them. However, the cumulative impact assessment report remains a part of a living policy for water management and planning in the state of Uttarakhand. Its status was bolstered when the World Bank cited it in the safeguard sheet as the required environmental impact assessment for the dam [42].

12.3.4.2 Measuring Externalities

One method for measuring the externalities, that is, the negative or positive effects of such projects, is to calculate the total proposed land use changes and then plot out the geomorphological and hydrological effects to the river system and the carbon footprint generated from the carbon fluxes of land use and river system changes. Documentation of sediment loads and land and hydrological changes can be plotted using ground and satellite data. In addition, the values of biodiversity and human cultural practices must be added to the total ecosystem services for the river system. In terms of biodiversity, the Cumulative Impact Assessment report notes that hydropower facilities are located in the small streams that act as hatcheries for biological production at the first and second tropic levels of freshwater ecosystems. They are a rich source of aquatic biodiversity ([7]: 9). The report also notes that, "landslides and sediment movement as debris flow are frequent in these reaches" ([7]: 8). However in the end, the protection for feeder streams advocated in the Assessment report is undermined by the report's own recommendations to approve all projects; these are projects that will alter all regime flows, dry sacred confluences, and tamper with the fragile headwaters and ecosystems near glacial snouts. A holistic measure of climate change should account for these mitigation effects, for changes to river beds and land uses in the upper reaches that impact a range of biodiversity and cultural values essential to ecosystem services and species survival. These can be tallied into the final measure of carbon fluxes so important to climate policies [17].

This procedure can also be followed when measuring the energy returned on energy invested (EROI) for a particular hydropower project and especially for a series of projects [43]. The metric must list the carbon tonnage "saved" by switching from coal to hydropower and then subtract all the carbon fluxes ensuing from land uses and hydrological changes. Then the human losses associated with cultural practices related to the sacrality of the river, and the economic costs and benefits of tourism and pilgrimage can be included. All these human needs and losses can be added to the total evaluation of river water uses and trade-offs between various uses, to determine how to allocate water across a range of needs and interests in a way that is beneficial to the largest number of people and the supporting ecosystems. This ensures that the metric of EROI, the energy returned on energy invested, does not just calculate energy but all the ecological, hydrological, land use and human and social costs of ecological change. Even though energy is the highest public and private good in industrializing society, water weaves through all ecosystems and human communities as a non-substitutable element and cannot be compromised.

The environmental impact assessments (EIAs) produced for specific projects should calculate safe levels of ecological and hydrological change through the guarantee of minimum flows, the protection for biodiversity and eco-fragile regions, and the support for cultural practices vital to local economies. Climate change policies legitimize the push to exploit the power of flow and allow energy interests to trump other uses and priorities. Engineering models support this push when they downplay the effects of water storages and alterations on river beds and flow regimes. In this way water security is subordinated to energy demands at the highest levels and in all quarters. This is a complex threat for residents of these river basins.

12.3.5 Hydropower and Sacred Prayags in the Upper Basin

For Hindus, the land between the Bhagirathi and the Alaknanda tributaries of the great river Ganga is sacred, and so are the confluences or "prayags" that unite all the streams in this basin. Ancient and contemporary Hindu traditions are steeped in worship of these tributary rivers and prayags. At the town of Deoprayag (Devprayag) both rivers meet and from there the road follows the Alaknanda upstream to the towns of Srinagar, Rudraprayag, Pipalkoti and Joshimath. Srinagar is the site of

Image 12.1 Dam at Srinagar, Uttarakhand under construction (Photo by Kelly Alley [file name: Srinagardam1])

a storage dam that, when completed, will generate 330 MW (per hour of energy generated). Traveling upstream, dam sites are evident by their massive heaps of gravel, sand and crumbling rock. Just before the Srinagar dam site, fortified hills hold back the eroding sediment that threatens at any moment to become a dangerous landslide (see Image 12.1). Processes of erosion and sediment deposition affect the whole valley: daily life is determined by how and when the sediment moves into livable and workable spaces. Mountains are eroding into reservoir sites, and into tunnel entryways drilled and blasted through stable and ancient ravines.

The prayag of Karnaprayag lies at the confluence of the Pinder and the Alaknanda tributaries. Just upstream from this prayag a 400 MW run-of-the-river dam is in the planning stage. Farther upstream, Rudraprayag is marked by several sacred spots including the Koteshwar temple and a riverbank cave where Shiva meditated on his way to Kedarnath. Off the road to the sacred site and popular pilgrimage place of Kedarnath two run of the river dams are under construction by a private firm. The state government is developing several smaller dams nearby. The site for the proposed Vishnugad Pipalkoti dam near Pipalkoti is now marked by a vast complex of offices and residential housing for construction workers and project staff. This dam site and all the others in the upper reaches are set in narrow channels, where the storage ponds will submerge homesteads and farmland along the riverbank. The Tapovan Vishnuprayag dam is under construction just below the operational dam at Vishnuprayag, nested just downstream from a glacial snout (see Image 12.2). At the confluence of the Dhauli Ganga and the Alaknanda, the Lata Tapovan dam is

Image 12.2 Dam at Tapovan Vishnugad under construction (Photo by Kelly Alley [file name: tapovanvishnugaddam2])

planned. In each location the hydropower project drives extensive land use changes and alterations to river beds and flow regimes [12]. It is therefore expected that all the sacred prayag will be impacted by reduced flows and the sites of religious worship below these dam sites may dry up in many locations. Is this cost appropriate for the survival of time honored cultural and religious traditions?

12.3.6 Hydropower and River Basin Management

In the water and hydropower politics of the Ganga basin, the concept of integrated river basin management emerges as a policy ideal; in this the goal is to have all stakeholders at the table with a fair say on how to use the river basin resources for all. But in reality the coordination is a confrontation, a push and push-back that characterizes the evolution of decisions and subsequent water uses. Now the epistemic or decision-making community has expanded to include university scientists and extension specialists, nongovernmental organizations, religious institutions, finance and resource investors, Ministry of Environment and Forests regulators,

World Bank and Asian Bank project managers and civil society members in India and Bangladesh [44–47]. Science groups have the potential to bring more ecological and climate expertise into the planning and assessment process. The new IIT (Indian Institute of Technology) consortium charting out the Ganga Basin Management Plan is a good example of a group carrying out research and specific management and policy and bridging government agencies and civil society (see Gangapedia web site http://gangapedia.iitk.ac.in/?q=content/grbemp-reports-1). Expert committees are also formed by court orders to offer analysis and comment on resource intensive projects including feasibility and detailed project reports. However, experts are hired within and outside India by all parties with an interest in a water project, and some of these experts find their way onto policy committees. The decision-making table will be dynamic and should include the leaders and citizens who depend upon this river system for their agricultural, municipal, religious and cultural practices. These are practices that provide a vital link to the ongoing human necessity for living rivers.

12.4 Conclusions

This paper has outlined the developments, policies and assessments of hydropower development in the Ganga basin. In the process it has laid out some suggestions on alternative ways to measure impacts and externalities to understand the hidden and unintended costs and consequences of rapid hydropower development. The paper also suggests a more realistic method for measuring energy returned on the energy invested when creating a project. This metric must be comprehensive to account for all energy expenditures associated with construction and land use changes. The main aim of outlining these developments, policies and assessments is to find a way to develop energy facilities at a safe and beneficial level, while limiting the insatiable temptation to overbuild on the Himalayan landscape and withdraw water and alter river beds and flows in irreversible ways. Regulations and limits on rapid development are not present in the key policy reports and assessments reviewed here; the only pressures to ensure water for non-energy uses come from ad hoc initiatives by the Ministry of Environment and Forests when pushed by citizen resistances and court orders. Hydropower projects and the policies that justify the massive river system and land use changes for domestic and international agencies must be assessed together so that the thrust of energy interests and their consequences can be fully understood. When understanding that recent climate policies tend to support rather than critically assess rapid hydropower development, the calls of critics to find ways to scale back and in some cases fully close some plans and projects can be better appreciated. Limits and regulations can be achieved to protect eco-fragile regions, reduce the carbon footprint, and allow the river flows needed for other essential and non-substitutable ecosystem services and human needs.

References

1. Chellaney B (2011) Water: Asia's new battleground. Georgetown University Press, Georgetown
2. Immerzeel WW, van Beek LPH, Bierkens MFB (2010) Climate change will affect the Asian water towers. Science 328(5984):1382–1385
3. China Dialogue and Humanitarian Futures programme (2010) The waters of the third pole: sources of threat, sources of survival. Aon Benfield UCL Hazard Research Centre
4. Turral H., Jacob B, Jean-Marc F (2011) Climate change, water and food security. FAO water reports, Rome, p xxii
5. Malone E (2010) Changing glaciers and hydrology in Asia: addressing vulnerabilities to glacier melt impacts. United States Agency for International Development, Washington, DC
6. Hosterman HR, McCornick PG, Kistin EJ, Pant A, Sharma B, Bharati L (2009) Water, climate change and adaptation: focus on the Ganges River basin. Nicholas Institute for Environmental Policy Solutions working paper NI WP 09-03, Duke University, August
7. Alternate Hydro Energy Centre, Indian Institute of Technology, Roorkee (2011) Study on assessment of cumulative impact of hydropower projects in Alaknanda and Bhagirathi basins up to Devprayag, vols I and II. Ministry of Environment and Forests, National River Conservation Directorate, Government of India, New Delhi, March
8. Central Pollution Control Board, Ministry of Environment and Forests, Government of India. http://cpcb.nic.in/water.php. Accessed 18 July 2012
9. Alley KD (2002) On the banks of the Ganga: when wastewater meets a sacred river. University of Michigan Press, Ann Arbor
10. Rodell M, Velicogna I, Famiglietti JS (2009) Satellite-based estimates of groundwater depletion in India. Nature 460(7258):999–1002
11. Scott CA, Sharma B (2009) Energy supply and the expansion of groundwater irrigation in the Indus-Ganges basin. Int J River Basin Manag 7(2):119–124
12. Mustafa D, Wrathall D (2011) Indus basin floods of 2010: souring of a Faustian bargain? Water Altern 4(1):72–85
13. Bosshard P (2010) Down and out downstream: new study documents the forgotten victims of dams. Int River 27(1):14
14. Lahiri-Dutt K (2012) Large dams and changes in an agrarian society: gendering the impacts of Damodar Valley Corporation in Eastern India. Water Altern 5(2):529–542
15. Schwarzenbach RP, Egli T, Hofstetter TB, von Gunten U, Wehrli B (2010) Global water pollution and human health. Annu Rev Environ Resour 35:109–136
16. Mann M (2005) Ecological change in North India: deforestation and Agrarian distress in the Ganga-Jamna Doab 1800–1850. Environ Hist 1(2):201–220
17. Tian H, Lu C, Chen G, Xu X, Liu M, Ren W, Tao B, Sun G, Pan S, Liu J (2011) Climate and land use controls over terrestrial water use efficiency in monsoon Asia. Ecohydrology 4:322–340
18. Baviskar A (2005) In the belly of the river: tribal conflicts over development in the Narmada valley, Studies in social ecology and environmental history. Oxford University Press, Delhi
19. Dharmadhikary S (2005) Unravelling Bhakra: assessing the temple of resurgent India. Report of a study. Manthan Adhyayan Kendra, Badwani MP
20. Gilmartin D (1995) Models of the hydraulic environment: colonial irrigation, state power and community in the Indus basin. In: Arnold D, Guha R (eds) Nature, culture, imperialism. Oxford University Press, New Delhi, pp 210–236
21. Singh S (1997) Taming the waters: the political economy of large dams in India. Oxford University Press, Delhi
22. Wagle S, Warghade S, Sathe M (2012) Exploiting policy obscurity for legalising water grabbing in the era of economic reform: the case of Maharashtra, India. Water Altern 5(2):412–430

23. Bisht T (2009) Development-induced displacement and women: the case of the Tehri dam, India. Asia Pac J Anthropol 10(4):301–317
24. Drew G (2011) Ganga is 'Disappearing': women, development, and contentious practice on the Ganges River. Unpublished PhD dissertation, University of North Carolina, Chapel Hill
25. Kumar K (1996) Controversy over the Tehri dam and Sardar Sarovar: a case study of India's two most controversial hydro electric projects and the devastation they are feared to bring about. Farm Digest Publications, Delhi
26. Mawdsley E (2005) The abuse of religion and ecology: the Vishwa Hindu Parishad and Tehri Dam. Worldviews 9(1):1–24
27. McAnally EA (2012) Toward a philosophy of water: politics of the pollution and damming along the Ganges River. Denton. UNT Digital Library. http://digital.library.unt.edu/ark:/67531/metadc3643/. Accessed 27 July 2012
28. Yumnam J (2012) An assessment of dams in India's North East seeking carbon credits from Clean Development Mechanism of the United Nations Framework Convention on Climate Change. Citizens' Concern for Dams and Development, Manipur
29. Government of India (2010) Report of (IMG) Inter Ministerial Group: to evolve a suitable framework to guide and accelerate the development of hydropower in the North Eastern Region, New Delhi, February
30. Menon M, Kohli K (2005) comp. Large dams for hydropower in Northeast India: a dossier. South Asia Network on Dams, Rivers and People and Kalpavriksh, New Delhi
31. Vagholikar N, Das P (2010) Damming Northeast India. Kalpavriksh, Aaranyak and ActionAid India, Pune/Guwahati/New Delhi
32. Dharmadhikary S (2008) Mountains of concrete: dam building in the Himalayas. International Rivers, Berkeley
33. Dharmadhikary S (2010) The protocol of vested interests. India Together. www.indiatogether.org/2010/nov/opi-protocol.htm. Accessed 3 June 2013
34. Sreekumar N, Dixit S (2010) Electricity for all. India Together, 17 September 2010. www.indiatogether.org/2010/sep/env-zforall.htm
35. Drew G (2012) Ecological change and the sociocultural consequences of the Ganges River's decline. In: Johnston B et al (eds) Water, cultural diversity, and global environmental change: emerging trends, sustainable futures? UNESCO-IHP, Paris, pp 203–218
36. Tripathi PS (2011) Damning audit. Frontline 27 (Nov. 20–Dec. 03, 2010). http://www.frontlineonnet.com/fl2724/stories/20101203272404900.htm. Accessed 10 May 2011
37. Alley KD (2011) The disappearing rivers of India. Asia Pacific Memo, 15 November 2011. http://www.asiapacificmemo.ca/the-disappearing-rivers-of-india. Accessed 4 June 2012
38. Government of India (2010a) Draft National Water Policy, 2012. www.scribd.com/doc/133562462/Draft-National-Water-Policy-2012
39. Government of India (2011) Ministry of Environment and Forests Notification for Eco-Fragile Zone on Upper Bhagirathi, New Delhi, 18 December 2012. http://moef.nic.in/assets/2930.pdf. Accessed 3 June 2013
40. Strang V (2011) Fluid forms: owning water in Australia. In: Strang V, Busse M (eds) Ownership and appropriation. Berg, Oxford, pp 171–196
41. Seth BL (2012) National water policy, silent on priorities. Down to Earth. www.downtoearth.org.in/content/national-water-policy-2012-silent-priorities
42. World Bank (2011) India – Vishnugad Pipalkoti hydro electric project. The World Bank, Washington, DC. http://documents.worldbank.org/curated/en/2009/11/11427475/india-vishnugad-pipalkoti-hydro-electric-india-vishnugad-pipalkoti-hydro-electric-project. Accessed 3 June 2013
43. Lerch D (2010) Making sense of peak oil and energy uncertainty. In: Heinberg R, Lerch D (eds) The post carbon reader: managing the 21st century's sustainability crises. University of California Press, Berkeley, pp 207–210

44. Alley KD (2004) The making of a river linking plan in India: suppressed science and spheres of expert debate. India Rev 3(3):210–238
45. Bhaduri A (2012) Citizens voice alarm over recent Supreme Court judgement on interlinking of rivers: the directive urges the Government to begin work on the proposal. India Water Portal. http://www.indiawaterportal.org/post/26045.
46. Feroze AM, Ahmad QK, Khalequzzaman M (eds) (2004) Regional cooperation on transboundary rivers: impact of the Indian river-linking Project. MatiarManush. alt, Bangladesh PoribeshAndolon, Dhaka
47. Zawahri NA, Hensengerth O (2012) Domestic environmental activists and the governance of the Ganges and Mekong Rivers in India and China. Int Environ Agreem Polit Law Econ. Springer 12(3):269–298. doi:10.1007/s10784-012-9179-9

Maneri Bhali II (600MHz) dam (8.5 km east of Uttarkashi) (Courtesy Rashmi Sanghi)

Chapter 13
3Ps (*Population, Poverty and Pollution*) and the Pious Poor Ganga

Niraj Kumar

Abstract The river Ganges occupies an unique position in the cultural ethos of India. It is a life-line, a symbol of purity and virtue for millions of people living on its bank, representing their ethos, socio-scientific culture and identity in every forum, not just in India, but in the entire world. But the holy river Ganges's existence is itself under threat, due to enormous load of population and subsequent pollution year by year. The poverty further enhances the grim situation by adding several unforeseen problems. All these have been discussed in detail in this article with reference to the efforts being made by the governmental and non-governmental agencies; and peoples participation.

Keywords Multidimensional Poverty Index (MPI) • Nature-Man-Science Complex • Maladaptive interaction between population & poverty

13.1 An Epilogue

From times immemorial, the Ganges has been India's river of faith, devotion and worship. No other river in the world is so closely identified with the cultural ethos of a country as the Ganges is with India. It is a life-line, a symbol of purity and virtue for millions of people living on its bank, representing their ethos, socio-scientific culture and identity in every forum, not just in India, but in the entire world. Legend says that the river has descended from heaven to the earth as a result of the long and arduous prayers of King Bhagirath for the salvation (***Moksha***) of his deceased ancestors. But the holy (***mokshadayani***) river Ganges's existence is itself

N. Kumar (✉)
The National Academy of Sciences, India (NASI), 5, Lajpatrai Road, Allahabad, Uttar Pradesh 211002, India
e-mail: nasi.allahabad@gmail.com

R. Sanghi (ed.), *Our National River Ganga: Lifeline of Millions*,

DOI 10.1007/978-3-319-00530-0_13,

under threat, due to enormous load of population and subsequent pollution year by year. Further, the role of poverty as a catalyst leading to manifold increase in the dimension of this menace (both the uncontrolled population and pollution) cannot be ignored or overlooked.

13.2 Factual and Problematic Details

The Ganges*: *Gangotri is an opening site of the river Ganges in the mountainous gorge at the foot of the Himalayas. ***Bhagirathi***, one of the tributaries, originates from the ***Gaumukh*** snout of Gangotri glacier, whereas the Alaknanda, the other main tributary emerges from the ***Bhagirathi Khark*** and ***Satopanth*** glaciers, east of Chaukhambha peaks. After flowing in opposite directions from their respective snouts they meet at Deoprayag and henceforth collectively known as the river Ganges.

***Tributaries of Ganges*:** The Ganges drains the young Himalayan fold mountains in the north, the stable northern part of the peninsular region in south and Gangetic alluvial plane in the centre. It's path is about 2,525 km long and its basin is spread over about 8,61,404 km^2 area, draining about one-fourth area of the country. The river system covers cool upland streams and warm water stretches, including deltaic habitats. It has many tributaries, both in the Himalayan region before it enters the plains at Haridwar and further downstream before its confluence with the Bay of Bengal. The important tributaries and sub – tributaries are: Yamuna, Ramganga, Gomti, Ghaghara, Gandak, Kosi, Kali/Sharda, Chambal, Sindh, Betwa, Ken, Tons, Sone, Punpun, Damodar and Kangsbati Haldi. Rainfall, subsurface flows and snow melt from glaciers are the main sources of water in river Ganga. Surface water resources of Ganga have been assessed at 525 billion cubic meter (BCM). Out of its 17 main existing tributaries Yamuna, Sone, Ghagra and Kosi contribute over half of the annual water yield of the Ganga. These tributaries meet the Ganga at Allahabad and further downstream. Tables 13.1 and 13.2 depict the relevant data about the geo-hydro status of the river Ganges.

13.2.1 Population Stress

The Ganga basin was so fertile, that the people living around depended for their livelihood by cultivating a variety of food grains; and with every harvest they used to offer their share just to pay regards to Ganges on whom they were dependent for their livelihood. Gradually these became the mark of their rituals taking shape in many celebrations, puja, purva and festivals. The main festivals are: Kumbh Mela, Sawan Mela, Magha Mela, Kartik Poornima, Ganga Dusshera, Chhath Puja and Ganga Sagar Mela. The nearby cities along the river Ganga grew into trade centres as navigation along the Ganga was the main source of transportation in ancient India, giving

Table 13.1 Catchment area, annual yield of water and mean flow of tributaries of Ganga Basin*

Sr. No.	Sub-basin	Mean annual flow (Billion Cubic Meter BCM)	Percentage contribution (%)
1.	Ramganga	17.789	3.39
2.	Yamuna (excluding Chambal)	57.241	19.90
3.	Chambal	32.554	6.20
4.	Tons-Kararmnasa	10.609	2.02
5.	Gomti-Ghaghra	113.511	21.62
6.	Sone-East of Sone	44.144	8.41
7.	Gandak-Burhi Gandak	58.967	11.23
8.	Kodi-Mahananda	81.848	15.59
	Total (Tributaries)	416.663	79.36
9.	Ganga main stem	84.980	16.19
10.	Evaporation* (attributable to ground water)	23.380	4.45
	Total Ganga (up to Indian border)	**525.023**	100.00

Source: *CWC [1]

Table 13.2 Other features of river Ganges

Total length	2,525 km
Uttar Pradesh and Uttarakhand	1,450 km
Boundary between Uttar Pradesh and Bihar	110 km
Bihar	445 km
West Bengal	520 km
Geographical area of India	3.28 million Sq. km
Reported area – River Basins	3.05 million Sq. km
Catchment area – Ganga Basin	8,61,404 km^2 (26.4 %)
Average annual discharge	**4,93,400 Million cubic meter**
East Longitude of Basin area	73°30′ and 89°
North Latitudes of Basin Area	22°30′ and 31°30′
Surface water resource potential	**525 billion cubic meter**
States of India in the Indo Gangetic Basin	11
Average population density	**Over 550 persons per Sq. km**

Source: CWC [1]

rise to association and fusion of different trades and culture. The outcome in terms of celebrated products and traditions are: Varanasi Paan and Thandai, Madhubani Art and Murshidabad Painting, Shehnai, Akharas and Ganga Arti.

But all these traditions and trade forced the population to grow and concentrate on the bank of the river; thereby increasing the load by leaps and bound. The below Table 13.3 clearly depicts the present and future state of population in the Ganga river basin. The proposed figures (as shown in the Table 13.3) clearly give an estimated growth of population in Indo-Gangetic region by the year 2050. This will lead to the formation of water stressed area having low per capita water availability.

Table 13.3 Population status

	Total population (millions)			
Basin	1991	2001	2025	2050
Indus	39	46	61	56
Ganga	362	445	684	868
Basin total	401	491	745	924
% of India	48	49	54	58
India	844	1,007	1,389	1,583

Source: Amarsinghe et al. [2]

Presently, more than 500 million people living in its basin are directly or indirectly dependent on the river Ganges. The river system is conventionally and dynamically the source of livelihood for the people involved in the following occupation/trades/tradition-

- Agriculture
- Fisheries
- Power generation
- Navigation, Transport and Tourism
- Poultry and Live-stock Farming
- Sand-mining
- Pottery
- Art and Craft
- Apiculture, silviculture, jute industry, leather industries, pharmaceuticals, electronics, textile, paper, tanneries, fertilizer and oil refineries.
- Priest hood snake charmers and magicians

By the year 2050, the population residing in this region will be almost double, leading to acute inter and intra competition among occupation holders, making the situation bad to worse. Not only this, per capita water availability will also reduce; and it is estimated [3] that it would go down to the level of <1,700 m^3, further deteriorating the situation. As far as water use is concerned, 90 % of the annual water withdrawals in the Ganga is still being diverted for agriculture, as about 70 % population of Gangetic basin mainly depends on agriculture and related activities. The scarcity of water will therefore pose threat not only for regular consumption but also for agricultural production, thus almost 1,000 million people will have to face grave consequences by the year 2050. This may even lead to conformational changes in the social dynamics, as the stressed population would have no option but to either change their habitat or shift to some hydro-independent occupation. More dependence on ground water, leading to its increased exploitation for the agriculture and other hydro dependent livelihood would make the situation grim in next few decades, i.e. by the end of this century.

Apart from intra-specific struggle, the increase in human population will also lead to inter specific and extra specific struggle for food and water; further eliminating several species which have become endangered, in turn leading to imbalance in the food chain and food web. Finally the consequences will be beyond control, if proper measures are not taken with immediate effect by governmental and non-governmental agencies as well as by each and every individual.

13.2.2 Poverty Effect

The new measure, called the Multidimensional Poverty Index (MPI), was developed and applied by the Oxford Poverty and Human Development Initiative with UNDP support (Website) [4]; an analysis by MPI creators reveals that there are more 'MPI poor' people in eight Indian states (421 million in Bihar, Chhattisgarh, Jharkhand, Madhya Pradesh, Orissa, Rajasthan, Uttar Pradesh, and West Bengal) than in the 26 poorest African countries combined (410 million). Is this not enough to alarm us about the poverty status of our Gangetic basin?

Out of these eight Indian states, part of four states are within the Gangetic basin; and approx. % of population below poverty line in main five states (in Gangetic basin) of India is- Bihar (41.4 %); Jharkhand (40.3 %); Uttar Pradesh (32.8 %); Uttarakhand (39.6 %); and W.B. (24.7 %), as per Press Note dated March 2007, Press Information Bureau, Government of India. This is much higher (except for W.B.) than overall poverty (27.5 %) in India. There are several socio-economic causes for this; but this poverty status is a serious cause of deteriorating condition of the river Ganges. The majority of this population depends on the river for their livelihood, in turn increasing the exploitation of river water and other river-resources (as sand, soil, vegetation and forest resources, etc.) in absence of alternate source of their livelihood. The changing social dynamics further forces the said population to depend more and more on these resources, as evolution of backward elites in last one or two decades has further dragged the majority of poor to "have-nots" state, having no option but to depend totally on natural resources only. A thousand of cases have been reported on such type of exploitation of river resources. The case studies done by our group in the Institute of Applied Sciences, Allahabad; also published (Rizvi et al. [5, 6]), and reported about the sharp decline in the fish population in and around Allahabad, due to over exploitation by the deprived poor, as one of the major cause of diminishing fish population.

Similarly, it applies to other river-resources, such as sand, soil, vegetation and other animal species. For example, dolphins are other major cause of concern, as their population has also been declining making them endangered.

Not only this, hundreds of living and non-living resources are on the verge of extinction; thanks to over exploitation. Actually, we have become callus to our mother earth for our endless need and greed. Our great Indian tradition was based on "**Nature-Man-Spirit Complex**", but gradually the rapid industrialization/urbanization/modernization eventually replaced the "spirit (supernatural powers)" factor by "science" factor, making it "Nature-Man-Science Complex"; but unfortunately, loss of moral values allowed injudicious use of scientific tools for unethical exploitation of nature and natural resources for our comforts and ease, finally making the survival difficult. Now, galloping population and poverty have increased the problem manifold, in fact beyond control.

13.2.3 Pollution Menace

The final blow to the problem has been given by the increasing load of pollution, which is nothing but the effect of maladaptive interaction between above two factors (i.e. population and poverty). In the year 2007, the river Ganges was placed among the five most polluted rivers of the world (greendiary) [7] with fecal coliform level estimated as more than hundred times the prescribed limits, near Varanasi city. Pollution poses threat not only on humans, but also on more than 140 fish species, 90 amphibian species and the dolphins [8]. The Ganga Action Plan (GAP), an initiative to clean the river Ganges in India, was proved failure [9–11] due to lack of will and technical expertise [12], lack of sound (environmental) planning [13], our age old traditions and beliefs [14] and lack of support from religious corners [15].

As per World Bank estimates, the health cost due to water pollution alone in India is equal to 3 % of India's GDP [16] and about 80 % of the total types of illnesses in India and one-third of deaths be attributed to water-borne diseases [15]. What more is needed to declare India a sick country?

Ironically, once upon a time, the same water (i.e. Gangajal) was having pristine properties and it was treated as sacred due to its purifying effects; thus linked to several rituals and sanskaras of Hindu tradition. But then it was different, as the extraordinary self-purifying properties of Gangajal were because of-

- its high content of dissolved oxygen (DO)
- extraordinary high rate of re-aeration, long DO-retention abilities
- the very fast assimilation of the putrefiable organic matter
- the presence of sulphur and radioactive traces
- the naturally occurring extracts of medicinal herbs
- the low temperature gradient at the origin
- the high flux density and flow
- the presence of beneficial algal and diatom population
- the abundance of predatory fauna; and over all and above
- the long duration of sunshine, along the Gangetic region, having its impact (electro-magnetic radiations) on microbial population.

But the overload of pollution has maligned the purity of Gangajal. The principal sources of pollution of the Ganga river can be characterized as follows [17]:

- Domestic and industrial wastes. It has been estimated that about $1.4 \times 106\ m^3\ day^{-1}$ of domestic wastewater and $0.26 \times 106\ m^3\ day^{-1}$ of industrial sewage are going into the river.
- Solid garbage thrown directly into the river.

- Non-point sources of pollution from agricultural run-off containing residues of harmful pesticides and fertilizers.
- Animal carcasses and half-burned and unburned human corpses thrown into the river.
- Defecation on the banks by the low-income people.
- Mass bathing and ritualistic practices.

The river Ganges and its basin have been treated by people as a huge reservoir and a logical place for the quiet disposal of the tremendous volume of waste on its banks for past several decades. Dumping of plastic containers, plastic bags, cans, mineral water bottles, rubbers and plastics, faeces, soaps and detergents, medical waste and several other articles in the river has defaced the Ganges. The Human impact is robbing the pristine values of the river-water and its sacredness. A collective action from all the corners of the society, i.e. stakeholders, people and, the state, is needed to bring Ganges to its past glory.

13.2.4 Government's Role

Looking into shortcomings of the GAP it was felt that a pragmatic and holistic approach be adopted. Accordingly, the Government of India has given Ganga the status of a National River and has constituted the National Ganga River Basin Authority (NGRBA) on 20th February 2009 under Section 3(3) of the Environment (Protection) Act, 1986. The NGRBA is a planning, financing, monitoring and coordinating body of the centre and the states. The objective of the NGRBA is to ensure effective abatement of pollution and conservation of the river Ganges by adopting a river basin approach for comprehensive planning and management.

The NGBRA has both regulatory and developmental functions. The Authority is supposed to take measures for effective abatement of pollution and conservation of the river Ganges (as per its details available on the website), by adopting following main measures-

- Development of a river basin management plan;
- Regulation of activities aimed at prevention, control and abatement of pollution in river Ganges to maintain its water quality, and to take measures relevant to river ecology and management in the Ganga basin states;
- Maintenance of minimum ecological flows in the river Ganges;
- Measures necessary for planning, financing and execution of programmes for abatement of pollution in the river Ganges including augmentation of sewerage infrastructure, catchment area treatment, protection of flood plains, creating public awareness;
- Collection, analysis and dissemination of information relating to environmental pollution in the river Ganges;
- Investigations and research regarding problems of environmental pollution and conservation of the river Ganges;

- Promotion of water conservation practices including recycling and reuse, rain water harvesting, and decentralized sewage treatment systems;
- Monitoring and review of the implementation of various programmes or activities taken up for prevention, control and abatement of pollution in the river Ganga;
- Issue directions under section 5 of the Environment (Protection) Act, 1986 for the purpose of exercising and performing these functions and for achievement of its objectives.

Sustainable Development of River Ganges Ecosystem could be achieved only if we look into the problem with Bio-geo Pragmatic approach. Scientifically, sustainable development of any ecosystem depends on Conservation of Environmental gradients, Socio-economic development, Geo-political measures and over all and above people's participation. Therefore, sustainable development of river Ganges needs aforesaid holistic approach based on the matrix of Nature-Man-Science Complex, powered by pragmatic action, which may be achieved by adopting the following-

- Probabilities of Community Treatment Plants and recycling of the wastewater be explored for economically feasible solutions.
- Accurate quantity and quality of waste water (e.g. sewage, industrial effluents and laboratory chemicals etc.) generated at Varanasi, Allahabad and Kanpur be assessed for long term environmental planning.
- Small scale and cottage industries (including motor workshops and garbage disposal) located in vicinity of the cities be identified and suitable treatment plants be suggested.
- Proper quantification and quality examination of solid wastes generated in each zone of the city be assessed for their recycling and other utilizations.
- Suitability and economic viability of the new Sewage Treatment Plants be ensured before their establishment.
- For economic saving old and existing STPs be upgraded/modified in light of their past drawbacks and recent advances in Wastewater Treatment Technologies.
- Land use patterns, fertility level of land and upstream/downstream locations for new STPs be examined so that no STP be constructed on highly fertile lands.
- Electric crematorium on respective cremation grounds, incineration plants for cattle's and separate washing ghats at downstream be established.
- Unauthorized constructions near the banks and its catchment areas be stopped immediately to maintain the natural flow.
- Production activities (e.g. fishery, aquaculture, navigation etc.) of Ganga-water be encouraged.
- Mass awareness programs through popular lectures, workshops, documentaries, *Nukkad natak*, quizzes be organized at school/college level.
- Development and adoption of micro-cosmic approach to understand the gravity of problems of 3Ps (Population, Poverty and Pollution) is urgently required.
- Rain Water Harvesting, its storage, multiple uses at micro community level and Ground Water Recharging practices be encouraged to maintain the ground water status and save Ganga-water.

- Forestry be intensively recharged for mitigating the problems of flood and soil erosion.
- Dam construction be in harmony with the ecosystem; and required only when no other option is left.
- Over all and above, man and machinery be not allowed now to play mischief with the mother nature.

13.2.5 NASI's Initiative

The Council of the National Academy of Sciences, India, the oldest Science Academy of the country, has taken concrete steps by establishing a magnificent Ganga gallery (idea mooted by Hon'ble Prof M S Swaminathan, Prof M G K Menon and Prof Manju Sharma; all are Past Presidents of the Academy), in order to make the people aware about the socio-scientific, economic, ecological and cultural aspects of the holy river Ganges, so that they could realize its vast importance and take steps for its preservation and conservation. On 15th April 2011, Union Science and Technology Minister Hon'ble Shri Pawan Kumar Bansal inaugurated the scientifically installed magnificent and aesthetic Ganga-gallery (Images).

The Gallery was established with technical help provided by the National Council of Science Museum, Kolkata, to fulfill the following broad objectives-

1. To create 'The Ganga Model' in a large hall indicating different pollution sources from Gaumukh to Ganga Sagar and its impact on the water quality with the help of relevant data and photographs.
2. To establish a library with the literature on the origin; geological structure; geographical distribution; flow of water through different states; physico-chemical and biological properties; pollution sources and other degradational forces; bio-diversity and endangered species; cultural, religious, economic, socio-scientific and ecological significance of the river Ganges.
3. To institute fellowships and start lecture series on the Ganges in different states specially where those the river flows.
4. To prepare CDs and documentary films, arrange photography and essay competition on the Ganges and involve NGOs, Universities, Colleges, Schools, Electronic and Print Media etc. for public awareness and action programmes.
5. To create a website for wider publicity and disseminate the knowledge at global level about the Ganges.
6. To encourage researchers of various disciplines and develop indigenous technologies for the treatment and recycling of waste waters to solve the problems of the Ganga Water Pollution.
7. To develop trees, shrubs and grasses plantation strategy to control the impact of pollution, silt load and maintain the ecological integrity of the river ecosystem.
8. To organize seminar, symposia, workshop and training programmes time to time on the issues related to the river Ganges.

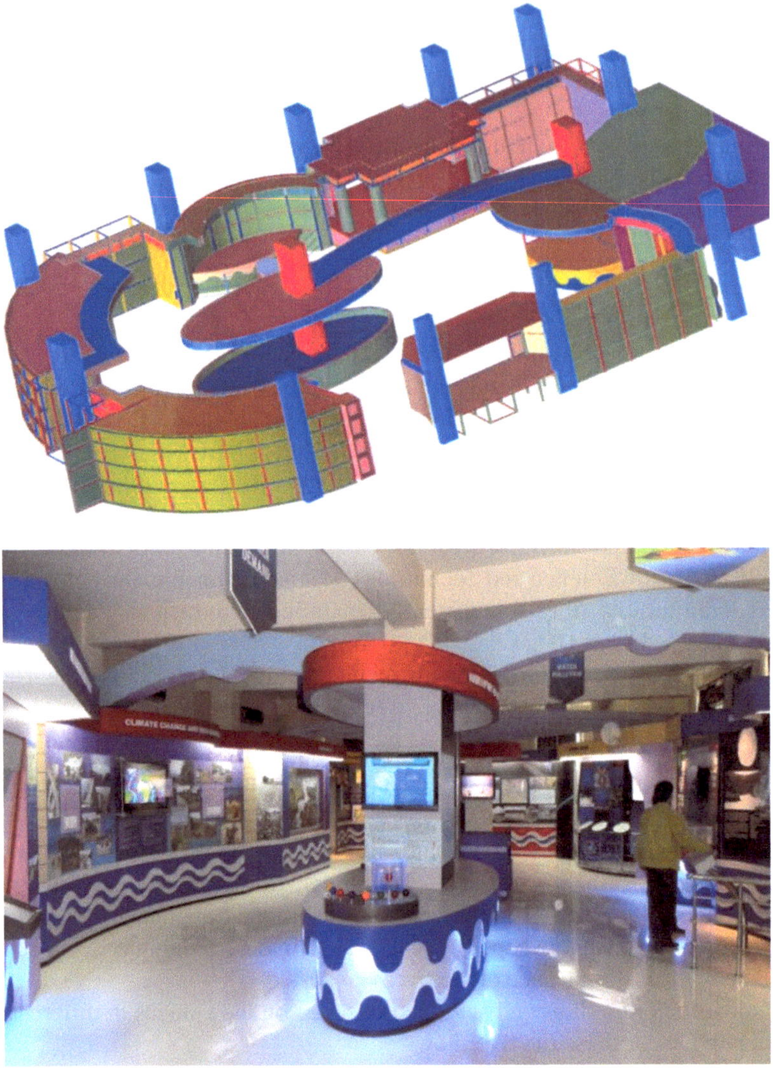

The virtual and real pictures of the gallery

And at last, the following 13 actions be prohibited in holy River Ganges, as per Brahmananda Purana 800 AD:

- Defecation
- Ablutions
- Discharge of wastewater
- Throwing of used floral offerings
- Rubbing of filth

- Body shampooing
- Frolicking
- Acceptance of donations
- Obscenity
- Offering of inappropriate praises or even hymns in a incorrect way
- Discarding of garments
- Washing of cloth
- Swimming across, in particular

Acknowledgments Author is grateful to Prof (Mrs.) Manju Sharma, Former Secretary to the Govt. of India, for always inspiring him for innovative approaches and endeavors. He is also thankful to Prof Krishna Misra, General Secretary (HQ), NASI, Sri Samarendra Kumar, Regional Science City, NCSM, Lucknow and all the learned members of the NASI-Council, who have given him their valuable advices and regular support. Thanks are also due to Dr. Chitranjan Kumar for helping in editing this article.

References

1. CWC. Publication No.30/88, New Delhi, India, pp 1–53. http://www.igu.in/9-3/6sarma.pdf
2. Amarsinghe UA, Shah T, Turral H, Anand BK (2007) India's water future to 2025–2050: business as usual scenario and deviations. International Water Management Institute, Srilanka, p 47
3. Sharma BR, Amarsinghe UA, Sikka A (2008) Indo-Gangetic river basin: summary situation analysis, New Delhi
4. http://articles.timesofindia.indiatimes.com/2010-07-12/india/
5. Rizvi AF, Amitabh CD, Singh KP (2012) Growth pattern of Labeo calbasu (Ham.) Affected by socio-geographic condition of the Riverine Habitat. Curr Res J Biol Sci 4(3):250–257
6. Rizvi AF, Singh KP, Kumar N (2011) Population dynamics of non-commercial fishes in rivers around Allahabad and possibility of their commercialisation to enhance socio-economic status of fishermen. Natl Acad Sci Lett 34(11and12):437–448
7. http://www.greendiary.com/entry/ganga-is-dying-at-kanpur/
8. India and pollution: up to their necks in it. The Economist, 27 July 2008. www.webline.co.in/gap_final/view_fullarticle.php?logauth=7
9. Clean up or perish. The Times of India, 19 March, 2010. www.webline.co.in/gap_final/view_fullarticle.php?logauth=3
10. Gardner G (2003) Engaging religion in the quest for a sustainable world. In: Bright C et al State of the world. W. W. Norton and Company, New York, p 256, pp 152–176. ISBN 0-393-32386-2
11. Haberman DL (2006) River of love in an age of pollution: the Yamuna River of northern India. University of California Press, Berkeley, p 277. ISBN 0-520-24790-6
12. Sheth JN (2008) Chindia rising. Tata McGraw-Hill Education, New Delhi, p 205. ISBN 0-07-065708-4
13. Singh M, Singh Amit K (2007) Bibliography of environmental studies in natural characteristics and anthropogenic influences on the Ganga River. Environ Monit Assess 129:421–432. doi:10.1007/s10661-006-9374-7
14. Tiwari RC (2008) Environmental scenario in India. In: Dutt AK et al (eds) Explorations in applied geography. PHI Learning Pvt. Ltd, New Delhi, p 524. ISBN 81-203-3384-5

15. Puttick E (2008) Mother Ganges, India's sacred river. In: Emoto M The healing power of water. Hay House Inc, Carlsbad, California. p 275, pp 241–252. ISBN 1-4019-0877-2
16. Bharati RK (2006) Interlinking of Indian rivers. Lotus Press, New Delhi, p 129. ISBN 81-8382-041-7
17. Helmer R, Hespanhol I (eds) (1997) Water pollution control – a guide to the use of water quality management principles. United Nations Environment Programme, the Water Supply and Sanitation Collaborative Council and the World Health Organization, WHO/UNEP, London. ISBN 0 419 22910 8

Ferrying is a crucial livelihood activity in the villages on the banks of Ganga. Men returning to villages at Chyaser, near Kannauj, UP (Courtesy Julian Crandall Hollick)

Part V
Livelihood and Tourism

Pallaze agriculture in sand island (Courtesy Sandeep Behera)

Chapter 14
Living Out of *Ganga*: A Traditional Yet Imperiled Livelihood on Bamboo Post Harvest Processing and Emerging Problems of *Ganga*

Dipankar Ghorai and Himadri Sekhar Sen

Abstract River *Ganga* with its tributaries provides livelihood and support for some 450 million people in India. Never in the history of mankind had a river of medium length such colossal importance ranging from social, cultural, economic and even spiritual motives. From its beginning at *Gangotri* to its ending at *Gangasagar*, in the entire stretch of 2,510 km, it provides means for livelihood of diverse nature and forms to groups of people. Post harvest processing of bamboo for its endured use utilizing *Ganga* water is one such livelihood. Considering the wide scale and multifarious utility of bamboo, existence of this group is utterly indispensable and requires attention discussed in the article. The present article is an attempt to put focus on their life and times with the changing *Ganga* scenarios which are actually two pronged – drying up and pollution facets. While the reasons for drying up of the river affecting the above and all other forms of livelihood are partially climatogenic and geogenic, but mostly anthropogenic in nature, the reasons for pollution are entirely anthropogenic. *Ganga* is the face of the nation and as such awareness and efforts are needed in all possible quarters to keep the face enlivened before it is too late.

Keywords Livelihood • Bamboo • Post harvest management • Traditional living

---

D. Ghorai
Subject Matter Specialist, Krishi Vigyan Kendra, CRIJAF (ICAR),
Bud Bud, Burdwan, West Bengal, India

H.S. Sen (✉)
Director, Central Research Institute for Jute & Allied Fibres (ICAR)-Retired,

Present address: 2/74 Naktala, Kolkata 700 047, West Bengal, India
e-mail: hssen.india@gmail.com; hssen2000@hotmail.com

R. Sanghi (ed.), *Our National River Ganga: Lifeline of Millions*,
DOI 10.1007/978-3-319-00530-0_14, © Springer International Publishing Switzerland 2014

14.1 Exordium

GANGA – 'The Ganges' in anglicized form– epitomizes purity, sanctity, divinity, motherliness and godliness to Indians. To them, it is not just another river; but, an ecclesiastical embodiment of life flowing through the sacred land of India – *a la* blood runs through veins. To them, '*Maa Ganga*' (Mother Ganges) is omnipresent and omnipotent – present in their chants, in their hymns, in their prayers and potent in relieving their pains, their sins, and their sorrows. To them, one dip in *Ganga* expunges sins of ten births (*Dasahara*), one drop of holy water from *Ganga* refurbishes all adulterated drinks, one voyage along its' length from source to sea is the ultimate *Shangri-la.* Quintessentially, *Ganga* symbolizes the rich and age old cultural and religious heritage of a nation – a nation where mythology and cultural tradition is entwined with the sacred topography, where rich biodiversity is entangled with affluent linguistic diversity, where socio-economic well being of millions is gyrated around one particular river – in the Ganges.

14.2 The Ganges, Enlivening and Myth

The Ganges and livelihood is connected – reality apart – even in our rich mythological fables. Keen readers and readers unfamiliar with Hindu mythology may find interest in this section. The Ganges is so deeply revered by the Hindus that every cults and sects – be it the *Vaishnavas,* or *Shaktas,* or *Shaivas* and so on – of Hinduism has their own mythical tale to tell about the origin and development of the Ganges in their respective religious anthologies with every one of them varying from the other to some degree. But the most widely popular one is that depicted in one of their epics, the *Ramayana*, according to which *Ganga* was created as a result of liquefaction of the energies of Lord *Bishnu* in heaven and was known as *Alakananda.* Later on, some mischievousness of Lord *Indra* led to carnage and calcination of 60,000 odd sons (their mythical ancestors) of King *Sagar* by the Sage *Kapila.* It was *Bhagiratha*, great-great-grandson of *Sagar,* who conciliated Sage *Kapila* and was advised to bring *Ganga* to earth for revivification of his ancestors. His rigorous asceticism placated heavenly Gods and they aided *Bhagiratha* in *Ganga's* decent. *En route, Bhagiratha* faced stiff challenges from heavenly mountains where *Ganga* lost her way and subsequently *Airabata* (the omnipotent mythical white elephant) came into her rescue using his tusks and thus *Bhagiratha* gave new lease of life to his forefathers.

14.3 The Ganges, Livelihood and Reality – An Overview

In reality, *Ganga* has its origin in The Garhwal Himalayas at an altitude of 4,100 m through liquefaction of G*angotri* glacier and curves its' way through the mountains in Uttaranchal and in Dev Prayag meets her twin sisters, *Mandakini* and *Alkananda*, wherefrom the conjoint stream gets its name as The Ganges. During its passage it cleanses the states of Uttar Pradesh, Bihar and West Bengal within Indian territory *en route* to Bay of Bengal, where it meets the sea and in conjunction with her myriad tributaries serves millions of people for their livelihood – livelihoods those are diverse in nature and form; livelihoods those are often indiscernible yet indispensable; livelihoods those have immense social significance yet economically trifling.

In its passage through the plains, the Ganges supports as much as 450 million of people – dependent on her water, directly or indirectly. Never in the history of civilization has a single river, of medium length, been of such colossal importance as the Ganges. Agriculture, of course, is the prime livelihood that is directly linked to the Ganges. In truth, the Gangetic alluvium is one of the most fertile tracts in the world apart from the fact that the Ganges with its irrigation canals is the major source of irrigation in the Gangetic plains. Two main canals from the river are the Upper *Ganga* Canal, which begins at *Hardwar*, and the Lower *Ganga* Canal, which begins at *Naraura*. Other forms of livelihood include fisheries, navigation, tourism, etc.

Continuing with livelihoods associated with *Ganga*, let us give a momentary look on two such livelihoods, fishery and idol making, which are of immense economic and social significance, respectively, yet each of them imperiled to some extent due to emerging problems of *Ganga*. An estimated 13 million people get directly livelihood out of the rivers in the entire *Ganga* basin upto an altitude of 1,200 m above MSL. As much as 161 different species of fishes are found in the middle and lower segments of *Ganga*. Unfortunately, the dam building spree along with augmented pollution in the river is steadily truncating the fish density putting a question mark on livelihood security of this group of people. The proportion of major carps in the fishery declined from 43.5 to 29 % by 1972–1976 and 13 % today [1]. Anadromous Hilsa (*Tenulosa ilisha*) is a traditional Bengalee delicacy and the fishermen used to profit notably during rainy season which is now on the wane, mostly on account of decreasing water flow particularly in the downstream to Farakka Barrage and partly due to decreased rainfall. Pollution in the river has also added to the woes by diminishing the Biological Oxygen Demand (BOD) and thereby decreasing the fish density.

The fact that Hindus are traditionally idol worshipers has opened another avenue of livelihood in idol making using the river mud. The lower segment of *Ganga* having high loads of fine silt and clay has let this group of people make easy procurement of primary raw material for their purpose. The mud is used over a structure of straw and bamboo to give shapes to various Gods and Goddesses of Hindu mythology. There is a well known place near Calcutta, namely *Kumartuli*, where many of these idol makers have congregated thereby rendering the place a unique identity in the world map. Although, these people are not really bothered by the gradual drying up of *Ganga* they are finding increased difficulty to collect quality clay from the river bed for the beds are often being loaded with all kinds of debris, waste materials and pollutants.

Apart from these there are other livelihoods like navigation and tourism. Since the Ganges does not have adequate number of bridges across it, the usual way of transportation across the river is the local boats or *ferry* which has given the opportunity to scores of people in generating an effective livelihood. Along the stretch there are numerous places having historical/social/religious/cultural importance on either bank of the river. This has opened an efficacious avenue for tourism and millions of people earn their livelihood through it.

These are obvious livelihoods and are discernible to all. But, there are livelihoods which, although being traditional, are often ignored yet invaluable for their social significance. Let us look into the interiors of such a livelihood – their lives, their times, social significance of their very existence, and the woes and threats they are facing with the passage of time.

14.3.1 A Traditional Living Out of Ganga

Bamboo is a forest product that is probably unfamiliar to none. But what most people are unfamiliar to is the fact that some very painstaking and fortitude toiling is involved in post harvest fortification of bamboo for its endured use. There is a group of people – not huge in number but traditionally present – residing on either bank in the lowest stretch of the Ganges (known as *Bhagirathi* or *Hoogly* river) in West Bengal who have embraced this post harvest processing of bamboo as their livelihood. These people basically congregate in particular places along the banks with their locale being familiar under the name of *Bansgola* (Bamboo warehouse). Places in the vicinity of Kolkata and its suburbs are these people's abodes. Collectively, number of people directly engaged in this profession may not be more than 20,000 but, considering the extremely broad spectrum of use of bamboo, they are indirectly instrumental in satisfying social, economic, cultural, even religious need of scores of people in the country, apart from the governmental need for production of paper & paper pulp for making currency notes, although the need for the latter is being argued at various fora.

Before we go into the lives of these people let us have a quick glance on bamboo – its post harvest preservation and end uses – to properly expound the service these people are rendering in a traditional and eco-friendly manner.

14.3.1.1 Bamboo Uses

Bamboo can hardly be regarded as 'poor man's wood considering its myriad uses in almost every sphere of life – be it urban, suburban, or village life. Every conceivable aspect of our life – social, economic, cultural, and even religious – requires bamboo in someway or the other. Bamboo means so much for our society that shortage in supply of durable bamboo can impair our everyday life.

14.3.1.2 Bamboo Post Harvest Management

A major drawback of bamboo is that it is not durable against wood degrading organisms. Bamboo, being a lingo-cellulosic material, is thus easily prone to biological degradation. The presence of large amounts of starch makes bamboo highly susceptible to attack by brown-rot fungi and powder-post beetles. Enhanced durability of bamboo, thus, requires post harvest treatment which can be mechanical, chemical or traditional. While fresh bamboo is treated by steeping, sap displacement, diffusion and Boucherie processes, dry bamboo is processed by either soaking in water soluble preservatives, hot-cold process or pressure treatments. Most of these processes are not cost-efficient. One technique that has been in existence in India and elsewhere, like Malyasia, Vietnam and African countries, is the traditional method of water soaking of bamboo in running water – a method that is cost-efficient as well as fairly effective in its purpose. Traditionally, bamboo is soaked in water for 90–100 days which causes starch to ooze out and get dissolved in water thereby making it less prone to wood-mite attack. Added to this, prolonged soaking in water causes the brown rot fungi like *Fusarium* spp., *Trichoderma* spp., *Aspergillus* spp., *Alternaria* spp., etc. to denature a portion of hemicellulose which prolongs the durability of bamboo further.

14.3.1.3 An Eco-Friendly Practice

Having had this enlightenment on bamboo let us now go back to where we left, i.e., the traditional eco-friendly livelihood out of *Ganga* on post harvest management of bamboo and have an introspection into their lives and times – their social subsistence, their economic constrictions, their physical afflictions, and the emerging problems those threaten their very existence. But before we proceed, it needs to be mentioned that whatever information or data that is given subsequently has been gathered through interviewing these people in person by the authors at different places in the vicinity of Kolkata and its suburbs possibly for the first time. The authors had to accept the facts on their face value for the reason that their life has not, probably, been articulated elsewhere, or at least, authors could not find any. There may be location-specific variations, in places other than those mentioned in this critique, in the lifestyle elsewhere in India where, and if, this particular profession is practised, but the overall structure should remain the same. It should also be taken note of that the views expressed and the inferences drawn in this article are authors' personal understanding of the prevailing state of affairs. Reader's discretion is invited in that this piece is basically a realization in form of a narrative – *a realization of the appalling antipathy by concerned corners toward the Ganges putting at risk livelihood of thousands, illustrated by the livelihood under discussion.*

As has been said earlier, these people congregate in one particular place and in West Bengal are found on the either bank of the Ganges in districts of 24-Paraganas (N), Howrah and Hoogly. Each place has around 10–20 traditional families who run the business and according to their financial capability each of them has control over specified area on the river banks and in the river water which they would use for post harvest management of bamboo. Intruding in others' area is regarded culpable and the guilty one is penalized according to their set protocols.

The division of labour in each *Gola* (warehouse/depot, Picture 14.1) has three separate groups where 2–3 men – owners and managers – manage the entire operations; one group – the transporters – engaged in transportation of bamboo

Picture 14.1 One typical *Bansgola* (Bamboo warehouse) (© Authors)

through rivers; and the labours who are engaged in the seasoning process as well as transportation of seasoned bamboo to the places of requirement. Let us have a close look on the living of these groups of people individually.

The transporters: The principal source of bamboo that these people operate with comes from the districts of Murshidabad, Midnapore and Nadia. Traditionally, around 10,000–15,000 bamboos are closely tied with rigid nylon ropes into a pattern known to them as *Chali* (roofing pattern; Picture 14.2). This they would tie with a boat that is manually rowed (for reducing transport cost) and embark on their voyage – and quite a voyage! One onward journey from Murshidabad via river *Jalangi* or *Churni* into *Ganga* to Kolkata takes as long as one and half month or more! They would adjust to the alternating tides and ebbs and anchor accordingly and favourable conditions prevailing they would start afresh. This particular way of transportation is also practised elsewhere in India and few other countries like Malaysia, Vietnam, Myanmar, Thailand, etc.

Presently, the group depending on the size of the *Chali* usually consists of 7–9 people out of which two are *Majhi* (Boatman) to guide the boat; one who owns the rope (for the ropes are mighty costly and owners more often than not prefers to hire the ropes instead of keeping one of their own); and the rest to manage the *Chali*. To them, '*Boat thy name is home*'. They live, they eat and drink, and they sleep, and they dream – dream for a better tomorrow – all in that same boat!

***Hurdles during the passage*:** Often, it is not a smooth sailing for them as they have to confront inclement weather frequently and even from human perpetrators known as *Jal Chor* (River pirates) who threatens them with their lives for money! Still they risk their

Picture 14.2 Bamboo culms tied into a *chali* (© Authors)

lives in doing this to provide two square meals to their families. These privations notwithstanding, there is emerging an added paradigm to their toil due to ill management of the Ganges resulting in augmented silting of river beds which often results in sand-beds flourishing above water level. Since the boat that guides the *chali* is manually rowed, often they fail to dodge the sand-bed effectively and the *chali* gets stuck in the beds and they have no other options than to manually tow the *chali* out of those sand-beds. This is easier said than done for the bamboos in the *chalis* are so tied as to just float smoothly in water and any kind of coercive force externally applied can unloose the bamboo culms out of the *chali*. Then they would tie them afresh which further prolongs their voyage. Increasing river pollution is another trouble that they are finding hard to cope up with. As they are often compelled to use river water for cooking, drinking and other routine purposes; polluted sewage effluents, toxic industrial runoffs, increasing amount of suspended particles in water, stomach-churning smell of human and animal corpses floating are rendering them increasingly disease prone.

***Hard labour and the earnings*:** Now, let us examine their earnings from such toil. The group is paid on the number of bamboo in the *chali*. For each bamboo they would receive presently Rs. 1.5–2.5, depending on the length of the bamboo. The total sum when divided equally among all the group members comes to a paltry sum of Rs. 2,000–3,000 per month which, considering the flagrant inflationary trend of all kinds of food materials and living accessories, is no less than ridiculous! So, what do you call it – a living or subsisting? A very melancholy KHOKON BALA (thus earning livelihood for 30 years) answers, "*We were living satisfactorily 15 years back, but this price hike and the emerging problems of Ganga is making our lives no less than hell. We are finding it hard to make ends meet.*"

Picture 14.3 The group of labours (© Authors)

The labours: In comparison to the transporters, the other group who puts in physical exertion – and the maximum actually –, i.e., the group of labours (Picture 14.3), in keeping this particular tradition of bamboo post harvesting alive, are a touch more solvent. It entails mentioning that a lion share of their physical exertion is not directly linked with the Ganges or the changes in it with time and hence a discussion on their livelihood may seem gratuitous, but the very fact that as a part of the whole system their presence is a *sine qua non* in maintaining this tradition justifies that a few lines be spoken on them here.

Most of the people of this group hail from the states of Bihar, Jharkhand and Uttar Pradesh while few of them only are Bengalees (due to well-known distaste of Bengalees to hard physical labour!). Except transportation through river, they do all the physical labouring needed in making the bamboo culms reach its end users. After cleansing the culms from mud, they dry those in vertical or horizontal stacks. Then as per requirement they load and take those to the customers in hand drawn carts. Few of their co-workers (and mostly Bengalees!) prefer cycle-van over hand drawn carts, but are vastly outnumbered by the latter. Most of the workers from outside states do not have a permanent shelter over their head. They eat, they rest, and they sleep in a *khatia* (a bamboo and rope cot) or whatever vantage position they can grab to (Picture 14.4). During daytime they usually have nominal work to do and they are used to spend the day by resting and sleeping. The actual toiling starts after midnight when they have to take the cart-load of bamboo culms to various places within the city and its outskirts for authorities prohibit their movement during daytime due to traffic. Often they have to traverse 20–40 km, from one end of city to the other, and mostly on bare feet! Cities being interspersed with multiple

Picture 14.4 Labours retiring places (© Authors)

overbridges, many of which are steeper than usual for space constriction, they cannot generate enough frictional force with the slippers they use in ascending up or descending down such bridges with a load of around one tonne or more! Often they are so sapped of their energies that they do not have the power to speak for an hour long when, engulfed in perspiration all over, they reach their destination. Now, what else task is Herculean if not this!

Now let us see their earnings from such an ordeal. They used to get 12–15 days – or nights to be precise – in a month to work and get paid accordingly, for beyond 15 days a month their physical endurance is stretched to the limit. They are paid on the number, length/weight of culms and the distance they carry, and on an average they would get between Rs. 4,000 and 6,000 per month presently. But then though they do not have to counter the emerging problems of *Ganga*, their occupation is more risky than the other group. Now and then, in the dead of the night, they are slain by an intoxicated truck driver driving in a devil-may-care manner or a heedless car. Hundreds have died in such 'accidents' in the past 10 years or so. They do not get any kind of security either for their owns or for their family left behind back home. So, why are you risking your life? One philosophical DHARMENDRA YADAV (a 52 year old man, originated in UP, in this profession for 22 years) returns in a remorse yet candid manner, "*We are born to give physical labour. We know that there is no one to take care of us but we have to take care of our family. Earlier it was good money and less risk but now the situation has changed. Yes, risks are there but life must go on.*"

The owners**:** Now the remaining group, i.e. the group of owners, who, although much more solvent than the previous groups, are much more apprehensive as well in regard of future prospects. Each *gola* is run by a separate family, while some of these families are in the same trade for as much as ten generations long. But unlike

Picture 14.5 'Office' of one *Gola* (© Authors)

a profitable business, their dealings are waning with dwindling water flow in the *Ganga* and pollution facets as well as the misanthropic attitude of government people. A good number of warehouses has ceased to run or has changed hands with the passage of time owing to non-sustainability.

***Sustainability and eco-friendliness*:** Each of the warehouses has one room roofed with mud tiles which they call their 'office' (Picture 14.5) from where they coordinate the proceedings. One of the principal uses of bamboo is in construction works which now has in many places been replaced with hollow iron poles and as such bamboo requirement as well as market price has not kept pace either with time or with population. The river transportation mode of bamboo is gradually calling it a day for the emerging problems of *Ganga* and as such they have no other option left than to shift to road transportation mode which is far more expensive. Still more, the authorities often harass and extort them with money in the name of *Ganga* pollution while the crux of the matter is that, pollution remaining a falsehood; this practice is actually eco-friendly – more than many other means of living associated with the Ganges.

The starch material secreting out of the bamboo culms, while in water, actually multiplies the phytoplankton density in the vicinity of the *chali* which augment fish density – a fact that has a direct and traditional bearing on fishermen's locale being adjacent to theirs' (It has to be mentioned that authors' keen observation along with these peoples' experiences form the basis of this fact. No reference to such works is however available).

Ironically, it is the escalating river pollution by other means which is causing them headache increasingly. Now and then, multiple staining chemicals are discharged into *Ganga* from the dyeing and other chemical factories located on the banks and as such the bamboo culms get stained thereby trimming down the market value. Apart from this the increasing density of detrimental bio-agents are biodegrading the culms more than required thereby decreasing the durability. So, why are you continuing with this profession? A dejected looking PANCHANAN HALDER (an elderly gentleman from a family continuing in this trade for seven generations running) presents an indifferent eye – an eye that has more apprehension than aspiration – and somewhat soliloquizes, "*I am just living on the tradition of my forefathers. What remains are only shadows from the past*".

14.4 Dwindling Water Flow and Increasing Pollution

So, there goes the chronicle of their life and livelihood – unbiased, unabridged and a bit emotion provoking, too! There are many a KHOKON or DHARMENDRA or PANCHANAN making a living of some kind out of the Ganges, and whose livelihoods are under threat, courtesy of inordinate insensitivities – social, regional and political – towards the Ganges, which the authors have tried to articulate, solemnly yet categorically, through their example. But, before emotions get better of senses, let us contemplate over the problems which are imperiling them. The crisis is basically two pronged – dwindling water flow and burgeoning water pollution in the Ganges. While the latter can well be coped with, the former has to be lived with unless the Government is careful enough to control indiscriminate construction of dams in the river course and diversion of water otherwise, particularly in the upstream, harmful effect due to global warming notwithstanding. Let us discuss these in brief and excogitate the means to attenuate the problems intended not only to revitalize the issue on bamboo post harvesting discussed above, but also to restore the livelihood security on various fronts along the entire stretch of flow.

14.4.1 Drying Up of Ganga

There are, actually, two separate facets behind the dwindling water flow in *Ganga*, both of which are anthropogenic – one affecting directly while the other indirectly. The indirect cause is global warming, most of which stemming from anthropogenic enrichment of Green House Gasses (GHGs) in the atmosphere inducing melting of glacial ice caps with glacier *Gangotri* being no exception. Earth's temperature is rising and could rise between 0.3 and 6.4 °C by 2100 depending on various resources utilization scenarios [2]. This could spell doom for glacial ice caps like *Gangotri* and others. In one study by Intergovernmental Panel on Climate Change [2] it was shown that tongue of the glacier has severely shortened in the past 200 years or so. In one report named "Gangotri glacier melting rapidly" the snowmelt run-off of the Gangotri glacier had been recorded at a huge 57.45 m^3 annually within just 5 years since 1999 [3]. According to Professor Syed Iqbal Hasnain, an expert on glaciers, the Hindukush-Himalayan-Tibetan glaciers can see a 43 % decrease in the glacial area by 2070 and a 75 % decrease by the end of the twenty-first century at the current rate. Although this augmented melting has increased the total water flux in the upper *Ganga* segment, the suspended sedimentation that the snowmelt run-off of the Gangotri glacier carries with it has increased to around 17.78 lakh tons a year which, apart from endangering some vitally important dams like Tehri Dam, is raising the river bed, causing enhanced lateral water flow through spill channels, and thereby reduces the flux through the main channel. Although dams are an effective and cheap source of energy, there is limit in number beyond which building of these through their direct influence can seriously impair the natural riverine ecosystem. In case of *Ganga*, these are actually the principal anthropogenic undoing which is decreasing the water flux in the middle and lower segments of *Ganga* thereby jeopardizing livelihood of millions. The northern states, and especially Uttarakhand, are on a dam building frenzy to harness hydro power. So much so, that the Upper *Ganga* System alone, including Bhagirathi and Alkananda Rivers and their tributaries, till their confluence at Dev Prayag, has more than 130 large and small hydro power dams either planned [4], commissioned, or under construction. This flawed, shoddy, biased, unacceptable and unprecedented cascade of dams has some serious negative implications for the fragile river system.

Continuing with dams, let us now concentrate on one particular dam in Farakka Barrage which has altogether altered the hydrologic cycle of entire Bengal basins. It is located at Farakka in Murshidabad district of West Bengal over *Ganga* and was commissioned in 1975 with a principal objective to resuscitate Kolkata (erstwhile Calcutta) Port. Calcutta being the epicenter of operations during British rule, construction of one port was planned near the city and got commissioned in 1890 without consideration of proper technical and hydraulic factors [5]. But the neo-tectonic movements during sixteenth to eighteenth century having had tilted the Bengal basin eastward, when *Ganga* got bifurcated into twin flows near Mithipur, 40 km south-east from Farakka barrage – Bhagirathi within the Indian territory and Padma in Bangladesh – the Bhagirathi was rendered into mostly a spill channel flow. Then one undoing, in building the Calcutta port itself, was endeavoured to counterbalance with another – in devising the Farakka Barrage – to divert enough water towards the Bhagirathi to flush out the sediment load that was impairing Calcutta port in regard of navigational aspects! But the arithmetic hydrology that worked in favour of the barrage was subsequently proved inadequate to bring about any positive impact either in flushing out the sediment load, or to share dry-season flow between the two countries for their mutual benefits, the very purposes for which it was conceived and thus given rise to an array of problems – in within and trans-border – like severe erosion in the upstream resulting in large scale devastation of natural and human capital, augmented river salinity rendering it less potable than ever, ethnic conflicts and serious tensions between neighboring countries, etc. Since then much water has flown through the Ganges but the problems and tensions persisted when in 1996, sense of international brotherhood prevailing, it was decided by the then Indian Government to divert an assured 33,000 cusec of water towards

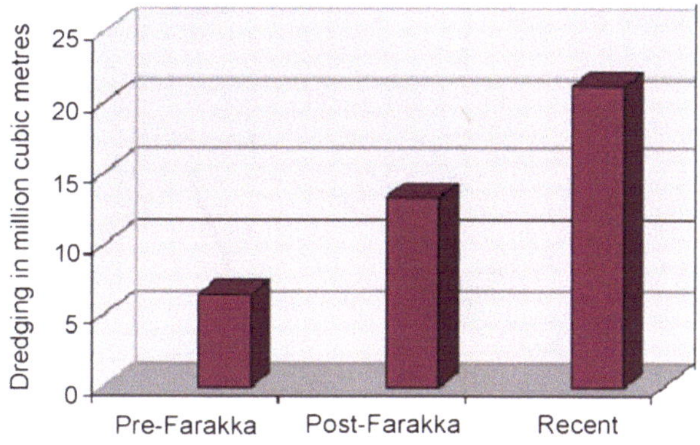

Fig. 14.1 Effect of Farakka barrage in augmented silting in the Ganges (Source: Rudra [6]; reproduced with permission)

the Padma during lean seasons between January and May. Neither of the countries eventually gets the required quota of their share due to the above direct and indirect influences, and for the Indian part in particular increased sedimentation along the Bhagirathi (Fig. 14.1) has been causing colossal erosion of the river banks, damaging the livelihood and eco-system, and finally threatening the very existence of the Calcutta Port.

14.4.2 Pollution in Ganga

In its different stretches the divine river *Ganga* encounters different challenges ranging from the deprived environmental flow in the upper stretches by virtue of plethora of dams to enormous pollution loads discharged by developing cities and industrial clusters in the middle and lower segments that has eventually transformed the sacred water to nearly sewage water in many places. According to the Central Pollution Control Board 42 % of the 12,690 km length of the river *Ganga* including its tributaries is moderately or severely polluted (BOD – biochemical oxygen demand greater than 3 mg l^{-1}) and hence unfit for bathing and drinking. This is almost 40 % of India's polluted riverine length. While *Ganga* accounts for 28 % share of India's total riverine length, its corresponding share of the total polluted riverine length is 36 % (Fig. 14.2) – the most by far.

In one study on monitoring of fecal coliforms in *Ganga* water, it was found that excepting Rishikesh all other monitoring stations has coliforms population in water much higher than that is suited for bathing or drinking purposes with monitoring stations in West Bengal being on the pinnacle (Fig. 14.3).

Apart from mass scale urbanization leading to burgeoning pollution, superstitions and religious significance of *Ganga* also adds on to it. Most of us, Indians with our pious mindset, do not seem to realize the value of this pristine river and as such the *Ganga* Action Plan (GAP) initiated by the Government is a failure in many places.

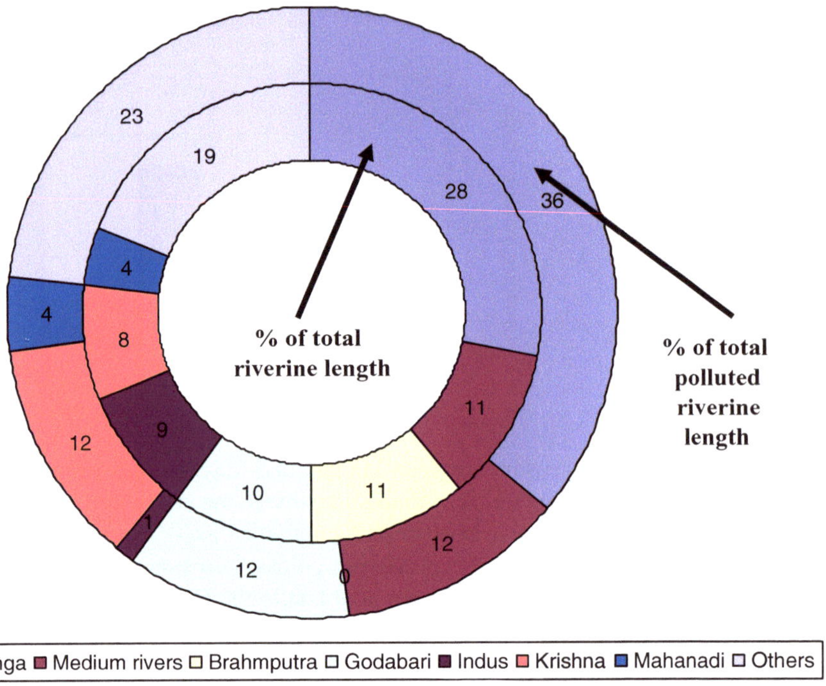

Fig. 14.2 Pollution in Indian rivers (Drawn based on data source: http://www.old.cseindia.org/misc/ganga/state_pollution.pdf)

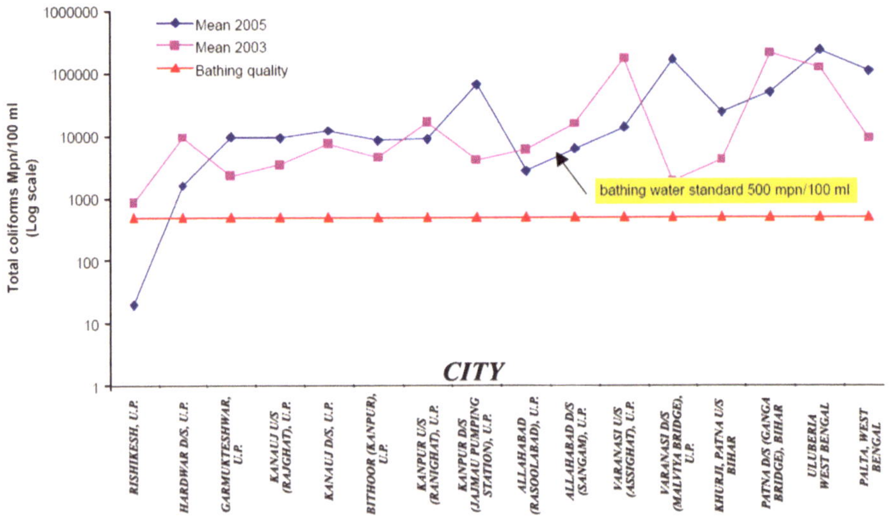

Fig. 14.3 Fecal coliforms in water at different monitoring places on Ganga (Source: http://www.old.cseindia.org/misc/ganga/state_pollution.pdf; reproduced with permission of R.C.Trivedi, CPCB)

14.4.3 Towards Attenuation of the Emerging Problems: Abating the Abuse

Sans mass scale realization of the enormous value of *Ganga*, its abuse cannot be abated. Any amount of time, energy and money spent in resolving the issue would have Achilles' heel without holistic appreciation of the true value of the Ganges by authorities as well as common people. Here authors would like to take the opportunity to suggest some measures for improved livelihood security which should be pondered upon by the appropriate authorities.

1. There is need for a study to regulate water flow through construction of structures and diversion of water at strategic points along the river systems upstream in order to ensure that minimal required water flows to and through the Farakka barrage.
2. Forthwith stopping of the proposed dams in the upstream for any future action plans for improvement will be futile if the upstream regulation is not viewed seriously. Hopefully, the government of India through a decision of an authority chaired by the Prime Minister has decided that no hydropower projects will be built on the initial 135 km stretch of *Ganga*.
3. The shortcomings in the planning and execution of the much-hyped Ganga Action Plan should be carefully studied.
4. In dealing with this issue a total *Ganga* Basin Management Plan has to be formulated taking the basin in its totality and micro level issues being at the heart of the planning with active participation of peoples.
5. Appropriate interventions may be developed for (i) higher productivity in agriculture, aquaculture, forestry, etc. under favourable soil and water conditions, and (ii) reduced hazard due to flooding of low lands and erosion of river banks [7].

14.5 Dénouement

Swami Vivekananda used to preach, "*Gita and Ganga constitute the essence of Hinduism; one in theory and the other in practice*." Indeed. *Ganga* is so inextricably woven into the social, economic, cultural and religious fabric of Indians and penetrates so deep into common Indian psyche that often our practices on or with *Ganga* turns into malpractices, ours' remaining oblivious and ignorant. *Ganga* is a river probably second to none on this planet which means so much for the life and livelihood of people around it. *Ganga* is the FACE of India and Indians' salvation lies in enlivening the FACE. The apparent antipathy of concerned authorities in conjunction with absence of realpolitik measures and appropriate laws is slowly, but surely, assisting *Ganga* into the oblivion. THEY are continuing to devise faulty plans to solve the issue in piecemeal rather than holistic manner and without participation of the people living around it. THEY are even tormenting their grey cells in chalking elaborate – and expensive as well – plans to decorate *Ganga* and her banks to beautify her and give a facelift to the cities without realizing the defacement of nature!

Bhagiratha had his salvation in resuscitating his ancestors by carrying *Ganga* to earth, while WE, ***Bhagiratha***s of modern times, bear the accountability of regulating steady passage of ***Ganga*** so that our descendants may continue to live. ***Indra***s are bound to be there imperiling the right of millions to live, but WE have to defeat their vicious intentions and invoke ***Airabata***s, wherever necessary, to make HER way. It is time that we must realize that THE MOTHER is dying. Listen carefully to her sighs! SHE is groaning for breathing space and so are the KHOKONs or DHARMENDRAs or PANCHANANs and thousands like them living out of *Ganga*. SO, GIVE EAR!

References

1. Payne AI, Sinha R, Singh HR, Huq S (2002) A review of the Ganges basin: its fish and fisheries. www.fao.orgdocrep007ad525ead525e0f.htm
2. IPCC (2007) In: Parry ML, Canziani OF, Palutikof JP, van der Linden PJ, Hanson CE (eds) Impacts, adaptation and vulnerability. Contribution of working group II to the fourth assessment report of the intergovernmental panel on climate change. Cambridge University Press, Cambridge, pp 1000
3. Knapp S (2010) Disappearance of the Ganga. http://www.stephen-knapp.com/disappearance_of_the_ganges.htm
4. Anonymous (2011) Dams, rivers & people. www.sandrp.in/drp/June_July_Aug_2011.pdf
5. Sinha A (1986) A victim of political indecision. In: Sarma SS (ed) Farakka – a gordian knot. Ishika, Kolkata
6. Rudra K (2006) Report, centre for development and environment policy. Indian Institute of Management, Kolkata, pp 59
7. Sen HS (2010) The drying up of River Ganga: an issue of common concern to both India and Bangladesh. Curr Sci 99(6):725–727

Another usage of traditional boats, i.e. ferrying the construction material, in places where the bridges are non-existent or far off. Crushed stones loaded near Rajmahal (Jharkhand) for transport down to Kolkata (Courtesy Julian Crandall Hollick)

Chapter 15
E-Flows Related Livelihood in the Ganga River: A Case Study of Tourism

Murali Prasad Panta

Abstract River Ganga contributes to livelihood in India. She is personified as a Goddess and holds an important place culturally, among Hindus. According to Hindu mythology, it is believed that bathing in the river water erases sins and facilitates the attainment of salvation or nirvana. Similarly, immersion of ashes of deceased ones in the river water not only keeps the soul of deceased in peace but places it in heaven. The River is also known for unique landscape and habitat for species which promote livelihood. Over the time period, exploitation of river resources by the inhabitants has imbalanced the flow related livelihood. Moreover, the river resources at certain stretches are at the verge of extinction. The proposed study focuses on the relationship between the river resources and livelihood. The study reveals that the Mega Pilgrimage Sites in the states of Uttarakhand and Uttar Pradesh, during 2006–2010, have attracted 236.78 million pilgrims and generated revenue of Rs. 98,921.07 crores. The empirical study also provides suggestions to increase environmental flows (e-flows) in the river to promote livelihood. The primary survey indicates that the present flow conditions for tourism in Uttarakhand (Haridwar-Rishikesh) is 'Maximum Sustainable Yield (MSY)' and in Uttar Pradesh (Kachhla Ghat-Bithoor) is 'Livelihood activities are not economically feasible'. Hence, the study suggests that MSY shall be maintained in the desirable state to preserve river resources and promote livelihood.

Keywords River Resources • Livelihood • E-flows

M.P. Panta (✉)
Department of Humanities and Social Sciences, Indian Institute of Technology
Kanpur, Kanpur, 208016, UP, India
e-mail: pmprasad@iitk.ac.in

R. Sanghi (ed.), *Our National River Ganga: Lifeline of Millions*,
DOI 10.1007/978-3-319-00530-0_15, © Springer International Publishing Switzerland 2014

15.1 Introduction

The write up is divided into five parts. Part-I refers an over view of livelihood, Part-II deals with livelihood specific analysis on Tourism, Part-III examines the e-flow analysis and Part-IV A short film and Part-V summary and conclusion.

15.2 Livelihoods – An Overview

The Ganga (or Ganges), the national river of India, contributes to the social capital, manmade capital and natural capital. The river resources provide livelihood for approximately 500 million Indians. Hence, the wellbeing of riparian community is dependent on the health of river ecosystem. Similarly, we assume that the stakeholders' environmental consciousness will preserve river resources.

The Ganga has traditionally been regarded to be an inexhaustible gift of nature, and also the symbol of Hindu values. There are numerous pilgrimage sites, bathing Ghats, ashrams and religious trusts situated on the banks of river Ganga. According to the Hindu mythology, it is believed that a holy dip in the river liberates one from all the sins and facilitates the attainment of salvation or *nirvana*. Hindus immerse ashes of their deceased one's in the river waters. They believe that the deceased one's soul will be sent to heaven or release from the mundane cycle of life, that is, death and rebirth. In addition, the river water is considered as *teerth* (holy water) and it is believed that sip a drop of it will cure diseases. The Hindus' carry the holy water to their homes for performing special *poojas* such as the *Sathyanarayan Vratam*, and is also offered to a dying person to obtain *moksha* (liberation from the cycle of rebirth).

The Ganga also promotes livelihood activities pertain to beach camping, water rafting, ferry, fisheries, mangroves, water falls, bird sanctuaries, landscapes, and medicinal plants. The river water is used for domestic, industrial, irrigation and power (hydro, thermal & nuclear energy) generation activities. In addition, she needs a few resources for herself too for her survival. The relationship between the river resource and livelihood can be explained with the following Chart 15.1.

The below Chart 15.1 Ganga River Related Livelihood indicates that there are numerous livelihood avenues promoted by the river for peoples' health and wealth. If it is so, why is there apathy among the people to preserve river resources?

15.3 Livelihood Analysis: Tourism

The Ganga with its characteristics of culturalism, predominantly promotes tourism-related livelihood. Tourism activities on river Ganga are predominantly nature-based, and can be broadly classified into religious tourism and other forms (recreational/adventurous/eco-friendly) of river tourism. The pilgrimage sites and recreational spots provide incentives to the tourists to rejuvenate their physical and spirituality

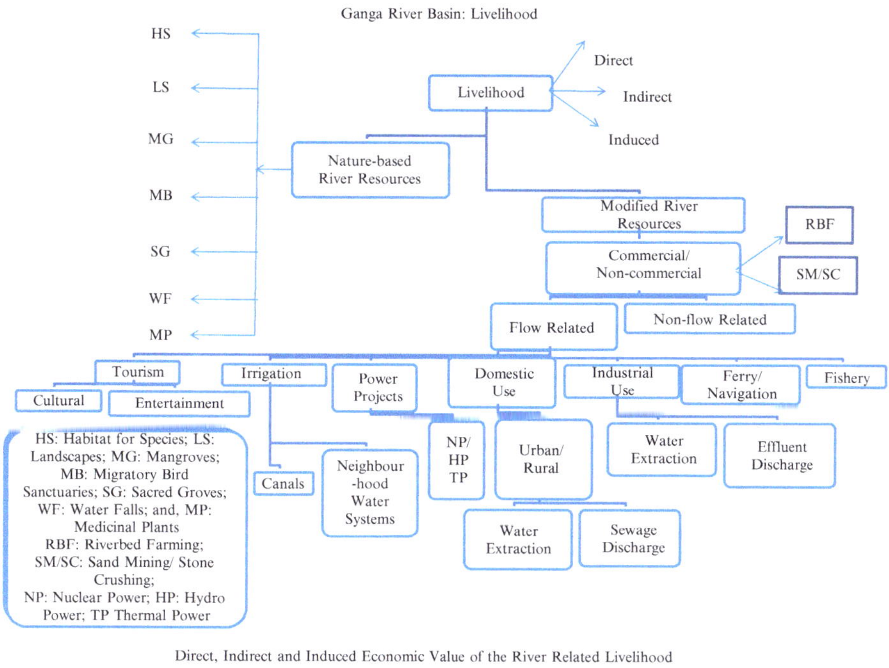

Chart 15.1 Ganga river related livelihood. Direct, Indirect and Induced Economic Value of the River Related Livelihood

which enhances their confidence levels to improve the quality of life. Moreover, economic theory states that a rational consumer (tourist) expects equal satisfaction against his expenditure. Hence, we assume that the amount of money spent by the tourist implies that he will obtain equal worth of satisfaction/benefit. Accordingly, the total benefits of the tourists indicate the economic value of resources at the site.

The river Ganga possess unique characteristics namely, International River, Interstate River, Gangatva, and promotes livelihoods. The characteristics of the river indicate that the river resources are ‘quasi-public good’ in nature. This has intensified the problem of survival of the river, with or without state intervention. The users of the river resources perceive that they are there to use and not for preserve river resources. The myopic behaviour of the users enhances the marginal cost of their efforts to harvest the river resources for livelihood. Moreover, the users of the river resources perceive that it is the states’ responsibility to promote livelihood and preserve resources.

The theory of renewable resources [1] indicates that the gap between demand for and supply of river resources like fisheries may impel the resources into critical depensation. This can be explained with the help of following graph.

The Graph 15.1 Growth of Renewable Resources indicates that, at X (max), the resources are at maximum and no more will be added. The reason for this is that

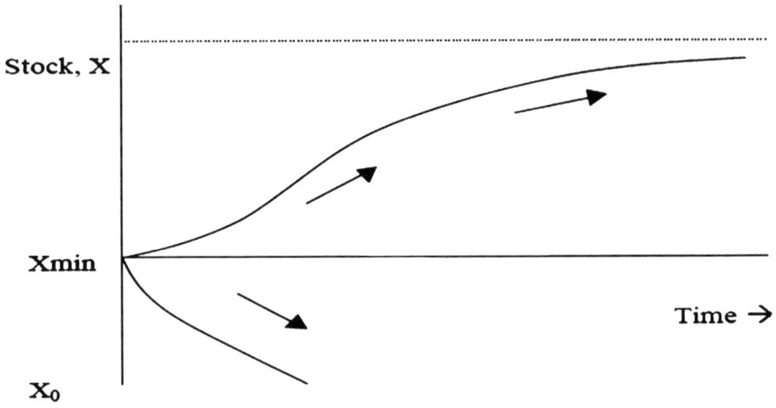

Graph 15.1 Growth of renewable resources (Source: David Pearce work on RRs Tourism data (15.1): Compilation based on the information received from the Department of Tourism, Government of U.P. & U.K.)

resources will reach its upper carrying capacity. However, at X (min) there is a real risk that the resources are at the verge of extinction. Finally, at X_0 resources are completely vanished. When the resources fall below the X min then that would be called as critical depensation. Hence, the harvest of river resources should be without depleting the stock of resources (i.e. for Maximum Sustainable Yield-MSY). However, the MSY is unlikely to be equal to Economically Sustainable Yield (ESY). This can be illustrated with the following diagram.

The Graph 15.2: Pattern of Renewable Resources Use indicates that the more effort, E, there is, the lower the stock. If there was no effort, the stock would be maximised at Xmax. If there is a massive effort and all the fish are caught, the stock would go to X_0. Effort would be measured by something like the number of trawlers weighted by their engine capacity, or number of boats, or number of fishermen. Fish in the river resources are free of cost. However, 'Effort' is clearly an input and if we multiply effort by the price of effort, we get cost. If we multiply output by the price of the fish sold to the market we will have total revenue.

The diagram also reveals that Private Property Regime (PPR) may provide incentives to the owner of resources to maximize their livelihood. However, in reality most of the river resources have scantily defined Property Rights. Hence, study would seek to find out the psychological behavior of users of river resource. It is obvious that as long as there are profits in the activity involved then each user makes more and more efforts for their better off. Consequently, a large number of users exploit the river resources for their greedy. In given such conditions, it is very hard to find out that users have any incentives to stop the over use of river resources at ESY. Therefore, users will continue to exploit the resources until total revenue (TR) gets equal to total cost (TC). The point below the intersection of TR and TC will represent as negative returns to the users of resources. In case, they exploits the resources further with the help of government subsidies may push the resources to critical depensation

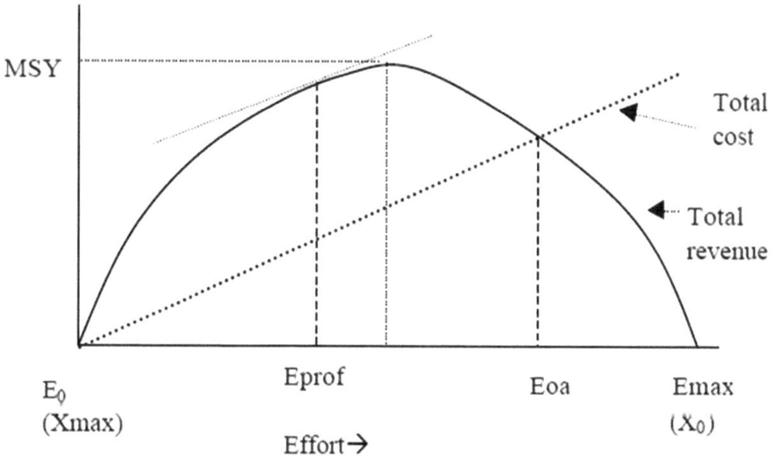

Graph 15.2 Pattern of renewable resources use (Source: David Pearce work on RRs Revenue data (15.2) is based on the field survey conducted by the researcher in a study on "Livelihood" as part of the WWF-India and HSCB study on "Living Ganga Programme for E-Flows Assessment")

People are depending upon the river resources for their livelihoods and the river in turn is depending upon the people for preservation of resources in a mutual dependence framework. However, the conduct of their transaction will depend upon the degree of assets that can be redeployed outside the relationship between the people and river. In this context, one can ask:

Whavt are the major economic, socio-cultural, institutional, and environmental factors that influence stakeholder's behavior towards the river? What are the most effective ways to gain community participation towards conservation of the river? What are the requirements and expectations of various stakeholder groups from the river Ganga resources? What kind of intervention do citizens anticipate from the state to sustain environmental flows in river Ganga?

We will work, *tentatively*, on the following prepositions to promote the relationships between people and river:

1. River resources as Common Property Resources (CPRs) and the users shall create their criteria for promotion of livelihood and preservation of resources;
2. Market negotiations with fully specified contract among user of the river resources; and
3. State intervention with less abusiveness [2].

The paper examines the generation, utilization and preservation of Ganga river resources in ideal (past), present and desirable (future) conditions to promote livelihood. In order to estimate the value of river resources in terms of tourism there is a need for the data on cost of the tourist (per trip). As a matter of fact, we do not have ready-made data on the cost of tourist at pilgrimage sites. The acquired information has been analyzed with the help of theoretical model of renewable resources, environmental evaluation (revealed and stated preference) methods and the *Building Block Methodology* (BBM) for estimation of e-flow based livelihood's economic value.

Secondary data is obtained from the Ministry of Environment and Forest, The Ministry of Water Resources, The Ministry of Tourism, Parliament Library, and few academic institutions of Kolkata, Varanasi, and Patna. The secondary data contributed to the information on the historical importance of River Ganga, and the status of river resources in the past and the present. We have collected primary data based on field survey in and around *Haridwar- Rishikesh* (UK) and in *Bhithoor-Kanpur* (UP) stretches of Ganga. The river stretch in UK predominantly promotes cultural and entertainment livelihoods. The field survey was conducted in and around Haridwar (including Rishikesh) on three occasions i.e., During 26th September-2nd October, 2009, 14th January, 2010, 2nd–6th March, 2010 in order to obtain opinion of the respondents about the status of the river Ganga on generation, preservation and utilization of resources. The researcher has interviewed 235 respondents based on open ended (structured) questionnaire.

The *Kumbh Mela* is celebrated in accordance with the Indian astrological calculations. It is held at a range of intervals. The dates are calculated in advance with respect to the zodiac position of the Sun, Moon and Jupiter at the various pilgrimage sites. The regular *Kumbh Mela* takes place once in 3 years. The site of *Kumbh Mela* observance rotates between four pilgrimage destinations of Haridwar (in Uttaranchal, on the banks of river Ganga), Allahabad (in Uttar Pradesh, on the banks of river Ganga), Nasik (in Maharashtra, on the banks of *Godavari*) and Ujjain (in Madhya Pradesh, on the banks of river *Shipra*). The *Purna* (full) *Kumbh Mela* takes place in every 12 years, whereas the *Adrh* (half) *Kumbh Mela* occurs in 6 years. The *Maha Kumbh Mela* (held at Allahabad) is the most spectacular fair and occurs after 12 *Purna Kumbh Melas* or 144 years. The *Kumbh Mela* also enhances culture-based tourism at the various *Kumbh Mela* destinations, particularly on the days following the fair.

The Ganga's uncommon environmental diversity, (with impressive scenery, and outstanding possibilities for adventurous and wilderness experience) have a strong influence on the length and quality of recreational tourism. The river offers some of the best and safest rafting opportunities such as bump, swirl and ride from grade I to V rapids. Riverside camping and water rafting (one of the most popular adventure sports) are the fastest growing tourism segments on the river Ganga.

The tourism analysis is limited to the state of Uttarkhand (UK) and Uttar Pradesh (UP) due to the paucity of data on religious tourism in Bihar and West Bengal. The total number of tourists (cultural and recreational) includes both domestic and international tourists. However, the official statistics accounts only up to 40 % of the total tourists' volume. The analysis also estimated the economic value of the tourist visitation at the pilgrimage sites during 2006–2010.

In the states of Uttarakhand and Uttar Pradesh, 8 and 11 pilgrimage sites are located on the banks of the Ganga were considered for the analysis. The majority of the sites in Uttarakhand are within the purview of ecologically sensitive zones. The following table explains the number of pilgrims and total economic value generated by cultural/entertainment tourism at these sites.

The Table 15.1 Pilgrimage Sites in UK and UP: Pilgrims and Revenue indicates that the total number of tourists in UK and UP were 81.47 and 187.43 million respectively, during 2006–2010. In the state of UK, Haridwar (including Rishikesh) has received the maximum number of tourists that is 81.59 % of the total number of

Table 15.1 Pilgrimage sites in UK and UP: pilgrims and revenue

Pilgrimage sites	Pilgrims (millions)						Field survey		Revenue at the Mega Pilgrimage Sites (Rs. in millions)					
	2006	2007	2008	2009	2010	Total	Respondents	Expend./benefit (pppt in Rs.)	2006	2007	2008	2009	2010	Total
Uttarakhand														
Haridwar	10.15	11.78	11.93	12.66	19.97	**66.48**	235	4,177.50	42,392.64	49,223.71	49,840.29	52,870.17	83,405.43	**277,732.24**
Pauri	0.08	0.08	0.09	0.09	0.10	**0.44**								
Srinagar Garhwal	0.20	0.22	0.20	0.25	0.22	**1.10**								
Rudra prayag	0.50	0.59	0.59	0.42	0.40	**2.51**								
Gopeshwar	0.23	0.28	0.39	0.40	0.38	**1.69**								
Tehri	0.83	0.87	1.08	0.89	0.91	**4.58**								
Uttarkashi	0.51	0.62	0.66	0.68	0.60	**3.07**								
Gangotri	0.26	0.33	0.33	0.38	0.31	**1.61**								
Total	**12.77**	**14.78**	**15.27**	**15.77**	**22.89**	**81.47**								
Uttar Pradesh														
Garhmukteshwar	0.81	0.89	0.98	1.03	1.10	**4.82**								
Bijnor	0.03	0.04	0.04	0.05	0.05	**0.22**								
Narora	0.63	0.70	0.77	0.80	0.85	**3.75**								
Kannauj	0.07	0.07	0.08	0.08	0.08	**0.39**								
Bithor	0.10	0.10	0.11	0.13	0.14	**0.59**								
Kanpur	0.22	0.23	0.25	0.26	0.31	**1.27**								
Allahabad	17.90	20.14	23.79	28.43	30.38	**120.63**			74,769.28	84,135.47	99,376.31	118,764.77	126,899.04	**503,944.87**
Mirzapur	0.71	0.71	0.74	0.74	0.74	**3.64**								
Vindyachal	5.95	6.06	6.08	6.10	6.11	**30.30**			24,868.66	25,308.97	25,413.21	25,464.70	25,514.16	**126,574.70**
Varanasi	3.36	3.68	3.89	4.09	4.36	**19.37**			14,025.30	15,360.48	16,259.69	17,075.16	18,208.36	**80,928.98**
Gahamar	0.49	0.53	0.55	0.58	0.60	**2.75**								
Total	**30.27**	**33.14**	**37.30**	**42.30**	**44.73**	**187.73**			**113,663.23**	**124,804.91**	**141,054.21**	**161,304.63**	**170,621.56**	**711,448.54**

Source: Data compiled based on the reports of the Department of Tourism: Tourism Statistics, Department of Tourism, Governments of Uttarakhand and Uttar Pradesh

Note: According to concerned authorities in the Department of Tourism, Uttarakhand, the registered pilgrims account for 40 % of the total pilgrims visitation

Haridwar (includes Rishikesh), Rudraprayag(excluding Kedamath), Uttarkashi(excluding Gangotri and Yamunotri), Kanpur (urban & rural), Vindyachal (includes Astabhuja)

Field survey during 26th September- 2nd October, 2009, 14th January, 2010, 2nd–6th, March, 2010

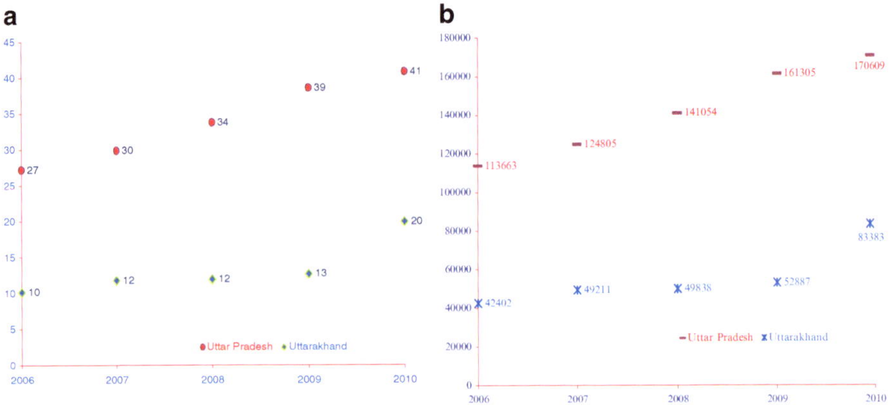

Fig. 15.1 (**a**) and (**b**) Tourism and revenue generation at the Mega Pilgrimage Sites

tourists. The number of international tourists accounts for 0.26 % of the total number of tourists. Similarly, the pilgrimage sites Allahabad, Vindhyachal (including Astabhuja) and Varanasi have received maximum number of tourists i.e., 90.72 % of the total number of tourists. Moreover, Allahabad, Vindhyachal and Varanasi account for 70.83, 17.79 and 11.37 % (of the 90.72 %), respectively. The combined share of international tourist arrivals at these sites is 0.72 %. The pilgrimage sites at Haridwar (including Rishikesh), Allahabad, Vindhyachal (including Astabhuja) and Varanasi are denoted as Mega Pilgrimage Sites (MPS) in UK and UP.

The inference based field survey helped us to estimate the average cost per pilgrim per trip. Accordingly the expenditure/benefit per pilgrim per trip is Rs. 4,177.50. We have applied the cost/benefit of the pilgrim to the number of tourists visited at the Mega Pilgrimage sites in order to calculate the economic value of the sites. The tourist visitation and revenue generation trends at these sites are explained using Fig. 15.1a, b.

It is evident from Fig. 15.1a, b: Tourism and Revenue Generation in the Mega Pilgrimage Sites reveals that both tourist volume and tourism revenue was remarkably high at the Mega Pilgrimage Site (Haridwar) in UK, 2010. This increase could be due to the *Kumbh Mela* held in 2010. The Site has attracted 66.48 million pilgrims and has generated revenue of Rs. 27,773.22 cores during 2006–2010. Similarly, the sites in UP namely Allahabad, Vindhyachal and Varanasi together have attracted 170.30 million pilgrims and have generated revenue of Rs. 71,144.85 crores. In other words, the MPS in the states of UK and UP have attracted 236 million pilgrims and have generated Rs. 98,921.07 cores during 2006–2010.

It is apparent from the tourism trend analysis that pilgrims' arrival and pilgrimage earnings are much significant at the MPS. However, analyzing the number of employment generated, nature of jobs, pay scale, local entrepreneurship in tourism, and so on will provide further insights on tourism-related livelihoods, and its impact on the sustainability of the Ganga. Mathieson and Wall [3] suggests that tourism engenders three types of employment opportunities: **direct,**

indirect and **induced**. Jobs (in establishments) that cater mainly to the tourists are referred to as direct employment. For instance, in the Ganga river basin job opportunities created in the accommodation establishments serving the tourists, river rafting companies, travel agencies, and souvenir shops can be treated as direct employment. Indirect employments are those jobs which are at times dependent on tourism. An example could be a fish dealer in the Ganga river basin who supplies fish occasionally to the restaurants. The additional job opportunities resulting from multiplier effects are termed induced jobs. In other words, induced employments are generated when direct and indirect tourism employees spend their wages in other economic sectors. For instance, a hotel manager spending salary to construct a house generates induced employments in the construction sector. The total tourism employment generated in a destination is the sum of direct, indirect and induced employments. The tourism analysis clearly indicates that the river promotes livelihood. However, the people who depend up on the river resources for their livelihood will really protect the river characteristics? We will assess the status of the river resources in UK and UP with the help of e-flow objectives.

15.4 E-Flows

> *Environmental Flows describe the quantity, timing, and quality of water flows required to sustain freshwater and estuarine ecosystems and the human livelihood and well-being that depend on these ecosystems. [4]*

E-Flows are required for:

1. Maintaining river regimes
2. Making it possible for the river to purify itself
3. Maintaining aquatic biodiversity
4. Recharging groundwater
5. Supporting livelihood
6. Allowing the river to play its role in cultural and spiritual lives of people.

The Building Block Methodology (BBM) has been considered to be one of the most advanced and refined methodology. The methodology for assessment of E-Flows can be explained with the help of following chart:

Chart 15.2: Building Block Methodology provides guide lines to carry out the study on Ganga River Resources based livelihood. The study has developed a method on environmental flow conditions with the help of Building Block Methodology (BBM) [5] and theory of renewable resources. The conditions are as follows:

A: Ideal State where resources are abundant;
B: Economically Sustainable Yield (ESY)

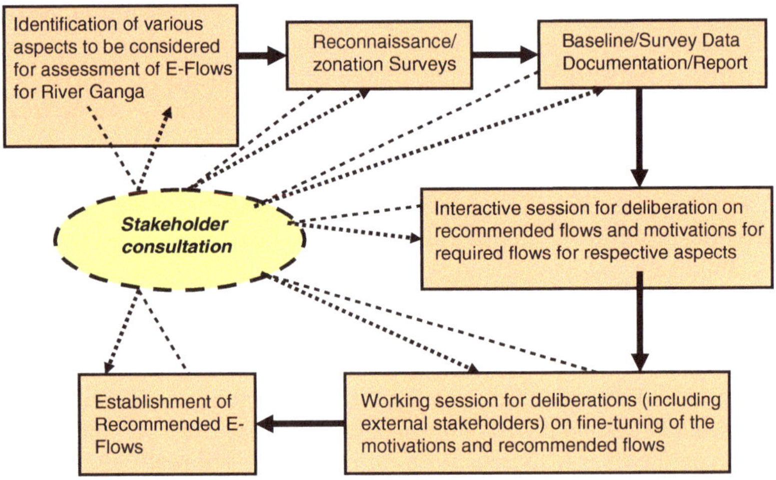

Chart 15.2 Building Block Methodology (Source: WWF Document Ganagapedia)

C: Maximum Sustainable Yield (MSY)
D: Total Revenue = Total Cost (TR = TC)
E: Economically not feasible; and
F: Critical Depensation (Extinction).

The field survey on livelihood in the upper, middle and lower stretches with the application of BBM provides us an opportunity to set the e-flow objectives. This can be explained with the help of the following Table 15.2.

The Table 15.2 Environmental Flow Object vies indicated that the conditions of the river related livelihood in the present and desirable conditions. It also emphasizes on collection of data from the stakeholders. The field survey observations indicate that the inundation of Ghats during floods hampers the activity of cultural tourism. Similarly, the torrents in the river as well as prohibitory order from the state reduce the activities of river rafting and beach camping. In addition, the prevalence of drought (lean e-flows) conditions too has its impact on livelihood activities such as water rafting and beach camping. In other words, the imbalance in e-flows that is over flows (floods) and lean flows (drought) limits the livelihood activities. The status of river resources in the present and in desirable condition will be explained with the help of following Table 15.3.

The Table 15.3 Status of E-flows in UK and UP stretches of Ganga River reveals that, given the data, the study has come up with e-flows recommendations for the three sites i.e., Kaudiyala (UK), Kachhla Ghat and Bithoor (UP). The proposed tentative e-flows were estimated with the help of Biodiversity, Geomorphology, Hydraulics and Hydrology groups. The recommended e-flows are based on the flow requirements of livelihood for different scenarios can be explained with the help of the following table.

The Table 15.4 Recommended E-flows in UK and UP Stretches of Ganga River indicates that the suggested e-flow of water resources during maintenance year and the drought year will provide sustainable livelihood. However, the imbalances in

Table 15.2 Environmental flow objectives

	Zone I	Zone II	Zone III
	Upper stretch (Gangotri to Haridwar)	Middle stretch (Haridwar to Farrukhabad)	Lower stretch Farrukhabad to Kanpur)
Present Livelihood Conditions (PLC) EFC	PLC- EFC: C	PLC EFC: E	PLC EFC: E
Vision or Desired Future State EFC	DFS- EFC: C	DFS EFC: C	DFS EFC: C
Desired Future State (DFS)	The environmental flow of the river Ganga should provide resources to meet the sustainable livelihood practices	The environmental flow of the river Ganga should provide resources to meet the sustainable livelihood practices	The environmental flow of the river Ganga should provide resources to meet the sustainable livelihood practices
General Flow Objectives (GFO)	Sufficient flow and quality of water to carryout sustainable livelihood practices (MSY)	Sufficient flow and quality of water to enable to carryout sustainable livelihood practices (MSY)	Sufficient flow and quality of water to carryout sustainable livelihood practices (MSY)
Component Objectives (CO)	Flows in terms of: cultural tourism, entertainment (water sports/adventure) tourism, Ferry, and Fishery	Flows in terms of: fishery, waterways, and tourism	Flows in terms of: fishery, waterways, and tourism
Indicators of threshold s of potential concerns (TPC)	Total revenue = Total cost Tourism (cultural and entertainment); ferry services	Total revenue = Total cost Seasonal fish catch; tourism in relation to cultural and spiritual events; ferry services	Total revenue = Total cost Seasonal fish catch; tourism in relation to cultural and spiritual events; ferry services
Data requirements	Collect the information on river based livelihood in the past and present	Collect the information on river based livelihood in the past and present	Collect the information on river based livelihood in the past and present

Table 15.3 Status of E-flows in UK and UP stretches of Ganga River

Flow related livelihood	Purpose of E-flows	UK Present status	UK Desirable status	UP Present status	UP Desirable status
Cultural tourists	Carrying holy water to home Drinking (as Teerth) Holy dip Immersion of ashes	C	C	E	C
Entertainment tourists	Beach camping White-water rafting	C	C		

Table 15.4 Recommended E-flows in UK and UP stretches of Ganga River

		Maintenance year		Drought year	
Stretch	Flows	January (Driest)	August (Wettest)	January (Driest)	August (Wettest)
UK (Kaudiyala)	Discharge in Cumecs	305	3,250	290	2,702
	Depth in meters	8.70	21.05	8.28	19.84
	Average velocity in m/s	0.58	1.15	0.57	1.08
UP (Kachhla Ghat)	Discharge in Cumecs	250	601	120	500
	Depth in meters	3.31	4.50	2.68	4.40
	Average velocity in m/s	0.36	0.34	0.29	0.45
UP (Bithoor)	Discharge in Cumecs	330	1,500	151	431
	Depth in meters	3.81	6.13	2.89	3.90
	Average velocity in m/s	1.03	1.51	0.82	1.11

e-flows i.e. over flows (floods) and lean flows (drought) will impose implications on livelihood. Hence, there is a need for management of sustainable e-flows to preserve river resource and promote livelihood.

15.5 A Short Film on River Based Livelihood

We intended to capture the respondents' opinion on the pattern of river resources and livelihood. Moreover, the unique attributes of river Ganga and the actions of the users of the river resources were captured in the film. This will help the viewers to interrelate the theory with the real world situation on the relationship between livelihood and e-flows. The film is based on the interaction with respondents in and around Haridwar-Rishikesh during September-October 2009 and March 2010. The film (1.83 GB) will be available on demand.

15.6 Summary and Conclusion

The Ganga River provides cultural, livelihood and ecological wealth to Indians. However, the status of river resources at present motivated us to assess the economic value of livelihood. This will make a case for preservation of the resources. The study has focused on livelihood namely, tourism. In order to estimate the economic value of tourism we need information about the cost/benefit of river based pilgrimage. There is no readymade data available so we have conducted a field survey in and around Haridwar-Rishikesh in UK and Kachhla Ghat-Bithoor in UP. The field survey indicates that on an average per pilgrim spends 4,177.50 per trip (2 days) to visit Haridwar-Rishikesh. We have applied the average cost/benefit to the available pilgrimage data on Allahabad, Vindhyachal and Varanasi in UP. The Mega Pilgrimage Sites namely Haridwar, Allahabad, Vindyachal and Varanasi all together have attracted 88.05 % of the total pilgrims that is 268.90 millions and generated revenue of Rs. 98,921.07 crores during 2006–2010.

The survey also analyzed the status of river resources (e-flows) with the help of Building Block Methodology. The approach of e-flows provides us information about the present status of river resources and promotion of livelihood in UK and UP. The survey reveals that the present flow conditions for tourism in Uttarakhand (Haridwar-Rishikesh) is 'Maximum Sustainable Yield (MSY)' and in Uttar Pradesh (Kachhla Ghat-Bithoor) is 'Livelihood activities are not economically feasible'. Hence, the study suggests that MSY shall be maintained in the desirable state to preserve river resources and promote livelihood. It also helps to set the desirable flow conditions (tentative) to promote livelihood and preserve river resources.

Acknowledgements Author express special thanks World Wide Fund – India (WWF) for its support to carry out a project on "Livelihood Objectives for Maintaining Sustainable Environmental Flows", Living Ganga Programme. Grateful thanks are due to Prof. V Tare, Coordinator, GRBMP (IITs Consortium) MoEF, GoI. Thanks are also due to project associates at IITK and IITM. Special thanks to the students of ECO 341: Environmental Economics and Policy (Y6 and Y7 batches: M.Sc. Economics Programme and Students of ECO 747: Environmental Economics, Legislations and Social Impact (M. Tech: Environmental Engineering and Management Programme) who have carried out field work in Haridwar-Rishikesh along with the Project Associate.

References

1. Pearce D Renewable resources: the fishery and wildlife (Unpublished notes). http://www.ucl.ac.uk/~uctpa15/B48_fish.pdf
2. Oliver W, Elinor O (2009) Nobel laureates
3. Mathieson A, Wall G (1982) Tourism: economic physical and social impacts. Longman, New York
4. Brisbane Declaration (2007) Environmental flows: definitions. http://www.eflownet.org/download_documents/brisbane-declarationenglish.pdf
5. Tharme RE, King JM (1998) Development of the Building Block Methodology (BBM) for in stream flow assessments, and supporting research on the effects of different magnitude flows on riverine ecosystems. Water research commission report no. 576/1/98, p 452

General Readings

1. Department of Tourism (Govt. of Uttar Pradesh) Tourism Statistics, Department of Tourism, Government of Uttar Pradesh
2. Department of Tourism (Govt. of Uttarakhand) Tourism Statistics, Department of Tourism, Government of Uttarakhand
3. Freeman RE (1984) Strategic management: a stakeholder approach. Pitman, Boston

Pilgrims on their way to kedarnath (Courtesy Geeta Bansal)

Ganga as Perceived by Some Ganga Lovers

Mother Ganga's Rights Are Our Rights

Pujya Swami Chidanand Saraswati

Nearly 500 million people depend every day on the Ganga and Her tributaries for life itself. Like the most loving of mothers, She has served us, nourished us and enabled us to grow as a people, without hesitation, without discrimination, without vacation for millennia.

Regardless of what we have done to Her, the Ganga continues in Her steady flow, providing the waters that offer nourishment, livelihoods, faith and hope: the waters that represents the very life-blood of our nation.

If one may think of the planet Earth as a body, its trees would be its lungs, its rivers would be its veins, and the Ganga would be its very soul.

For pilgrims, Her course is a lure: From Gaumukh, where she emerges like a beacon of hope from icy glaciers, to the Prayag of Allahabad, where Mother Ganga stretches out Her glorious hands to become one with the Yamuna and Saraswati Rivers, to Ganga Sagar, where She finally merges with the ocean in a tender embrace.

As all oceans unite together, Ganga's reach stretches far beyond national borders. All are Her children.

For perhaps a billion people, Mother Ganga is a living goddess who can elevate the soul to blissful union with the Divine. She provides benediction for infants, hope for worshipful adults, and the promise of liberation for the dying and deceased.

Every year, millions come to bathe in Ganga's waters as a holy act of worship: closing their eyes in deep prayer as they reverently enter the waters equated with Divinity itself. In their palms may be flowers, a tiny cup, or an invisible prayer. Reaching with loving fingers, they capture a tiny scoop of Her flow. Arms stretch upwards to the sun. Ancient mantras are uttered, and an age-old ritual of the worship of nature is carried forth time and time again.

At dusk, when the sky prepares to enrobe in its glorious dress of stars, worshippers congregate once more on Ganga's shores. Seeming to emulate the cosmos, they light oil lamps small and large, which they wave while chanting affectionate hymns in adoration of the river they call, "Mother."

One cannot imagine such a vision on the shores of the Thames or Mississippi Rivers, nor on the banks of the Yangtze, Danube or Hudson. Yet, here in India, the

R. Sanghi (ed.), *Our National River Ganga: Lifeline of Millions*,
DOI 10.1007/978-3-319-00530-0, © Springer International Publishing Switzerland 2014

land of rishis, sages and saints, Her worship is seen as normal and natural. "Of course the Ganga is worshipped by so many," a devotee may say. "When has She not been?"

The Sacred Origins of the River Ganga

In times of ancient lore, there was a mighty ruler named King Sagara. Proud and strong, he wished only to become even mightier. Given the protocol of the time, he arranged for a horse yajna, in which a healthy steed is set forth to freely roam. For as far as the horse may journey, the King may claim the land as his own. Potentially, he could thus become Emperor of all.

It is said that the God, Indra, didn't like this plan. Hence, he stole the horse.

King Sagara, in his state of distress, dispatched all of his 60,000 sons to find the animal. Having been snatched away by a heavenly being, the stallion was nowhere to be found. Yet, being dutiful sons, far and wide they searched until they reached the underworld hermitage of Sage Kapila.

"Of all sages," says Lord Krishna in the Bhagavad Gita, "I am Kapila." Yet, the sons saw only a meditating derelict – a potentially felonious one at that. For right there, next to the sage, was tied their father's missing horse.

From the mouths of 60,000 men came tremendous shouts of insult. So loud did they raise their cries of outrage that the deeply-meditating sage opened his eyes for the first time in many, many years. With one powerful glance, all sons were immediately reduced to heaps of smoldering black-gray ashes.

Their bodies were gone, but their souls were trapped to wander without peace until the proper funeral rituals could be performed – rituals which would need a river within which their ashes could be submerged. Such a river, at this point in history, did not exist in this world.

Learning of this, the King became inconsolable, as would any loving parent who loses one child, much less 60,000. His boundless sorrow was such that it crossed the generations, compelling a lineage of descendants to try era after era to bring down a river from the heavens, so that their ancestors could finally rest in peace.

Centuries later, King Bhagiratha ascended the throne. By now, his land was tormented by natural disasters that were attributed to the fact that his ancestral uncles yet roamed as tormented ghosts. Abdicating his royal post, he set forth to the Himalayas, where he practiced ground-shaking spiritual practices for 1,000 years. At the conclusion of the millennia, Lord Brahma, the creator, appeared before the king to offer any boon.

"My Lord, my only request is that you bring the River Ganga down to Earth," said the saintly king, "so that my ancestors can finally be free, and our world can become safe once more from the disasters that plague it."

Touched by his sincerity, Lord Brahma said his request would of course be granted, but first, the king would need to please Lord Shiva. Lord Shiva was the only one who could slow down the descent of Mother Ganga's heavenly waters, so that She wouldn't inadvertently flood and destroy our planet at the same time.

Living only on air, King Bhagiratha performed more penance until he attracted the loving glimpse of Lord Shiva, who agreed to cushion Mother Ganga's descent in His free-flowing hair.

Thus, Ganga came to this Earth. Called forth by Lord Brahma, She became a creator in Her own right. Cushioned by the hair of Lord Shiva, She became an emancipator. With love knowing no bounds, She freed the princes' souls. She fed the lands. She became our Mother.

The Wisdom of the Sages

As Mother Ganga steadily flowed, lovingly nurturing planes, valleys and mountain sides, ancient teachers of ancient times saw Her intrinsic importance to the lives and well-being of all. As they sat alongside Her peaceful shores, wondrous visions would unfold, bringing them to blissful admiration of the Divine. With eyes held shut, they saw. And they realized the Godly bonds that forever fused together nature and humanity as one. Where there is nature, they realized, and they wrote, there is God. Where there is humanity, there is Divinity. A man. A flower. The wind. The sun. All are the same: all are glorious embodiments of the one loving, nurturing God who creates and pervades everything.

Understanding this, they implored the masses to see nature as a holy gift from and of God. Rituals thus were developed entailing the worship of plants, rivers, the sun, the wind. Everything of nature was seen for its holiness and meditated upon for its grace.

Even today, one can witness remnants of such ancient traditions: in the Tulsi plants that are worshipped every morning in millions of courtyards; as the threads of prayer that are tied to holy trees; as the lamps that are illuminated and then floated on the Ganga for Her nightly embrace.

Such love for nature established a pattern in life. In Vedic times, only what was needed was taken. What was taken was replaced. Thus, without realizing it, the ancients established a sustainable lifestyle that resulted in plentiful yields of crops and sparkling, ever-present waters. In such a culture, famines and water-borne diseases would have been rare indeed. Nature which is revered is not simultaneously exploited.

Sadly, history changed. Mother Ganga and nature didn't.

The Changing Times and the Paradox of Today

The population grew, and in its haste for development, ancient ways were discarded. Factories, in their enthusiasm to manufacture, poured foul toxins into our waters. Tiny creatures, such as insects and indigenous plants, became pests to be destroyed by chemical warfare. Forests and mountain tops were leveled by mighty machines so that building materials and fuel could be harvested. Our lives became more and more isolated, as open doors were replaced with walls of glass and steel.

Thus, our natural resources began to sicken and disappear, so slowly at first that it was barely perceptible. Then suddenly, history sped up, and people started to die in enormous numbers.

Every day, about a billion liters of chemicals are poured into the Ganga alone. Some substances, which issue forth from industry, are so toxic that they can cause convolutions, shock and death. Agricultural run-off furthermore carries the pesticides and chemical fertilizers of numerous farms and fields into our sacred River.

Just recently, a study determined that those living alongside portions of the Ganga are some of the most prone in the world to contracting certain cancers. Much of this is simply due to the chemicals that are used by humanity to make the items we believe will make our lives better. Instead, these items are killing us.

Adding to the burden, approximately two billion liters of sewage pour into Ganga's waters every day. In the holy city of Varanasi, where people have come for millennia to take their last breaths and to scatter the ashes of their loved-ones, Ganga's waters are darkened by human waste. Nearly 70 % of the people who use Ganga's waters will become sickened by waterborne diseases caused by the sewage upstream.

There was a time in which the Ganga had the ability to self-clean, to remove many of the toxins that threaten and take millions of lives. Sadly, much of that ability has been removed. Upstream, hydroelectric dams and barrages channel Her waters away. Sometimes, they are diverted for over 20 km into dark underground tunnels while the natural riverbed aboveground is left dry. Like the ashes of King Sagara son's, the lives that depend on Mother Ganga are once again crying for drops of water.

Making matters more difficult, Ganga's waters are also diverted so that they may quench thirsty cities and farms. Yet, sadly, much of this water is lost to evaporation and to poor public infrastructure.

Without enough water, the Ganga, quite simply, can't wash away the tons of waste heedlessly dumped by blinded humanity. Thus, Her waters are becoming more and more polluted, toxic and sparse, day by day, bringing more sorrow, more disease and more pain.

Can India Think Like Ecuador?

In 2008, the South American nation of Ecuador ratified its constitution, and included the first injunction in the world granting inalienable rights to nature. Said the New York Times, 'the Constitution includes a novel set of articles that appear to be the first in any Constitution… One passage says nature "has the right to exist, persist, maintain and regenerate its vital cycles, structure, functions and its processes in evolution."'[1]

According to the Christian Science Monitor, "No other country has gone as far as Ecuador in proposing to give trees their day in court, but it certainly is not alone in its recalibration of natural rights. Religious leaders, including the Archbishop of Canterbury, the Dalai Lama and the Archbishop of Constantinople, have declared

[1] Andrew C. Revkin, New York Times, "Ecuador Constitution Grants Rights to Nature," September 28, 2008.

that caring for the environment is a spiritual duty. And earlier this year, the Catholic Church updated its list of deadly sins to include polluting the environment. Ecuador is codifying this shift in sensibility. In some ways, this makes sense for a country whose cultural identity is almost indistinguishable from its regional geography."[2]

From its scriptures to sites of worship to places of retreat, the history and culture of India are similarly indistinguishable from its geography. Yet, India's natural resources, such as the historic, life-giving, and holy Ganga River, are in serious peril. One may suggest it is time that India begin to think like Ecuador, and consider a similar constitutional injunction or legislation before its natural treasures are forever lost.

It is as simple as this: if Ganga dies, India dies. If Ganga thrives, India thrives.

Traditional Environmental Laws Regulate the "Use" of Nature

Today, the Ganga River Basin is managed under environmental laws similar to traditional environmental laws found around the world. These are designed to regulate how we use nature, legalizing environmental harms by regulating how much pollution or destruction of nature can occur under law. Under such law, natural resources, such as Mother Ganga, are seen as property that can be exploited under management.

As this type of environmental law has spread across the globe over the past four decades, by almost every criteria, the condition of our natural environment has grown worse. We need look no further than the Ganga to see that this structure of law is not able to protect nature, or us, from the harmful activities that cause pollution and rob our children of their resources and heritage.

Before it is too late, fundamental change is needed.

Recognising Rights of the Ganga River Basin

As natural resources are legally viewed as commodities, the Ganga River has no rights of Her own. Despite the tremendous harm that has been done, despite the fact that millions of people have died because of this harm, the people and governments of India are unable to defend the Ganga's right to be healthy and thrive.

If the rights of the Ganga are violated, all of our rights are violated. Ganga's rights are our rights.

For this reason, civil society is proposing a National Ganga River Act, which would establish a new paradigm to protect the Ganga. To join the cause, one may visit www.gangaaction.org.

---

[2]Eoin O'Carroll, Christian Science Monitor, "Ecuador constitution would grant inalienable rights to nature," September 3, 2008.

The Choices We Make

For every choice we make, we create a new future. Whether we are working within the highest levels of governance or are simply deciding where to throw our waste, every action has meaning. That is why we must do all we can, while we can, to protect and preserve our Mother Ganga.

If we respect and revere nature, our choices become clear: never should we place our trash in the Ganga. Never should we drain our chemicals and sewage into Her beautiful waters, nor should we tolerate this behavior from others. A good portion of Her water must be allowed to flow freely in the natural riverbed. The trees on Her banks, which offer fortification from run-off and landslides while also providing the moisture that brings replenishing rains, must be cultivated and safe-guarded.

We should look towards the ways of old to learn how we may be better stewards of our God-given resources: ways such as large-scale rainwater harvesting, organic farming methods and the cultivation of water-saving indigenous crops. And we should look towards new ways, so that we may use less water for irrigation, through methods such as drip irrigation. Through other new approaches, we may also more efficiently treat our sewage, using technologies such as Biodigester toilets, which dissolve all waste into clear water for fertilizer and biogas for fuel.

The fate of Mother Ganga hangs in perilous balance. The future is in your hands. Your ideas matter. Your voice matters. Your choices matter. Now, it is time to do what is right. Now is the time to be the change.

Photographer Geeta Bansal

India Lives, If Ganga Lives

Vandana Shiva

My home is the land through which the Ganga flows – where the Bhagirathi, the Bhilangana, the Mandakini, the Alaknanda join to form the river which has supported our culture and civilization, our agriculture and economy from times.

Today, a war has been declared on our sacred mother Ganga.

Our region in the Himalaya is called Dev Bhoomi, the Divine and sacred land. Today it has been defined as "Urja Pradesh" – the land of hydroelectric power. Mukesh Semwal in his article "Dam(ed) Development in Uttarakhand. Where will it lead to?" (India Age, July 2010, p 20) quotes a folk song by well known folk singer – Narendra Singh Negi

Dev – bhoomi ko nao badaly
Bhyali – bhoomi karayala ji
Uttarakhand ki dharti
Yun damun damayali ji
Surango nu gawadhali ji

The Ganga is India's ecological, economic, cultural and spiritual life line. The myth of the descent of the Ganga is in fact an ecological tale. The following hymn is a tale of the hydrological problem associated with the descent of a mighty river like the Ganges.

Ganga, whose waves in swarga flow,
Is daughter of the Lord of Snow.
Win Shiva, that his aid be lent,
To hold her in her mid-descent.
For earth alone will never bear
These torrents traveled from the upper air.

R. Sanghi (ed.), *Our National River Ganga: Lifeline of Millions,*
DOI 10.1007/978-3-319-00530-0,

H. C. Reiger, the eminent Himalayan ecologist, described the material rationality of the hymn in the following words:

> In the scriptures a realization is there that if all the waters which descend upon the mountain were to beat down upon the naked earth would never bear the torrents…In Shiva's hair we have a very well known physical device, which breaks the force of the water coming down,…the vegetation of the mountains.

That is why Chipko, the Movement to protect the Himalayan forests, was so important for India's ecological security. I started my ecological activism with Chipko. After nearly a decade of Chipko actions, logging was banned in the high Himalaya in 1981. After the 1978 flood in the Ganga, it became clear that water conservation was the first gift of the Himalayan forests.

The threats to our Mother Ganga – Ganga Ma – are many. Deforestation was a major threat to the catchment of the Ganga in the 1970s. The Ganga is also threatened at its very source at the Gangotri glacier with climate change which is leading to a decline in snowfall and an increase in the rate of melting of the snow. From 1935 to 1956 the retreat of the Gangotri glacier was 4.35 m/year. In the period 1990–1996 it is 28.33 m/year. The average rate of retreat is 20–38 m/year. If this retreat continues, the Ganga would become a seasonal river, with major ecological and economic consequences for the entire Ganges basin. This is why we need climate justice for water justice.

The Ganga's tributaries are threatened by dams and diversions in the upper reaches. Their dam has been an unmitigated disaster. The 260.5 m high Tehri dam has been built at Tehri on the confluence of the Bhagirathi and Bhilangana. It submerged the ancient capital of Tehri Garhwal, it destroyed the lush and fertile fields of Bhilangana and Bhagirathi valleys and it displaced 100,000 from 125 villages of which 33 were completely submerged. But the displacement due to the dam continues. 100 new landslides have been triggered by the dam threatening another 100,000 people. In the heavy monsoon of 2010, the disasters intensified.

At the shore of the reservoir people were flooded from below and above simultaneously. Fields and homes by the dam shore were submerged as the water level rose from 820 to 835 m. The Tehri hydro power plant authorities did not want to release excess water from the dam, even though the water levels were affecting the surrounding villages. From their point of view release through the slush gates is spillage. Mooni Devi, who lives at water level, says "This used to be such a great place with great farms, the dam builders have turned us all into beggars". A chain of hydroelectric projects have stopped the "aviral" flow of Ganga and in many stretches, the Ganga runs dry.

In the plains a big threat to the Ganga and Yamuna is pollution – both from industry and from sewage. And even as billions are poured into cleaning the Ganga and Yamuna through the Ganga Action Plan and the Yamuna Action Plan, the pollution of our sacred river increases because of a combination of corruption and inappropriate technologies.

Industrialization and urbanization have turned our sacred rivers into sinks for pollutants. The Yamuna is clean before entering Delhi. In 22 km of its journey through Delhi, it picks up 70 % of all the pollution of the river in its total length. All that the Action Plans have done is set up centralized Sewage Treatment Plants (STPs) that do not work, and 70 % of all sewage is dumped untreated into the river. The river dies because of pollution; the land dies because it is deprived of rich nutrients. As Sunderlal Bahuguna reminded me, Gandhi called this "golden manure". Intelligent zero waste sewage treatment systems like those evolved in IIT, Kanpur, by Dr. Vinod Tare would both clean the Ganga and fertilize the soil. We would not be wasting Rs. 130,000 crores on fertilizer subsidies and thousands of crores on river action plans. Organic farming can be a major action for cleaning the Ganga.

The final threat to the Ganga is privatization. Privatization of water reduces water to a commodity, makes giant corporations owners and sellers of water and ordinary citizens, buyers and consumers. The role of citizens and communities as conservers and caretakers is destroyed. The human right to water which was recognized by the UN in April 2010 is undermined. That is why when the Ganges water which has been brought to Delhi from Tehri was being privatized to Suez through a World Bank project, we built a Citizens Alliance for Water Democracy and told the World Bank and the Delhi Government that our "Mother Ganga is not for sale". The World Bank project was withdrawn and the privatization stopped.

Planners do not see our rivers as rivers of life. They see them as 20,000 MW of hydro power. In 1916, Rai Patiram Bahadur in his book *Garhwal, Ancient and Modern* wrote "We may say that there is no country in the world of the dimension of Garhwal which has so many rivers as a traveller will find in this land. The district has 60 rivers of different size, besides these; there are rivulets, rills, springs and fountains in hundreds, showing that nature has been especially bountiful to this land in the matter of its water supply." (quoted in Semwal, p21)

The 99 MW power project at Singoli – Bhatwari near Augustmuni, being constructed by Larson & Toubro, which will affect 60 villages, is one of the 12 Hydro electric projects all coming up on the Mandakini. My colleague and local coordinator of Navdanya Chandrashekhar Bhatt has been part of the movement resisting the 13 project. In 2008, our teacher and friend, Dr. G D Aggarwal, a former Professor of IIT Kanpur where I attended summer schools as a Physics student, undertook a fast into death at the age of 76 to stop the string of dams built on the Bkaguella including the 600 MW Loharinag pala being built by Natural Thermal Power Corporation. The Ganga would be put into a 17 km tunnel to generate electricity. Other dams on the Bhagirathi include the 480 MW Pala Maneri, 381 MW Bharan Ghati, (Ref http://www.asiantribune.com). On 14 January 2009, Prof. Aggarwal undertook a fast again because the government failed to stop the dam. After continued fasts by Dr. Aggarwal the Lohari Nappala dam was stopped. In addition the government announced the setting up of the National Ganga River Basin Authority (NGRBA).

Announcing the stopping of dams on the Bhagirathi, the Union Minister of Environment, Jairam Ramesh said "I have said in Parliament that India is a civilization of rivers, that it should not become a land of tunnels.... There are no two opinions. There is just one mass opinion that the projects proposed on the river Bhagirathi, named Pala Maneri and Bhaironghate projects will not be entertained any further by the government". (Ref Jairam Ramesh emphasizes the need for river conservation. One India (Greynuin Information Technologies, 10.2.2010))

Sacred waters carry us beyond the marketplace into a world charged with myth and stories, beliefs and devotion, culture and celebration. These are the world that enable us to save and share water, and convert scarcity into abundance. We are all sagar's children, thirsting for waters that liberate and give us life – organically and spiritually. The struggle over the kumbh, between gods and demons, between those who protect and those who destroy, between those who nurture and those who exploit, is ongoing. Each of us has a role in shaping the creation story of the future. Each of us is responsible for the kumbh – the sacred water pot.

"There is only one struggle left – the struggle for right to life", said Magsaysay Award-wining writer Maheshwata Devi. Eminent author Arundhati Roy and Dr. Vandana Shiva stressed the urgent need to take collective united action to defend people's rights to land, water and biodiversity.

"Mother Ganga Is Not for Sale": The Haridwar Declaration

Today, the 8th of August 2002, on the *eve* of the 60th Anniversary of the "Quit India Movement," we all have gathered here to pledge that: We will never let the river Ganga to be sold to any Multinational Corporations. Ganga is revered as a mother (Ganga Maa) and prayed to, and on its banks important ceremonies starting from birth till death are performed (according to Hindu religious practices). We will never allow our mother or its water to be sold to Suez-Degremont or any other corporations.

The sacred waters of the Ganga cannot be the property of anyone individual or a company. Our mother Ganga is not for Sale. We boycott the commodification and privatisation of the Ganga and any other water resources.

We pledge to conserve and judiciously use our regional water resources to save our environment and ecology, so that we would gift our coming generation a clean and beautiful environment as well as safeguard their right to water resources

We pledge and declare that the local community will have the right over the local water resources. It is the duty of the local community to conserve and sensibly utilize their resources. Anyone from outside the community whether an individual, an organization or a corporation have to take permission of the Gram Sabha for utilizing these resources

The river Ganga was brought upon the face of earth by Bhagirath through his yagna (prayers) to sustain the existence of life on Earth. The Ganga is now intrinsic to our culture and a part of our heritage and our civilization. Our life and progress over the millennia has been dependent upon the sacred waters of Ganga. We will fight any multinational company trying to take away our rights to life by privatising Ganga waters

The "Water Liberation Movement" will continue till we liberate the sacred waters of Ganga from the clutches of corporations, like Suez-Ondeo Degremont.

The movement to Save the Ganga and its "nirmal" (clean) and "aviral" (uninterrupted) flow is not just a movement to save a river. It is a movement to save India's troubled soul, polluted and stifled by crass consumerism and greed, disconnected from its ecological and cultural foundations. Our movement is based on the acknowledgement of the sanctity and sovereignty of the river Ganga.

If the Ganga lives, India lives. If the Ganga dies, India dies.

A devotee taking a holy dip in the Ganga (Courtesy Julian Crandall Hollick)

A Long Love Affair

Julian Crandall Hollick

I spent most of my childhood on or near the Thames in southern England. Of moderate width, its banks lined by weeping willows or gentle green meadows where cows went about their lives, while on the water proud white swans patrolled against unwanted bathers. A pastorale river. Not a goddess. Just a sweet gently-flowing river. Strangely my family always looked inland away from the river towards rolling hills where they could walk or ride.

So it's ironical that I have spent the past 30 years in or around Ganga, about as different from the Thames as it's possible to be. My love affair with Ganga took a lot of time to develop. I first saw her in 1986 at Varanasi, where I was taken out on the obligatory boat ride at dawn to view bathers at the ghats. I suppose I was expected to take photos, like a good tourist peering at a new animal species. I didn't because I was working and taking tourist snaps has always made me distinctly uncomfortable. I was anyway underwhelmed.

Fast forward to 1996 to Kanpur. In 1995 and 1996, I spent 3 months there making a documentary on the monsoon. It was strictly a working relationship. I also spent several days each week with farmers and fishermen in the village of Bhithur on the banks of Ganga. Bhithur calls itself the "Center of the Universe" because it was there that Brahma rested after creating the universe.

It was high summer, just before the monsoon was due to break. The river was very low, no more than a hundred yards across. We used to cross the pontoon bridge by car or motorbike, riding a quarter of a mile across sand dunes to the far bank where a friend had a farm. Fishermen were harvesting watermelons planted in the sand dunes in February when the river had begun to shrink. It seemed hard to imagine this modest, placid river could become a sea just a month later. But anyway, I was there to live the monsoon, not Ganga.

Five years later, I got to know a very different Ganga. I was in delta country in southwestern Bangladesh and West Bengal where land, river and sky hazily were bleeding into each other to form an abstract landscape. The river in the delta was so vast its other shore couldn't be seen. It was hard to equate this ocean with the muddy

R. Sanghi (ed.), *Our National River Ganga: Lifeline of Millions*,
DOI 10.1007/978-3-319-00530-0, © Springer International Publishing Switzerland 2014

sacred Ganga at Bhithur or Varanasi, let alone the fierce torrent in the Himalayas. Yet it was still the same Ganga!

I was hooked. How can the same river have such disparate personalities? I realized I knew next to nothing about the physical river Ganga. Indeed, not too many people did. Travel writers, novelists, even distinguished academics focussed either on high culture or on the obvious tourist sites – Varanasi, Allahabad, Rishikesh – and their picturesque mythologies. None of them attempted to make any real connections between mythology and the environment. Scientists, hydrologists or geographers each concentrated on their local reach, on their specialty. Nobody studied the river either as a whole, or indeed holistically. Why? Because Humanities and Sciences are not supposed to mix?

I was intrigued how a river can also be a goddess. For millions of Hindus the river is the liquid expression of the Goddess Ganga. Bathing in her waters is both a spiritual ritual and necessary ablution. But in the West we separate sacred from secular so the very notion of their being indivisible seems absurd. And if Indians worship Ganga as sacred how can they at the same time pollute her as off-handedly as they do?

The best way (indeed the only way it seemed to me back in 2001 – and it still does) to get to know Ganga was to travel the whole river. But this unearths two unanticipated wrinkles. First, Ganga has two quite distinct basic personalities. The religious dimension dominates from the river's source in the Himalayas down to Varanasi. This is the Ganga of mythology, the Ganga who came down to earth from the Milky Way in a lock of the god Shiva's hair. *Ganga-ma* is a generous Goddess who gives life to the fields, cleanses the body of disease, the soul of sins and opens the way to the next life. The spiritual benefits of Ganga jal – the water of Ganga – go hand in glove with the physical benefits of bathing and drinking the water. From Gaumukh down to Varanasi Ganga is worshipped everywhere at *tirthas*, sacred crossing points between the worlds of the Gods and the world of human beings. People in the north are generally less concerned with the physical river than the river in their minds.

But beyond Varanasi everything changes, Ganga becomes an altogether much bigger river and an overwhelmingly physical presence. The monsoon completes this transformation of Ganga into a destroyer of land and life. Reality in Bihar and West Bengal means dealing daily with the sheer physicality of Ganga, not her spiritual generosity. Of course, millions do still worship Ganga as goddess, but they usually pray with one eye open for the fury of floods and storms.

In reality there are three distinct rivers, each called Ganga, although the dominant source of water is quite different in each instance. This distinction into three rivers seems to me important because it reveals how each separate reach has quite different characteristics: the issues in *Himalayan Ganga* are very different from those of *Nepalese Ganga*. *Yamuna Ganga* almost doubles in size at Allahabad. Revitalised and no longer on life support, *Yamuna Ganga* flows on to Varanasi, then into Bihar, where it becomes *Nepalese Ganga*, its third avatar.

It completes this metamorphosis from mountain stream to a giant river thanks to at least four huge rivers – the Son from the Deccan plateau to the south, the

Ghaghara, Gandak and Kosi from the Nepalese Himalaya. By the time it reaches Farakka *Nepalese Ganga* is more than three times the size of *Yamuna Ganga* and eight times the size of *Himalayan Ganga* when she entered the north Indian plains at Haridwar!

Nepalese Ganga is huge, sometimes so broad you can't even see the opposite bank. In the monsoon its fury simply carries away fields, animals and humans, and leaves behind highly fertile silt. *Nepalese Ganga* takes away with one hand and gives back with the other, characteristic of Shiva who originally brought Ganga down to earth in the locks of his hair. Shiva is never the life and soul of a party at the best of times. But he stamps his mark on Ganga everywhere, whether as creator or as destroyer, or both at the same time. Supreme irony: only in Kolkata, where the river is basically too shallow for anything except commuter ferries, does the river finally assume a majesty befitting her status as both goddess and river. She is at last Ganga, flowing through the heart of an ex-capital city. Unfortunately too little too late. The young girl has finally become a mature woman.

But back to my original puzzle: how can Indians pollute Ganga, yet at the same time worship her as a goddess? How can so many millions take a 'holy dip' every morning to wash away their sins in a river that is polluted by so much waste, both human and industrial? Many Indians simply do not see any contradiction. It is just so obvious to them that a river can be both in the mind and a very physical reality.

This ability to live on multiple mental levels simultaneously may also be a huge part of the failure to date to really make any improvement in Ganga's state of health. The physical river may actually die in places because of apathy or indifference, no matter the best intentions of environmentalists, scientists or concerned Indians. In the minds and hearts of her devotees 'what's the problem if we pour our garbage into the river or suck her dry? Ganga can always survive and purify herself, precisely because she's a goddess.

I suspect Indians have also been spending far too much time looking at Ganga's problems through Western lenses, that successive Indian governments have chosen Western parameters and technologies, often unsuited to Indian realities. While the scientific method is universal, what a scientist chooses to examine may be determined by his or her culture. The danger is that the dominance of Western science within India has caused a corresponding loss of knowledge of and confidence in India's own scientific values and heritage.

If Ganga had a problem, what exactly was it? Back in the 1980s, Indian government scientists and their international consultants decided pollution was the main issue. The irony is that the consultants chosen were from the Thames River Water Authority. Back in the late 1950s the Thames was severely polluted, especially in London. The government ordered it to be cleaned up so that fish could freely swim upstream. Sure enough salmon were spotted in the river opposite the Houses of Parliament after only a few years. Problem solved.

So it was natural that Delhi would turn to them. But that choice in turn dictated the dominant parameters of the 1986 Ganga Action Plan – Dissolved Oxygen, Biological Oxygen Demand – chosen to measure success or failure. It also largely dictated the 'western' methods they would adopt to tackle the 'problem,' although there were and still are less expensive alternation and indigenous solutions – variations on low cost settling ponds that use the sun, another of India's great natural assets, to let Nature do much of the work at a fraction of the cost. It's not that the choice of parameters was necessarily wrong. They just may not have been the most sensible or urgent solutions for a river in a tropical country. Why would what worked on the Thames be expected to work on Ganga?

In any case, pollution may be secondary to the lack of water. For the real 'problem' today is **flow**, the actual amount of water in the river. India simply cannot go on taking so much water out of Ganga and expect her to somehow survive. It's always possible that wise heads will prevail and understand there has to be balance between supply and demand. But economic pressures may prove too great. How much water will ever-more intensive agriculture, urbanization, headlong modernisation require? Are yet more dams or river-linking really viable solutions? Are the sources of Ganga stable, sufficient or in decline? Global warming in the short term may mean more runoff into rivers such as Ganga. In the long term it may mean much less snow-melt, rivers facing a lot less water and the best-laid plans for the generation of hydro-electricity literally may simply dry up.

What does all this mean for Ganga? These are questions I think far too few scientists within India ask. Why? Because the solution is too large for them to even contemplate? Because it's safer to concentrate on more narrowly-defined questions? Because they expect the central government in New Delhi to always take the initiative?

There's the awful possibility that too many Indians will simply continue on regardless, ignoring meaningful conservation and sucking Ganga dry, drawing her down, convinced that somehow the goddess will find always ways to renew her liquid form. But what if *Ganga-ma* cannot and does not?

Photographer Deepak Kumar

What and Where Is Ganga?

Manoj Misra

River Ganga has symbolised India for ages. It is not only its largest river but is an integral part of its mythology, its geography and history. Sadly today the river is endangered and all past efforts to rejuvenate it have failed to produce the desired results. We understand that it is a lack of sound understanding about what and where is Ganga, which is behind the failure of past planning and actions, for it.

In the recent past the government of India has constituted the Prime Minister led NGRBA for the river. The NGRBA stands for National Ganga River Basin Authority. What does the term 'basin' in it convey?

To experts it of course makes sense as a larger area that collects water and conveys it to the river Ganga. But to a common man 'basin' is some technical term that is best left for the experts to grapple with!

Imagine if we had rather used the term National Ganga 'Rivers' Authority. Would it not still have conveyed the same thing and yet made easy sense to one and all.

No river, not even a small stream stands alone. Thus we make a fundamental mistake and unwittingly limit our view and vision of a river and its restorative needs by talking only of a river rather than the 'rivers' that constitute it.

Let us see how this plays itself out in reference to rivers, Ganga and Yamuna.

Ganga begins to be called Ganga only after Devprayag in the state of Uttarakhand, whence Alaknanda has mingled itself with Bhagirathi. Alaknanda itself is incomplete without Pinder as is Bhagirathi without Bhilangana. Later of course Ram Ganga, Gomti, Ghaghara, Yamuna and Son complete the Ganga within the state of Uttar Pradesh before it enters the eastern state of Bihar.

Few drops of *Ganga jal* (water) and tulsi leaf in the mouth of a dying hindu is the most preferred manner of death. The reason being that Ganga is not only a sacred river, but is one whose water has magical properties of retaining its freshness and vitality even after long period of storage. There are reports of stored precious *ganga jal* having been transferred down the generations as part of family heirloom. Not surprisingly bacteriological and other researches have gone into determining the how and why of this unique feature of Ganga *jal*. But the jury on it is still out.

R. Sanghi (ed.), *Our National River Ganga: Lifeline of Millions*,
DOI 10.1007/978-3-319-00530-0, © Springer International Publishing Switzerland 2014

Just consider that while the precious *ganga jal* is collected by the devout mostly at Rishikesh and its downstream, have we ever bothered ourselves of which Ganga water are we talking about? Is it Bhagirathi, Bhilangana or Alaknanda and that too before or after the Pinder has joined it. Is it the property of the water itself or a unique geological source underneath this or that glacier that endows this wonderful property to the Ganga waters? Or is it a particular plant in a particular tributary's catchment that is responsible? Most critically not withstanding all our efforts to rid the Ganga at say Hardwar, Kanpur, Allahabad or Varanasi of its pollution load, this unique property of its water might still remain threatened until the concerned tributary is rid of various threats facing its health.

Similarly, Yamuna does not become formidable until Tons joins it near Kalsi. And then Asan, Giri and Bata complete it before it enters the plains of north India after crossing the Shiwalik hills. Now while it is well known to most anglers that there was a time when Yamuna abounded in golden mahseer – a large sized fish which is an angler's delight – it has been little appreciated that it is actually the river Giri amongst all the Yamuna rivers that has the best habitat available for the fish. This is because the golden mahseer avoids very cold glacial waters and it is only the river Giri system that has the appropriate thermal conditions in its water, for it has largely non-glacial origins and also carries the requisite wilderness as a river to facilitate the fish's sustenance and growth. Thus if the future of Yamuna as the abode of golden mahseer is to be secured then it can only be done by securing the river Giri system. Unfortunately the insatiable drinking water needs of Delhi, the national capital that sits some 250 km away could jeopardise once for all the wilderness and suitable habitat conditions for golden mahseer in river Giri with a 145 m high dam planned on it to provide some 275 MGD of additional water to the national capital.

Like most Indian deities, Ganga and Yamuna too have respectively, *makar* (crocodile) and *kacchapa* (turtle) as their *vahans* (mount). While this fact is known to most, but how many appreciate that it is river Ramganga that perhaps has the most suitable habitat for magar crocodiles and river Chambal (a Yamuna tributary) that is known for carrying suitable sand banks for the egg laying and hatching of turtles.

Well, despite the above facts when we speak of rivers Ganga and Yamuna then how many of us ever visualize Pinder as Ganga, or Giri as Yamuna. So much so that even the NGRBA while acknowledging Yamuna as being the biggest tributary of Ganga is keeping a safe distance from Yamuna in its planning and action.

Thus the NGRBA shall make worthwhile sense only when it takes into account the conservation needs of each and every tributary of the system as if the future health of each of the latter determined the ultimate results of various proposed plans and actions enunciated by it.

In this context it is suggested that just like the fact that "a chain is as strong as its weakest link", *a river is as healthy as its sickest tributary.*

So shall we talk of Ganga 'rivers' each and every time? AMEN!

A sunset view of Ganga at Rishikesh (Courtesy Nitin Kaushal)

Appendices

Appendix 1: The 108 Names of the Ganges (Gangastottara-sata-namavali)

1. Ganga (गंगा)	Ganges
2. Vishnu-padabja-sambhuta (वषि्णु -पदाब्ज –संभूता)	Born from the lotus-like foot of Vishnu
3. Hara-vallabha (हरा -वल्लभ)	Dear to Hara (Shiva)
4. Himacalendra-tanaya (हमिाकालेंद्र –तनया)	Daughter of the Lord of Himalaya
5. Giri-mandala-gamini (गरिी -मंडला –गामनिी)	Flowing through the mountain country
6. Tarakarati-janani (तरकरती -जननी)	Mother of [the demon] Taraka's enemy (i.e. Karttikeya)
7. Sagaratmaja-tarika (सगारात्मजा –तारकिा)	Liberator of the [60,000] sons of Sagara (who had been burnt to ashes by the angry glance of the sage Kapila)
8. Saraswati-samayukta (सरस्वती –समायुक्ता)	Joined to [the river] Saraswati (said to have flowed underground and joined the Ganges at Allahabad)
9. Sughosa (सुघोसा)	Melodious (or: Noisy)
10. Sindhu-gamini (सन्धि-गामनिी)	Flowing to the ocean
11. Bhagirathi (भागीरथी)	Pertaining to the saint Bhagiratha (whose prayers brought the Ganges down from Heaven)
12. Bhagyavati (भाग्यवती)	Happy, fortunate
13. Bhagiratha-rathanuga (भागीरथ –राथानुगा)	Following the chariot of Bhagiratha (who led the Ganges down to Hell to purify the ashes of Sagara's sons)
14. Trivikrama-padoddhuta (त्रविक्रिमा –पदोद्धुता)	Falling from the foot of Vishnu
15. Triloka-patha-gamini (त्रलिोक -पथ-गामनिी)	Flowing through the three worlds (i.e. Heaven, earth and the atmosphere or lower regions)
16. Ksira-subhra (कसीर –सुभ्रा)	White as milk
17. Bahu-ksira (बहु –कसीर)	[A cow] which gives much milk

(continued)

R. Sanghi (ed.), *Our National River Ganga: Lifeline of Millions*,
DOI 10.1007/978-3-319-00530-0, © Springer International Publishing Switzerland 2014

(continued)

18. Ksira-vrksa-samakula (कसीर -वर्क्स –समकुला)	Abounding in [the four] "milktrees" (i.e. Nyagrodha (Banyan) Udumbara (glamorous figtree), Asvattha (holy-fig- tree), and adhuka (Bassia Latifolia)
19. Trilocana-jata-vasini (त्रलिोकान -जाता-वासनिी)	Dwelling in the matted locks of Shiva
20. Rna-traya-vimocini (रन -त्रय –वमिोचनिी)	Releasing from the Three Debts, viz. Brahma-carya (study of the Vedas) to the rishis, sacrifice and worship to the Gods, and procreation of a son, to the Manes
21. Tripurari-siras-cuda (त्रपिुरारी -सरिस–कदा)	The tuft on the head of the enemy of Tripura (Shiva). (Tripura was a triple fortification, built in the sky, air and earth of gold, silver and iron respectively, by Maya for the Asuras, and burnt by Shiva)
22. Jahnavi (जाह्नवी)	Pertaining to Jahnu (who drank up the Ganges in a rage after it had flooded but relented and allowed it to flow from his ear)
23. Nata-bhiti-hrt (नाता -भीती -हर्ट)	Carrying away fear
24. Avyaya (अव्यय)	Imperishable
25. Nayanananda-dayini (नयनानान्दा –दायनिी)	Affording delight to the eye
26. Naga-putrika (नागा -पुत्रकिा)	Daughter of the mountain
27. Niranjana (नरिंजना)	Not painted with collyrium (i.e. colourless)
28. Nitya-suddha (नति्य –सुद्धा)	Eternally pure
29. Nira-jala-pariskrta (नरिा -जल -परसि्क्र्ता)	Adorned with a net of water
30. Savitri (सावति्री)	Stimulator
31. Salila-vasa (सललिा –वासा)	Dwelling in water
32. Sagarambusa-medhini (सगाराम्बुसा –मेधनिी)	Swelling the waters of the ocean
33. Ramya (राम्या)	Delightful
34. Bindu-saras (बदिु –सरस)	River made of water-drops
35. Avyakta (अव्यक्ता)	Unmanifest, unevolved
36. Vrndaraka-samasrita (व्र्न्दारका -समस्रति़ा)	Resort of the eminent
37. Uma-sapatni (उमा –सपत्नी)	Having the same husband (i.e. Shiva) as Uma (Parvati)
38. Subhrangi (सुभ्रंगी)	Having beautiful limbs (or body)
39. Shrimati (श्रीमती)	Beautiful, auspicious, illustrious, etc.
40. Dhavalambara (धवालाम्बरा)	Having a dazzling white garment
41. Akhandala-vana-vasa (अखंडला -वन –वासा)	Having Shiva as a forest-dweller (hermit)
42. Khandendu-krta-sekhara (खान्देंदु -करता -सखारा)	Having the crescent moon as a crest
43. Amrtakara-salila (अम्र्ताकारा -सललिा)	Whose water is a mine of nectar
44. Lila-lamghita-parvata (लीला -लम्घति़ा -पर्वत)	Leaping over mountains in sport
45. Virinci-kalasa-vasa (वरिचिि-कलस –वासा)	Dwelling in the water-pot of Brahma (or Vishnu, or Shiva)

(continued)

(continued)

46. Triveni (त्रविणी)	Triple-braided (i.e. consisting of the waters of three rivers; Ganges, Yamuna and Saraswati)
47. Trigunatmika (त्रगिुणात्मकि)	Possessing the three gunas
48. Sangataghaugha-samani (संगताघौघा-समानी)	Destroying the mass of sins of Sangata
49. Sankha-dundubhi-nisvana (संख -दुन्दुभि-नस्वािना)	Making a noise like a conch-shell and drum
50. Bhiti-hrt (भीती -हर्ट)	Carrying away fear
51. Bhagya-janani (भाग्य -जननी)	Creating happiness
52. Bhinna-brahmanda-darpini (भन्नि -ब्रह्मंडा –दर्पनिी)	Taking pride in the broken egg of Brahma
53. Nandini (नंदनिी)	Happy
54. Sighra-ga (शीघ्र -ग)	Swift-flowing
55. Siddha (सद्धि)	Perfect
56. Saranya (सरांय)	Yielding shelter, help or protection
57. Sasi-sekhara (ससी -सेखारा)	Moon-crested
58. Sankari (संकरी)	Belonging to Sankara (Shiva)
59. Saphari-purna (सफारी -पूर्ण)	Full of fish (esp. Cyprinus Saphore a kind of bright little fish that glistens when darting about in shallow water – or carp)
60. Bharga-murdha-krtalaya (भर्ग -मूर्धा –क्र्तालय)	Having Bharga's (Shiva's) head as an abode
61. Bhava-priya (भाव –प्रयि)	Dear to Bhava (Shiva)
62. Satya-sandha-priya I (सत्य -संध –प्रयि)	Dear to the faithful
63. Hamsa-svarupini (हमसा –स्वरूपणिी)	Embodied in the forms of swans
64. Bhagiratha-suta (भागीरथ –सुत)	Daughter of Bhagiratha
65. Ananta (अनंता)	Eternal
66. Sarac-candra-nibhanana (सरक -केन्द्र –नभिानना)	Resembling the autumn moon
67. Om-kara-rupini (ॐ -करा –रूपणिी)	Having the appearance of the syllable Om
68. Atula (अतुल)	Peerless
69. Krida-kallola-karini (क्रीडा-कल्लोल-कारणिी)	Sportively billowing
70. Svarga-sopana-sarani (स्वर्ग -सोपाना -सारणी)	Flowing like a staircase to Heaven
71. Ambhah-prada (अम्भः -प्रद)	Bestowing water
72. Duhkha-hantri (दुह्ख -हन्त्री)	Destroying sorrow
73. Santi-santana-karini (सन्तिं-सैन्टाना-कारणिी)	Bringing about the continuance of peace
74. Darirya-hantr (दरर्यि -हंटर)	Destroyer of poverty
75. Siva-da (सविा -दा)	Bestowing happiness
76. Samsara-visa-nasini (संसार -वीसा -नासनिी)	Destroying the poison of illusion

(continued)

(continued)

77. Prayaga-nilaya (प्रयागा –नलिय)	Having Prayaga (Allahabad) as an abode
78. Sita (सीता)	"Furrow". Name of the eastern branch of the four mythical branches into which the heavenly Ganges is supposed to divide after falling on Mount Meru
79. Tapa-traya-vimocini (तप -त्रय -वमिोचनिि)	Releasing from the Three Afflictions
80. Saranagata-dinarta-paritrana (सरनागता-दनिार्ता परत्रिना)	Protector of the sick and suffering who come to you for refuge
81. Sumukti-da (सुमुक्ति-दा)	Giving complete [spiritual] emancipation
82. Siddhi-yoga-nisevita (सद्धि-योग-नसिवति)	Resorted to (for acquisition of successor magic powers)
83. Papa-hantri (पापा –हन्त्री)	Destroyer of sin
84. Pavanangi (पवानंगी)	Having a pure body
85. Parabrahma-svarupini (परब्रह्म –स्वरूपणिी)	Embodiment of the Supreme Spirit
86. Purna (पूर्ण)	Full
87. Puratana (पुरातन)	Ancient
88. Punya (पुण्य)	Auspicious
89. Punya-da (पुण्य -दा)	Bestowing merit
90. Punya-vahini (पुण्य–वाहनिी)	Possessing (or producing) merit
91. Pulomajarcita (पुलोमाजर्सति)	Worshipped by Indrani (wife of Indra)
92. Puta (पुता)	Pure
93. Puta-tribhuvana (पुता -त्रभिुवन)	Purifier of the Three Worlds
94. Japa (जपा)	Muttering, whispering
95. Jangama (जंगम)	Moving, alive
96. Jangamadhara (जन्गामधरा)	Support of substratum of what lives or moves
97. Jala-rupa (जला-रूपा)	Consisting of water
98. Jagad-d-hita (जगद -द –हति)	Friend or benefactor of what lives or moves
99. Jahnu-putri (जह्नु –पुत्री)	Daughter of Jahnu
100. Jagan-matr (जगन –मात्र)	Mother of Bhisma
101. Siddha (सद्धि)	Holy
102. Ramya (राम्या)	Beautiful
103. Uma-kara-kamala-sanjata (उमा-करा -कमल–संजाता)	Born from the lotus which dreaded Uma (Parvati) (presumably a poetic way of saying that they were sisters)
104. Anjana-timira-bhanu (अंजना-तमिरि–भानु)	A light amid the darkness of ignorance
105. Sarva-deva-svarupini (सर्व -देवा -स्वरूपणिी)	Embodies about the continuance of peace
106. Jambu-dvipa-viharini (जम्बू -द्वीप –वहिारणिी)	Roaming about or delighting in Rose- apple – tree Island (Siva)
107. Bhava-patni (भाव –पत्नी)	Wife of Bhava (Siva)
108. Bhisma-matr (भीस्म – मात्र)	Mother of Bhisma

Source: Jai Shree Mataji SAHAJA YO GA MANTRA BOOK (SHRI MATAJI NIRMALA DEVI)

Trek to gaumukh (Courtesy Rashmi Sanghi)

Appendix 2: List of Books on River Ganga as Compiled from Internet Sources

Title	Author	Publisher	Year
Delhi's Contribution in Yamuna River Pollution & Its Control Measures: Present State of Yamuna	Sameer Arora	LAP Lambert Academic Publishing	2012
Hydrogeochemistry in Chhoti Gandak River Basin, Ganga Plain	Vikram Bhardwaj, Dhruv Sen Singh	LAP Lambert Academic Publishing	2012
Dolphin Diaries: My Twenty Five Years with Spotted Dolphins in the Bahamas	Dr. Denise L. Herzing	St Martin's Press	2011
Ecology of River Ganga	D. R. Khanna, Gagan Matta	Biotech Books	2011
In Search of Yamuna: Reflections on a River Lost	Sarandha Jain	Vitasta Pub.	2011
Status of Riparian Floral Diversity Along River Ganga: Riparian Floral Diversity along River Ganga in Garhwal Himalaya of India	Radhey Gangwar, Triveni Gangwar	LAP LAMBERT Academic Publishing	2011
The Yamuna River Basin: Water Resources and Environment	Raveendra Kumar Rai, Alka Upadhyay, C. Shekhar P. Ojha, Vajay P. Singh	Springer	2011
Archaeology of the Ganga Plain: Cultural-Historical Dimensions	Purushottam Singh	Indian Institute of Advanced Study	2010
Ek thi Nadi Saraswati	Khadag Singh Valdiya	Aryan Books International	2010
Ganga and Yamuna: River Goddesses and their Symbolism in Indian Temples	Heinrich von Stietencron	Permanent Black	2010
Ganges	Books, LLC, General Books LLC	General Books	2010
The Ganges	Jen Green	Hodder Children's Division	2010
The Ganges: India's Sacred River	Molly Aloian	Crabtree Publishing Company	2010
The Lost River: On the Trail of the Sarasvati	Michel Danino	Penguin Books India Pvt. Ltd.	2010
Ganga: The Divine Beauty	Anant Pai, S.S. Havaldar	Amar Chitra Kath Pvt. Ltd.	2009
Journey of the Pink Dolphins: An Amazon Quest	Sy Montgomery	Chelsea Green Publishing	2009
River Pollution	A. K. Jain	APH Publishing	2009

(continued)

(continued)

Title	Author	Publisher	Year
The Ganga: Water Use in the Indian Subcontinent	Pranab Kumar Parua	Springer	2009
Ganga: The River that Flows from Heaven to Earth	Sperling Vatsa, Harish Johari	Inner Traditions	2008
Ganges	Jon Nicholson, Ian Gray, Sharmila Choudhury, Tom Hugh-Jones, Dan Rees	BBC Books	2008
Large Rivers: Geomorphology and Management	Avijit Gupta	John Wiley & Sons	2008
The Holy Ganga	Kaushal Kishore	Rupa & Co.	2008
Witness to Extinction: How We Failed to Save the Yangtze River Dolphin	Sam Turvey	Oxford University Press	2008
The Five Theories of River Management	U.K. Choudhary	Ganga Scientific and Technical Council The Ganga Research Center I.T., B.H.U.,Varanasi	2008
Ganga: A Journey Down the Ganges River	Julian Crandall Hollick	Island Press	2007
Archaeological Geography of the Ganga Plain: The Upper Ganga (Oudh, Rohilkhand and the Doab)	Dilip K. Chakraborti	Munshiram Manoharlal Publishers Pvt Ltd	2007
Limnological Modeling	D. R. Khanna, R. Bhutiani	Daya Books	2007
Living on the River Ganges: World Cultures	Louise Spilsbury, Richard Spilsbury	Heinemann-Raintree	2007
People and Legends of Himalaya and the Ganga	M. P. Kuksal	Kalpaz Publications	2007
River Pollution	L.B. Singh, P N Pandey, Bhola Mahto	APH Publishing	2007
Ganga Pollution	Manish L. Srivastava	Bookwell Publications	2006
Ganga: The Holy River	Shantilal Nagar	BR Publishing Corporation	2006
River of Love in an Age of Pollution: The Yamuna River of Northern India	David L. Haberman	University of California Press	2006
Moonlight on the Ganga	Claire Krulikowski	iUniverse	2006
Along the Ganges	Ilija Trojanow	Haus Publishers Ltd.	2005, 2011
Freud along the Ganges: Psychoanalytic Reflections on the People and Culture of India	Salman Akhtar	Other Press	2005
Ganges River Dolphins	Kristin Petrie	ABDO	2005

(continued)

(continued)

Title	Author	Publisher	Year
River Ganga: A Cartographic Mystery	Madan, P.L.	Manohar Publication	2005
Settlements of the Ganges River	Richard Spilsbury	Heinemann/Raintree	2005
The Ganges: Along Sacred Waters	Aldo Pavan	Thames & Hudson	2005
The Story of Ganga	Priya Krishnan	Tulika Publishers	2005
Mahakumbh: The Greatest Show on Earth	J. S. Mishra	Har-Anand Publications	2004
Microbial Ecology: A Study of River Ganga	D.R. Khanna	Discovery Publishing House	2004
The Archaeology of Middle Ganga Plain: Excavations at Agiabir	Ashok Kumar Singh, Purushottam Singh	Indian Institute of Advanced Study	2004
The Ganges Water Diversion: Environmental Effects and Implications	Ema. Manirula Kādera Mirjā, M. Monirul Qader Mirza	Springer	2004
The Sacred Ganges	Charnan Simon	Child's World	2004
Thus Flows the Ganges	Sudhir Kumar Karan	Mittal Publications	2004
New Discoveries about Vedic Sarasvati	Ravi Prakash Arya	Indian Foundation for Vedic Science (Regd.)	2003
Ecology of Indian Rivers: A Review	K. Sankaran Unni	International Book Distributors	2003
River Pollution in India and Its Management	Krishna Gopal, Anil Agarwal	APH Publishing	2003
Sacred River: The Ganges of India	Ted Lewin	Houghton Mifflin Harcourt	2003
The Ganges	Rob Bowden	Heinemann/Raintree	2003
The Ganges	Raghubir Singh	Thames & Hudson Ltd	2003
The Ganges	Michael Pollard	Evans Publishing Group	2003
Living with the politics of floods: The Mystery of Flood Control	Dinesh Kumar Mishra	Peoples Science Institute, Dehradun	2002
Mahakumbha: A Spiritual Journey	Jiwesh Nandan	Rupa & Co.	2002
On the Banks of the Gaṅgā: When Wastewater Meets a Sacred River	Kelly D. Alley	University of Michigan Press	2002
Saraswati: The River That Disappeared	K.S. Valdiya	Universities Press	2002
The Ganges	James Barter	Lucent Books	2002
The Saraswati Flows on: the Continuity of Indian Culture	Braj Basi Lal	Aryan Books International	2002
Archaeological geography of the Ganga Plain: the lower and the middle Ganga	Dilip K. Chakrabarti	Orient Blackswan	2001

(continued)

(continued)

Title	Author	Publisher	Year
Plant Diversity Along River Ganga	S. Kumar	Sai Publishers	2001
The Ganges in Myth and History	Steven G. Darian	Motilal Banarsidass Publ. pvt ltd.	2001
Swallowing the River Ganges: A Practice Guide to the Path of Purification	Matthew Flickstein	Wisdom Publications	2001
Cleaning-up the Ganges: A Cost-Benefit Analysis of the Ganga Action Plan	Anil Markandya, Maddipati Narasimha Murty	Oxford University Press	2000
From the Ganges to the Snake River: An East Indian in the American West	Debu Majumdar	Caxton Press	2000
Commercial Activities and Development in the Ganga Basin	V.K. Shrivastava (Ed.)	Concept Publishing Company (p) Ltd.	1999
Ganga Safari: Ganga River Boat Adventure With Jayapataka Swami	Mahamaya Devi Dasi	Holy Cow Books	1999
Ecological Imbalance of the Ganga River System: Its Impact on Aquaculture	Parmila Ray	Daya Books	1998
The Birth of the Ganga	Harish Johari	Inner Traditions India	1998
Water Pollution Control: A Guide to the Use of Water Quality Management Principles	Richard Helmer, Ivanildo Hespanhol	Taylor & Francis	1997
Ganga: A Water Marvel	Ashok Chandra Shukla, Vandana Asthana	Ashish Publishing House	1995
Ganges River Dolphin: A Monograph	Tej Kumar Shrestha	Steven Simpson Books	1995
Sharing the Ganges: The Politics and Technology of River Development	Ben Crow, Alan Lindquist, David Wilson	Sage Publications and Book Review Literary Trust, New Delhi	1995
Water and the Quest for Sustainable Development in the Ganges Valley	Graham Chapman, Michael Thompson	Mansell Pub.	1995
Environment Protection and Pollution Control in the Ganga	Pramod Kumar Agrawal	M.D. Publications Pvt. Ltd.	1994
The Ganges Delta and Its People	David Cumming	Thomson Learning	1994
Ecology and Pollution of Ganga River: A Limnological Study at Hardwar	Dev Raj Khanna	Ashish Pub. House	1993

(continued)

(continued)

Title	Author	Publisher	Year
Plant Wealth of the Lower Ganga Delta: An Eco Taxonomical Approach In 2 Vols	Kumudranjan Naskar	Daya Publishing House	1993
The Ganga: A Scientific Study	C. R. Krishna Murti	Northern Book Centre	1991
Aquatic snd Semi Aquatic Plants of the Lower Ganga Delta: Its Taxonomy Ecology and Economic Importance	K R Naskar	Daya Publishing House	1990
Can the Ganga Be Cleaned	Brojendra Nath Banerjee	B.R. Pub. Corp.	1989
Jaya Ganga: In Search of the River Goddess	Vijay Singh	Penguin Books, London	1989
Pollution of Ganga River (Ecology of Mid-Ganga Basin)	Ghose, N. C; Sharma, C. B	Ashish Publishing House, New Delhi	1989
A Walk Along the Ganges	Dennison Berwick	Dennison Berwick	1986
Ganga Pollution and Health Hazard	Upendra Kumar Sinha	Inter-India Publications	1986
Human Impact on Ganga River Ecosystem	S.B. Chaphekar, G.N. Mhatre	Naurang Rai Concept publishing company, New Delhi	1986
Lower Ganga-Ghaghra Doab: A study in Rural Settlements	Minati Singh	Tara Book Agency	1983
The Ganges	Gina Douglas	Rcl Benziger	1978
Clan Settlements in the Saran Plain (Middle Ganga Valley): A Study in Cultural Geography	Rana P. B. Singh	National Geographical Society of India, Banaras Hindu University	1977
The Lower Ganga-Ghaghra Doab: A Study in Population and Settlement	Ram Niwas Misra	Sahitya Bhawan	1977
Ganga	Sivaramamurti, C.	Orient Longman	1976
Population Problem in the Ganges Valley	Kripa Nath Varma	Shiva Lal Agarwala	1967
The Ganges Delta	Kanangopal Bagchi	University of Calcutta	1944
Ganges Canal: A disquisition on the Heads of the Ganges of Jumna canals, North-western Provinces	Cautley, Proby Thomas	London, Printed for Private circulation	1864
An Account of the Fishes Found in the River Ganges and Its Branches	Francis Hamilton	Edinburgh, Printed for A. Constable and company	1822

Pristine stretch in the upper Ganga region (Courtesy Deepak Kumar)

Appendix 3: List of Ganga Festivals Held on the Ganga River (Compilation from Various Internet Sources)

Month	Ganga festival	When and why celebrated	City
January	Makar Sankranti	Transition of the Sun from Sagittarius to Capricorn during the winter solstice in the northern hemisphere (Uttarayana)	Uttar Pradesh, West Bengal, (Tamil Nadu, Andhra Pradesh, Maharashtra, Assam, Punjab)
	Mauni Amavasya	Sun and Moon enter and the Capricorn sign, because of the transit, on this day. This day is celebrated as the birthday of Manu Rishi	Mainly – Haridwar, Kanpur, Allahabad, Rishikesh, Varanasi
	Ganga Sagar Mela		West Bengal (Kolkata)
Feburary	Basant Panchami Snan, Rath Saptami Snan, Bhisma Ekadashi Snan, Maghi Purnima Snan		Mainly – Haridwar, Kanpur, Allahabad, Rishikesh, Varanasi
April–May	Ganga Saptami or Ganga Jayanti	Celebrated on the 7th day (Saptami) of Sukla Paksa, or the waxing fortnight of the full moon in the month of Baisakhi	Varanasi, Haridwar, Uttar Pradesh
May	Buddha Poornima	Lord Buddha was born on the Full Moon day in the month of Vaisakh in 563 BC	Mainly – Haridwar, Kanpur, Allahabad, Rishikesh, Varanasi
	Yama Dvitiya	Sacred bath is performed to mark the auspicious occasion in the month of Kartik. Brothers and sisters bathe on Yama Dvitiya	From Gujarat to Bengal
June	Ganga Dussehra	Birthday or descent of mother Ganges – Ganga Ma. Throughout India this festival lasts 10 days beginning on the Amavasya (dark moon night) and going through to the dasami tithi (tenth phase of the Moon, the day before Pandava Nirjal Ekadasi)	Mainly – Varanasi, Allahabad, Garhmukteshwar, Prayag, Haridwar and Rishikesh

(continued)

(continued)

Month	Ganga festival	When and why celebrated	City
July–August	Kanwar Mela	Is annual pilgrimage of devotees of Shiva, known as Kānvarias, to fetch holy waters of Ganges River, Ganga Jal, which is later offered at their local Shiva temples. The Yatra takes place during the sacred month of Shravan	Haridwar, Gaumukh, Gangotri, Rishikesh in Uttarakhand
October	Chhat Puja	Is celebrated on the sixth day of Kartik Shukla Paksha	Mainly – Bihar, Uttar Pradesh
	Prabodhani Ekadashi	Eleventh lunar day (ekadashi) in the bright fortnight (Shukl Paksha) of the Hindu month of Kartik	Mainly – Haridwar, Kanpur, Allahabad, Rishikesh, Varanasi, Bihar
November–December	Kartik Poornima	Celebrated on the full moon day or the fifteenth lunar day of Kartik	Mainly – Haridwar, Kanpur, Allahabad, Rishikesh, Varanasi, Bihar
	Dev Deepavali	Festival of lights of the gods	Mainly – Varanasi, Bihar (Patna), Gujarat, Kanpur, Haridwar, Allahabad, Rishikesh
Every Month	Ekadashi	Eleventh lunar day (Tithi) of the shukla (bright) or krishna (dark) paksha (fortnight) of every lunar month in the Hindu calendar (Panchang)	Mainly – Haridwar, Kanpur, Allahabad, Rishikesh, Varanasi, Bihar
	Purnima	Full Moon	Mainly – Haridwar, Kanpur, Allahabad, Rishikesh, Varanasi
	Amavasya	New moon	Mainly – Haridwar, Kanpur, Allahabad, Rishikesh, Varanasi, Bihar, Rishikesh
Whenever occurs	Somvati Amavasya	Observed when the no moon day falls on a Monday in traditional Hindu Lunar	Mainly – Haridwar, Kanpur Allahabad, Rishikesh, Varanasi, Bihar
	Surya Grahan	In a solar eclipse the Moon comes between the Sun and Earth. During a total solar eclipse the disk of the Moon fully covers that of the Sun, and only the Sun's corona is visible	Mainly – Haridwar, Kanpur, Allahabad, Rishikesh, Varanasi, Bihar

(continued)

(continued)

Month	Ganga festival	When and why celebrated	City
	Chandra Grahan	In a lunar eclipse all or a part of the Moon's disk enters the umbra of the Earth's shadow and is no longer illuminated by the Sun. Lunar eclipses occur only during a full moon, when the Moon is directly opposite the Sun	Mainly – Haridwar, Kanpur, Allahabad, Rishikesh, Varanasi, Bihar
Every year	Magh mela	Jupiter is in Aries or Taurus, Sun and Moon in Capricorn (Allahabad)	Allahabad
Once in 3 years	Kumbh mela	When Jupiter is in Aries or Taurus, Sun and Moon in Capricorn	Allahabad
		When Jupiter and Sun in Leo (Simha Rashi)	Nasik
		When Jupiter in Aquarius, Sun in Aries (Mesha Rashi)	Haridwar
		When Jupiter in Leo, Sun in Aries, or when Jupiter, Sun & Moon are in Libra	Ujjain
Once in 6 years	Ardh Kumbh	Jupiter is in Aries or Taurus, Sun and Moon in Capricorn	Allahabad
		Jupiter in Aquarius, Sun in Aries (Mesha Rashi)	Haridwar

Picturesque view enroute Harshil (Courtesy Deepak Kumar)

Authors Biography

Dr. Rashmi Sanghi (rsanghi@gmail.com) is the recipient of CHEMRAWN VII (IUPAC) Prize for Atmospheric and Green Chemistry award for 2012. After obtaining her D. Phil. degree from Chemistry Department, University of Allahabad in 1994, she worked at the Indian Institute of Technology, Kanpur, India, in various capacities as a research scientist. She has also worked as visiting scientist at the Rutgers University USA and LNMIIT, Jaipur. She is passionate about environmental green chemistry with focus on microbial and nanoparticle research. Her research mainly focuses on the development of methods that can help in minimizing or eliminating the hazardous substances in the environment. She has contributed to improved waste-water remediation using green technologies based on biopolymers and has also engaged herself in a variety of outreach activities by generating awareness about the need to embrace environment friendly ways for a better and healthier future. She has numerous publications and patents to her credit. Currently she is working at IIT Kanpur as Consultant in the 'Ganga River Basin Environment Management Plan' (GRBEMP) project funded by Ministry of environment and forests.

R. Sanghi (ed.), *Our National River Ganga: Lifeline of Millions*,
DOI 10.1007/978-3-319-00530-0, © Springer International Publishing Switzerland 2014

Prof. Kelly D. Alley (alleykd@auburn.edu) is Professor of Anthropology at Auburn University. She has worked in the Ganges river basin of India for over 20 years, studying interpretations of the sacred river Ganga and problems with river pollution and wastewater management. She is the author of *On the Banks of the Ganga: When Wastewater Meets a Sacred River* (2002), and book chapters and articles on religion and ecology and environmental law and justice in India and the USA. She is now working on water governance in the Ganga-Brahmaputra-Meghna basin. This research is supported by the Center for Forest Sustainability and the College of Liberal Arts at Auburn University.

Dr. Sandeep Kumar Behera (sbehera@wwfindia.net) is working as Associate Director, River basin and biodiversity Programme in WWF-India, New Delhi. He is working on various aspects of ecology and biodiversity of rivers, especially Ganga and its tributaries in the Indian subcontinent for last 20 years. He has worked for many important research projects of National and International importance. He was awarded Ph.D. during 1996 for his work entitled "**Population Dynamics, Habitat Utilization and Conservation of Gangetic Dolphin**". He was nominated as the member of the River Dolphin committee formulated by Ministry of Environment and Forests (under the National Ganga River Basin Authority, NGRBA), Government of India. He developed the river dolphin conservation document "**Ganges River Dolphin Action Plan for 2010–2020**", for the Indian Federal Government which will help in conservation of river dolphins in India.

Professor (Er.) Dr. Devendra Swaroop Bhargava (dsbhargava@yahoo.co.in) earned B.E. (Civil), PGD & M.E. (Public Health Engineering), Ph.D. (Environmental Engineering), and is Roorkee University's retired Environmental Professor. After PGD (1960), he also taught at IIT, Kanpur, AIT, Bangkok, served at the Central Pollution Control Board, New Delhi, and was Uttranchal Government's Advisor (QCC). He has authored 470 publications and won 32 academic awards including UGC's most prestigious Pranavananda Award (1992). As jury member of World Cultural Council awarding the Albert Einstein award in science, he is included in the prestigious Marquis Who's Who in the world. His 38 years research on Ganga and its self-purification is well known.

Dr. Malavika Chauhan (malavikachauhan@gmail.com), a doctorate in Limnology leads the Himmotthan Society, Dehradun, which is working in rural development in the Central Himalayan regions of the country. She is also a Development Manager at the Sir Ratan Tata Trust, Mumbai (www.srtt.org). Previously Malavika worked with World Wide Fund for Nature, Wetland International – South Asia, the Institute of Social and Economic Change, Delhi, and was a DST Young Scientist, a Global Environment Facility fellow at the University of East Anglia, UK, and a Fulbrighter with the U.S. Geological Survey's National Wetlands Research Centre at Louisiana. She has numerous publications on ecological and environmental management issues to her credit, and has consulted extensively, including for the World Bank, the Asian Development Bank and WWF-International.

Subhajyoti Das (Subhajyoti_Das@hotmail.com), former Director of Central Ground Water Board, India, is a hydrogeologist now working with the Geological Society of India as an Editor. He also studied Integrated Water Resources Management in the International School of Water Resources in Colorado State University, USA. His fields of interest include coastal and hard rock hydrogeology, water quality and pollution, and water management. He is a prolific writer in science journals and has authored or edited number of widely acclaimed books.

Dipankar Ghorai graduated with Chemistry (Hons.) at the University of Calcutta and subsequently did his M.Sc. and Ph.D. in Agriculture with specialization in Soil Science from the same university. His Ph.D. thesis was focused on the effect of long term fertilizer application on soil, crop quality and agricultural sustainability in jute-based intensive cultivation practices. Dr. Ghorai is presently engaged as a Subject Matter Specialist (Agril.) with Krishi Vigyan Kendra (CRIJAF, ICAR), Burdwan, West Bengal, specializing mainly on agri-based rural livelihood systems with areas of interest in fertility, water and crop management practices, and developing market chain with focus on value added goods. He has considerable expertise in other social and economic activities towards ensuring improved livelihood security.

Dr. Brij Gopal (brij44@gmail.com) formerly Professor at the Jawaharlal Nehru University, New Delhi, is currently a Member of the Rajasthan State Biodiversity Board and the Wetland Advisory Committee of Bihar government, Vice President of the National Institute of Ecology, Member of the Board of Directors of the International Society for River Science, Executive Vice President of the International Society of Limnology (SIL) and a Member of the International Scientific Committee of the Brazilian National Institute for Wetland Science and Technology (INAU). He is the only recipient, outside Europe and North America, of the Naumann-Thienemann Medal (2004) of the International Association of Limnology (SIL) and only the second Asian recipient of the International Fellow award (1997) of the Society of Wetland Scientists. He was one of the lead authors of the chapter on Ecosystems for the 2007 Nobel prize-winning IPCC's Fourth Assessment Report (2007).

Julian Crandall Hollick has produced many radio documentaries for National Public Radio and BBC Radio 4 on India, including Passages to India (1989), Letters from Jitvapur (1992), Apna Street (1995–2002), Monsoon (1997) and Ganga (2007). In 2004/2005 he travelled the entire length of Ganga by county boat. He has also written two books: *Ganga*, Random House India, Delhi 2007, and *Apna Street*. He teaches an annual workshop at Xavier's College, Mumbai, and is doing a Ph.D. on Ganga Jal at Leeds University (UK). He lives with his wife in Provence, France.

Mr. Nitin Kaushal (nkaushal@wwfindia.net) is Senior Manager–River Basins and Water Policy with WWF-India. He has been working in the field of water resources management from over 12 years. He started his professional journey as a field researcher in Irrigation Management Transfer. Afterwards, he contributed to consulting assignments of the World Bank funded irrigation reform projects in northern and central India. From last half a decade, he has been actively engaged in furthering the concept of Environmental Flows, including the assessment of E-Flows for the Ganga river basin and coordination of various activities pertaining to E-Flows group established for the preparation of GRBEMP by the IITs Consortia. He has contributed to number of project reports, research papers and publications.

Dr. Niraj Kumar (nasi.allahabad@gmail.com) is the Founder Fellow and Hony. Scientist, The Institute of Applied Sciences, and Institute of Rural Health and Development, Allahabad. He is also the Bioethics Expert, Global Ethics Observatory–UNESCO, and Principal Investigator, DBT Govt. of India's Project on Natural Resources Awareness. Currently he is the Executive Secretary of The National Academy of Sciences, India (NASI) Allahabad.

Dr. Asheesh Shivam Mishra (shivam_a2000@yahoo.co.in) received his D.Phil. degree in Zoology from the University of Allahabad, Allahabad, for his work on the "Longitudinal distribution of benthic macroinvertebrate community in upper Ganga basin." He has worked as team member of the biodiversity group for e-flow assessment of the upper Ganga River (WWF-India). He has also worked as a Research Associate in DCFR (ICAR) Bhimtal (UK) and Project Scientist in the Ganga River Basin Management Programme in IIT Kanpur. He has published more than 20 research papers in the journals and books of National and International repute. He has also published one book *Benthic Macroinvertebrate Fauna in the Rivers of Central India*, Lambert, Academic Publication, Germany.

Manoj Misra, previously of the Indian Forest Service, IFS, has been researching, documenting and advocating for the restoration of the river Yamuna since early 2007. He is the Convener of Yamuna Jiye Abhiyaan, a Delhi based civil society consortium for the restoration of the Yamuna river. (www.peaceinst.org and blog at www.yamunajiyeabhiyaan.blogspot.in)

Dr. Prakash Nautiyal (pn.mahseer@yahoo.com) is Professor, Zoology Department, HNB Garhwal University. He contributed to the *in situ* conservation issues of the threatened fish Golden mahseer, habitat needs and hence the assessment of major components of aquatic and e-flows needed to maintain integrity of related ecosystems. He is a recipient of Gold Medal (2012) Indian Academy of Environmental Sciences, Haridwar; J.S. Datta Munshi Gold Medal (2009), International Society for Ecological Communication; Scientist of the Year Award (2009), Bioved Society; Promising Scientist Gold Medal (2000), and Zoological Society of India. He authored *The Golden Mahseer (A Threatened Fish of Himalaya)* (2011) Compendium on *Environmental Audit of HEP* for Sustainable Development (2006). He was associated with GAP I Integrated Research Programme. He led the biodiversity group for assessment of e-flows (Living Ganga Programme, WWF-India).

Professor T.V. Prabhakar (tvp@iitk.ac.in) is Professor at the Dept. of Computer Science and Engineering IIT Kanpur. He is interested in building systems and is still excited about science and technology. He has worked in multiple areas like Databases, Logic Programming, User Interface Design, Internet Technologies, Software Architecture, Knowledge Modeling, and Indian Language Technologies. He has worked with CMC R&D, Satyam Computer Services and has been a consultant to a large number of IT organizations and e-governance initiatives. He has been heading the communication group of Ganga River Basin Management River Plan (GRBEMP) whereby he designed and developed the www.gangapedia.com website.

Dr. Panta Murali Prasad did his Ph.D. on the "Economic and Legal aspects of Consumer Protection in India: With a Comparison of the German Experience" from University of Hyderabad in 1999, India. He is currently working as Associate Professor in Economics in the Department of Humanities and Social Sciences, Indian Institute of Technology Kanpur, India. His teaching interests are in the areas of "Law and Economics" and "Environmental Economics and Policy". He has published a book on "Business, Consumer and the Government: Economic and Legal Perspectives (India and Germany)" and a couple of papers on "Environmentalism". His initial work on "Livelihood Aspects as of the River" under the Ganga Living Programme (WWF-India & HSBC) motivated him to be part of "Socio-economic and Cultural Group" to work further on Livelihood under the MoEF, GoI (IITs Consortium).

Viveksheel Sagar is working with WWF-India in River Basin & Biodiversity Division as an Assistant Coordinator and has more than 6 years of working Experience in Biological Monitoring, education and awareness programmes for the fishermen/local people regarding the importance of indicator species and to minimize the threats for the species from the target people and implementation of the programme.

H.H. Pujya Swami Chidanand Saraswatiji is President and Spiritual Head of Parmarth Niketan Ashram, Rishikesh, India, and also the Founder of Ganga Action Parivar (www.gangaaction.org). Pujya Swamiji has been a leader of frequent world peace pilgrimages across the world and in numerous international, inter-faith summits and parliaments, including the Parliament of World Religions, the Millennium World Peace Summit of Religious and Spiritual Leaders at the United Nations, the World Economic Forum, the World Council of Religious Leaders at the United Nations, the World Conference of Religions for Peace, the Global Youth Peace Summit at the United Nations, the first Hindu-Jewish Summit in Jerusalem and the Hindu-Christian dialogue initiated by the Vatican. He is the recipient of innumerable awards, including the Mahatma Gandhi Humanitarian award, Hindu of the Year Award, Prominent Personality Award, by Lions Club, Best Citizens of India Award, the Uttaranchal Ratan (Jewel of Uttaranchal) Award and the Bharat Vikas Parishad first Utkrishtta Samman Award and many others.

Dr. H.S. Sen, Former Director, Central Research Institute for Jute & Allied Fibres (CRIJAF) did his M.Sc. and Ph.D., both from IARI, New Delhi. His field of specialization in research has been water management and soil salinity, especially for the coastal ecosystem. He was conferred with a national award by ICAR in 1999 for his contribution on rainwater harvesting under rainfed ecosystem. A significant landmark during his tenure at CRIJAF was launching of Jute Technology Mission at national level sponsored by GOI in 2006 for an all round improvement of jute and allied fibre crops from production to marketing. He has been in post graduate teaching for the last 25 years. For more details please visit his blog http://www.hssen-coastalmanagement.blogspot.com

Dr. Vandana Shiva is trained as a Physicist and later shifted to inter-disciplinary research in science, technology and environmental policy, which she carried out at the Indian Institute of Science and the Indian Institute of Management in Bangalore, India. She is the founder of Navdanya, a national movement to protect the diversity and integrity of living resources, especially native seed, the promotion of organic farming and fair trade. Among her many awards are the Alternative Nobel Prize (Right Livelihood Award, 1993), Order of the Golden Ark, Global 500 Award of UN and Earth Day International Award. Lennon ONO grant for peace award by Yoko Ono in 2009, Sydney Peace Prize in 2010, Doshi Bridgebuildep Award, Calgary Peace Prize and Thomas Merton Award in the year 2011.

Dr. Hari Singh (hsingh7wild@gmail.com) has completed his Ph.D. on aquatic mammals (dolphin & otter), from Jiwaji University, Gwalior, Madhya Pradesh, India. Currently, he is studying the underwater behavior of Ganges river dolphin by using acoustic technology with WWF-India. He has more than 6 years of research experience in biodiversity conservation. His main research interest is on habitat ecology, population dynamics and acoustic study. He has received meritorious student fellowship from UGC-SAP, New Delhi. Prior, to joining WWF-India he has also worked with Gharial Conservation Alliance (GCA).

Dr. Praveen K. Thakur is currently working as Scientist/Engineer 'SE' at Water Resources Department (WRD) of Indian Institute of Remote Sensing (IIRS), Indian Space Research Organization (ISRO)-Dehradun. He did his B.Tech. in Civil Engineering from National Institute of Technology (NIT) Hamirpur (June 2001), Master of Technology (M.Tech.) in Water Resources Engineering from Indian Institute of Technology (IIT) Delhi (December 2002) and Ph.D. from IIT Roorkee in Geomatics engineering (September 2012). He joined IIRS Dehradun in 2004. He has published more than 30 research papers in peer-reviewed international/national journals and conferences/symposiums. His current research interests are geospatial technology application in water resources, microwave remote sensing for snow, ice and floods studies and remote sensing data assimilation.

Prof. B.D. Tripathi (tripathibd@gmail.com) is the **Founder Coordinator of** Centre for Environmental Science & Technology, BHU and the expert member of State Ganga River Conservation Authority Govt. of U.P. and NGRBA. He was also the Member of the **Prime Minister's** and **U.P. Chief Minister's committee (2008/2009)** on Ganga and has been awarded the **Ganga Bhakt Puraskar** (2007) by Society for Science & Technology, Allahabad University, **Kashi Gaurav Puraskar** (2005) by SAAR Sansthan, **Bhagirath Puraskar** (1999) by Ganga Sewa Samiti, **Paryavaran Mitra Puraskar** (1991), SNTM foundation by Hon'ble President of India, and **Founder President** (1979) National Environmental Conservation Association (www.ecaindia.org) India. He has developed seven eco-friendly technologies in the field of Environmental Pollution Control Technologies, Bioremediation, Ecological Conservation, Biodiversity and Resource Management also.

Dr. Smriti Tripathi is an eminent scholar in the field of Environmental Science and technology. She has developed two technologies for wastewater treatment, published ten research papers in high ranking international journals and one book, published from Germany. She is recipient of a few gold medals and awards.

Dr. Ramesh Chandra Trivedi (dr_r_c_trivedi@rediffmail.com) presently working with National Green Tribunal as an Expert Member has served Central Pollution Control Board, Govt. of India for 31 years. He was involved in various aspects of regulation and monitoring of pollution in the country. He was also responsible for inventory and quantification of pollution from different sources including domestic, industrial and non-point. He was also responsible for development of guidelines, policies, manuals, laboratories, setting of standards, criteria and procedures on pollution regulation. He has authored and co-authored 6 books, 75 research papers, over 100 reports of Central Pollution Control Board and 65 reports submitted to Hon'ble Supreme Court of India, which formed basis for taking many important decisions of national interest.

Dr. Jyoti Verma (diatombuster@gmail.com), PDF (UGC) at Zoology Department, University of Allahabad, Allahabad, is project scientist in IIT Kanpur-Ganga River Basin Management Plan working on Diatom biodiversity – upper Ganga River Basin.

Pray for a healthy and mighty Ganga (Courtesy Tripti Singh)

Index

R. Sanghi (ed.), *Our National River Ganga: Lifeline of Millions*,
DOI 10.1007/978-3-319-00530-0, © Springer International Publishing Switzerland 2014

V

W

FSC
www.fsc.org
MIX
Papier aus verantwortungsvollen Quellen
Paper from responsible sources
FSC® C105338

If you have any concerns about our products,
you can contact us on
ProductSafety@springernature.com

In case Publisher is established outside the EU,
the EU authorized representative is:
Springer Nature Customer Service Center GmbH
Europaplatz 3, 69115 Heidelberg, Germany

Printed by Libri Plureos GmbH
in Hamburg, Germany